Targets for the Design of Antiviral Agents

Targets for the Design of Antiviral Agents

Edited by

E. De Clercq

Rega Institute for Medical Research
Catholic University of Leuven
Leuven, Belgium

and

R. T. Walker

The University of Birmingham
Birmingham, United Kingdom

Springer Science+Business Media, LLC

Proceedings of the NATO Advanced Study Institute on
Targets for the Design of Antiviral Agents,
held June 19–July 2, 1983,
at Les Arcs, France

Library of Congress Cataloging in Publication Data

NATO Advanced Study Institute on Targets for the Design of Antiviral Agents
(1983: Les Arcs, France)
Targets for the design of antiviral agents.

(NATO ASI series. Series A, Life sciences; v. 73)
"Proceedings of the NATO Advanced Study Institute on Targets for the Design
of Antiviral Agents, held June 19–July 2, 1983, at Les Arcs, France"—T.p. verso.
Includes bibliographical references and index.
1. Antiviral agents—Congresses. I. De Clercq, Erik. II. Walker, Richard T. III.
North Atlantic Treaty Organization. Scientific Affairs Division. IV. Title. V. Series:
NATO advanced science institutes series. Series A, Life sciences; v. 73. [DNLM:
1. Antiviral agents—Congresses. QV 268.5 N279t 1983]
RM411.N38   1983   615'.3   83-24627
ISBN 978-1-4684-4711-8   ISBN 978-1-4684-4709-5 (eBook)
DOI 10.1007/978-1-4684-4709-5

©1984 Springer Science+Business Media New York
Originally published by Plenum Press, New York in 1984
Softcover reprint of the hardcover 1st edition 1984

PREFACE

This publication contains the Review Lectures presented at a joint NATO Advanced Study Institute and FEBS Advanced Study Course held at Les Arcs, Bourg-Saint-Maurice, France, from the 19th June - 2nd July 1983. The Course, entitled "Targets for the Design of Antiviral Agents" was in some ways a sequel to the NATO-FEBS Course held at SOGESTA (near Urbino), Italy from the 7th - 18th May 1979 and published as volume A26 in this series. During the subsequent four years, we have witnessed the first of the "new generation" of antiviral compounds, which are more efficacious and less toxic than the "classical" antiviral drugs, reach the clinic and we felt that it was the right time to assess the future prospects of this very important and exciting field.

The vast majority of the drugs developed recently have proved active against various members of the herpesvirus family and elsewhere in this publication we learn that the cure for only rather few viral diseases, such as the common cold, influenza and herpes, promises the return on investment required by the pharmaceutical industry. However, the aim of this Course was for eminent virologists to identify possible targets among the various virus classes against which the chemists could then design suitable therapeutic agents.

Recent advances with antiherpesvirus drugs have shown that a far greater selectivity and therapeutic index can be obtained than was previously thought to be possible. Now that the mode of action of many of these antiherpes drugs is being resolved, it is hoped that the information gained from these serendipitously discovered drugs could be applied to other virus classes so that a more rational approach might lead to the development of the next generation of antiviral drugs. Many chemists have only a very hazy idea of the mode of virus replication and thus one of the intentions of the meeting was to acquaint the chemist engaged in drug synthesis, with the potential targets in the different virus groups upon which antiviral drugs could act. Such an approach should hasten a more rationalized drug design.

As before, the aim of the meeting was to gather together many of the experts in different scientific disciplines, to enable senior scientists in one discipline to learn from their colleagues working in other related and relevant disciplines, and to present the opportunity to younger scientists who have just entered the field to assimilate current knowledge and make personal contacts with the more senior people. In the design of new antiviral agents, a coordinated effort of virologists, chemists, pharmacologists, toxicologists and clinicians is required and such an effort can be afforded only by a strong cooperation between academic institutes, pharmaceutical industry and governmental organizations. This is precisely the kind of scientists which assembled for the meeting. An important feature was the interest shown by many pharmaceutical companies, as can be seen from the Acknowledgments.

This publication only contains the review lectures which cover in depth the possible potential targets for attack within the various virus classes. Other review articles are focussed on the antiviral compounds which are presently in use or being assessed. As the molecular biologists and virologists continue to unravel the mechanism of viral replication, the structure of the viral genomes and the enzymes for which the viral genes code, this will increase considerably our insight into the targets available; and for some viruses we already have a reasonable basis on which to plan an attack. One exception is the class of slow viruses, where it is still not known whether the infective agent contains nucleic acid, and although this class of viruses hardly constitutes a commercially viable target, the mechanism by which these infections are transmitted may be of vital concern to us all.

The initial euphoria which was apparent four years ago has already started to evaporate. Serendipity still rules, nothing of fundamental significance has yet been designed and our knowledge of the detailed mechanism of viral replication is woefully inadequate. On a molecular level, our understanding of the mode of action of the virus-specified enzymes is even less well known but at least a start has been made. Several relatively non-toxic and specific antiviral agents have been developed and some of these agents have reached the market and what holds for one class of viruses (particularly herpes) may well hold for others. It is unlikely that the antiviral "penicillin" will ever exist but the coming years will undoubtedly reveal many new antiviral agents with a useful spectrum of activity combined with low toxicity.

In conclusion, the Directors of the Course would like to acknowledge the help given by Bernard Croise and his staff of Seminarc and by Charles Arnould and his staff of the Hotel du Golf.

We were able to enjoy the superb scenery and remove the effect of
the excess of calories by trying to follow the national ski speed
champion Pascal Budin up and (very occasionally) down dale or by
using some of the many other sports facilities available without
being worried by organizational difficulties. More importantly,
we were able to concentrate on the science, renew contacts with
old friends and set up new collaborative projects, the results
of which should become apparent in the years to come.

We finish with a quotation used by one of our review lecturers
(J.S. Oxford) and taken out of the context in which it was originally
used 40 years ago but which perhaps encapsulates the current position
in the field of the rational design of antiviral drugs:

"This is not the end. It is not even the beginning of
the end. But it is, perhaps, the end of the beginning"

W.S. Churchill, 1943

July 1983   E. De Clercq

R.T. Walker

ACKNOWLEDGEMENTS

This Advanced Study Institute was sponsored by NATO and co-sponsored by FEBS (Advanced Course no. 83/07) and by NIAID (National Institute of Allergy and Infectious Diseases, USA), the Fogarty International Center, and PIRMED (Programme Interdisciplinaire de Recherche sur les Bases Scientifiques des Medicaments, France). The Course Directors also acknowledge the financial assistance of the following companies:

Antibioticos, S.A., Madrid, Spain
Astra Lakemedel AB, Sodertalje, Sweden
Bayer AG, Wuppertal, W. Germany
Beecham Pharmaceuticals, Epsom, Surrey, United Kingdom
Behringwerke Aktiengesellschaft, Marburg, W. Germany
Boehringer Mannheim GmbH, Mannheim, W. Germany
Bristol-Myers Company, New York, New York, USA
Burroughs Wellcome Co., Research Triangle Park, North Carolina, USA
Continental Pharma, Brussels, Belgium
du Pont de Nemours & Co., Wilmington, Delaware, USA
Hoffmann-La Roche Inc., Nutley, New Jersey, USA
ICI (Imperial Chemical Industries), Macclesfield, Cheshire, United Kingdom
IRL Press Ltd., Eynsham, Oxford, United Kingdom
Janssen Pharmaceutica, Beerse, Belgium
Merck Sharp & Dohme Research Laboratories, West Point, Pennsylvania, USA
Ortho Pharmaceutical (Canada) Ltd., Don Mills, Ontario, Canada
Pfizer Central Research, Sandwich, Kent, United Kingdom
P.L. Biochemicals Ltd., Northampton, United Kingdom
Rhone-Poulenc Sante, Cedex 29, Paris, France
Roche Products Ltd., Welwyn Garden City, United Kingdom

Sandoz Forschungsinstitut GmbH, Wien, Austria
Searle Research and Development, High Wycombe, United Kingdom
Smith Kline & French Laboratories, Philadelphia,
Pennsylvania, USA
Sterling-Winthrop Laboratoria, Brussels, Belgium
Syntex Research, Palo Alto, California, USA
Wellcome Research Laboratories, Beckenham, Kent, United
Kingdom

CONTENTS

CONCLUSION

OVERVIEW OF THE POSSIBLE TARGETS FOR VIRAL CHEMOTHERAPY

William H. Prusoff, Tai-Shun Lin, William R. Mancini,
Michael J. Otto, Scott A. Siegel and J.J. Lee

Department of Pharmacology
Yale University School of Medicine
New Haven, Connecticut 06510

The ideal target for prevention of viral infections is eradication of either the virus or the host, since both are required to produce an infection. For a variety of reasons, mostly personal, we would all agree to preferably eliminate the virus. Such an approach appears to have been accomplished with the world-wide elimination of the virus responsible for smallpox. This was achieved by a highly successful campaign of surveillance and containment by vaccination in order to prevent its spread, rather than by universal routine immunization.

A number of live attenuated vaccines are available not only for prevention of small pox, which has been eradicated, but more importantly for rubeola (measles), rubella (German measles), mumps, and poliomyelitis. Inactivated vaccines are also available for potential prevention of influenza, poliomyelitis, and rabies (1). In 1981 the U.S. Food and Drug Administration approved a new vaccine for hepatitis B which is produced from hepatitis B surface antigens (HB_5Ag) found in human blood (2).

Of interest is the development of a vaccine against foot and mouth disease in cattle, swine, sheep and goats caused by a picornavirus. By use of recombinant DNA techniques, the viral RNA coding for VP_3, a coat protein, was inserted into an E. coli plasmid, and the desired antigenic protein was produced and purified (3). Two injections of 0.25 mg of this protein produced protection in animals from a challenge dose of virus. However since there are many strains of this virus, about 15 different proteins are expected to be eventually included in order to prepare a comprehensive foot and mouth disease vaccine (3).

An exciting development is the recent chemical synthesis of the first synthetic vaccine to be produced. Peptides were synthesized which corresponded to several regions of the foot-and-mouth disease virus, and one of the peptides (20 amino acids) produced neutralizing antibodies which protected guinea pigs against a challenge with this virus (4). The potential to develop synthetic vaccines for other virus infections is obvious and studies are in progress for development of synthetic vaccines (5-7).

Unfortunately vaccines are not available for all viruses that infect humans and there are a variety of reasons for this. The rhinoviruses consist of over 100 different serotypes and therefore it is most unlikely that vaccination would be a successful procedure for prevention because of the specificity of the immune reaction. A number of compounds however are under consideration for potential therapy of the rhinoviruses and these include arildone, enviroxime, dichloroflavin and interferon.

The influenza A virus presents another problem, in that it continuously changes its antigenic composition, so that this year's vaccine may be less effective against next year's emerging virus. Although amantadine is available for prophylactic or early use against respiratory infections caused by the influenza A viruses, the primary approach is use of the vaccine.

The family of herpesvirus are responsible for a number of clinically important infections:

HSV-1: herpes keratitis, herpes encephalitis, mucocutaneous herpes and more recently genital herpes.

HSV-2: genital herpes, and possibly cervical cancer.

Varicella-Zoster: Varicella (chicken pox) in children, herpes zoster (shingles) in adults and in immunocompromised cancer and organ transplant patients.

Cytomegalovirus: Congenital CNS infections. In adolescents and adults an infectious-mononucleosis-like syndrome. In immunocompromised cancer and organ transplant patients it is a major cause of death.

Epstein Barr Virus: Infectious mononucleosis, nasopharyngeal cancer, Burkitt's Lymphoma.

The development of a herpes simplex virus vaccine is under investigation. Inactivated virus cannot be used because of potential oncogenicity. The use of protein-subunits, free of DNA has been evaluated, but they lack potency. An interesting approach, discussed in a workshop at the NIH concerned with genital herpes vaccines (8), was the use of recombinant herpesvirus, free of those genes believed to be responsible for oncogenicity or the induction and maintenance of latency.

In the absence of an effective vaccine for interaction with the target virus, we are dependent upon the development of antiviral drugs. Ideally the drug should exert its antiviral activity as a consequence of interaction with a target that is unique to the virus, and hence of little or no concern to the uninfected host tissues. The interaction could consist of an inhibitory event unique for the virus, or of a preferential activation of the drug in the virus infected cell.

What are some of the targets that one can consider as potential sites for antiviral drug development? What are some of the strategies that one might employ to take advantage of the availability of these targets? What are some of the approaches that one might take to develop antiviral agents for specific interaction with known target sites, as well as those target sites yet to be uncovered as a consequence of future acquired knowledge of the biochemistry of viral replication, the physiology of host cell-virus interrelationships, and the physical chemistry of drug-target interactions.

At this time we know very few targets that we have been able to take advantage of, and so it is not surprising that we have as of today very few clinically useful drugs. Only 5 drugs have been approved for use in humans by the Food and Drug Administration in the United States, and these include: (1) amantadine (Adamantanamine) for prevention or very early treatment of infections caused by the influenza A viruses, (2) 1-(2-deoxy-β-D-ribofuranosyl)-5-iodouracil (5-iodo-2'-deoxyuridine, Idoxuridine, IdUrd), (3) 1-(2-deoxy-β-D-ribofuranosyl)-5-trifluoromethyluracil (5-trifluoromethyl-2'-deoxyuridine, Trifluridine, F_3dThd), (4) 9-β-D-arabinofuranosyl-9H-purine-6-amine (Vidarabine, adenine arabinoside, ara-A) for topical therapy of herpetic keratitis, for systemic therapy of herpes encephalitis and pending is FDA approval for neonatal herpes, and (5) 9-(2-hydroxyethoxymethyl)guanine (Acyclovir, ACV) for topical therapy of primary genital herpes and non-life threatening mucocutaneous herpes, as well as for intravenous administration for therapy of severe cases of initial genital herpes and cutaneous and mucosal herpes-simplex (HSV-1 and HSV-2) infections in immunocompromised patients.

In Europe 5-ethyl-2'-deoxyuridine (Aedurid) and 5-iodo-2'-deoxycytidine (Cebe-Viran) are available for therapy of herpes keratitis.

A number of compounds which are in clinical trial, or look promising, or are of interest include E-5-(2-bromovinyl)-2'-deoxyuridine (BVdU), 1-(2-deoxy-2-fluoro-β-D-arabinofuranosyl)-5-iodocytosine (FIAC), phosphonoformate (foscarnet, PFA), 5-methoxymethyl-2'-deoxyuridine, 5-n-propyl-2'-deoxyuridine, (S)-9-(2,3-dihydroxypropyl)adenine ((S)-DHPA), 9-(1,3-dihydroxy-2-propoxymethyl)guanine (DHPG, BIOLF-62, 2'-NDG), 2-deoxyglucose, L-lysine, arildone, ultrasound + "Hepergon", enviroxime, zinviroxime, dichloroflavin, isoprinosine, interferon, etc.

Table 1. Viral Targets

---

1. Extracellular
   A. Antibody
   B. Enzymes
   C. Chemicals
      1. Organic Solvents
      2. Detergents
      3. Formaldehyde

2. Adsorbtion of virus to cell surface

3. Transport across cell wall

4. Uncoating of virus

5. Transport of genome into cytoplasm or nucleus

6. Transcription (RNA formation)

7. Methylation of RNA

8. Protein synthesis or processing

9. Enzymes

10. Maturation (assembly of macromolecules into a virion)

11. Release of virion from cell

Table 1 indicates various targets for which antiviral drugs have been demonstrated to exert an effect, or are potential targets for future drug development. We have already discussed the use of vaccines which interact with the target virus, when it is present in the blood stream, or during the phase of reversible adsorption to the cell prior to penetration.

Extracellular Target

The virus prior to penetration into the host animal is a target for inactivation by a variety of biological, chemical and physical agents and some of these have been of value in preparation of inactivated vaccines (9).

Proteolytic enzymes can decrease the infectivity of enveloped viruses by removal of their glycoprotein spikes, and phospholipases can inactivate by hydrolysis of phospholipids present in such viruses. Steele and Black (10) found 2-thiouracil decreased the infectivity of polio virus by a direct reaction of the oxidized form of this compound with the capsid sulfhydryl groups.

Chemical agents which inactivate the virus directly include kethoxal (3-ethoxy-2-oxobutyraldehyde), calcium elenolate, and certain dihydroisoquinolines references are cited by Smith et al. (11). The first two exert an antiviral effect on both RNA and DNA viruses and the latter against the influenza and parainfluenza viruses, but none have proven to be clinically useful. Retinoids (12-14), long-chain unsaturated monogylcerides and alcohols (15) also inhibit viral replication by direct interaction.

Anionic detergents (sodium dodecyl sulfate) and nonionic detergents (Triton X-100, etc.) inactivate viruses by solubilizing the viral envelope. Compounds such as guanidine, urea or phenol dissociate the polypeptides of the viral capsids into their individual components. Formaldehyde interacts with amino groups of nucleic acid bases as well as free amino moieties of proteins.

Some viruses such as myxovirus and oncogenic RNA virus, are sensitive to heat inactivation, whereas many others are not. X-ray, gamma-ray and ultraviolet radiations also may produce a loss of infectivity. Whereas X-ray and gamma-ray radiations produce single stranded breaks in the DNA, ultraviolet radiations produce butane dimers between DNA pyrimidines.

Viral Attachment and/or Penetration of Virus as Targets

Reproduction of a virus requires that the virus adsorb to the cell membrane and penetrate into the cell. This involves an initial reversible attachment of the virus to a receptor on the host cell membrane, which quickly undergoes a change such that the virus can no longer be dissociated from the cell by mild procedures. The nature of the conversion from reversible to irreversible binding is not clear, but it may involve an increase

in the fluidity of the cell membrane, which allows other binding
sites on the virion to react with additional cell receptors.

Different viruses may interact with various specific sites
on the cell surface, however, unrelated viruses may also share
the same cellular binding site. The site of attachment may be a
glycoprotein, a glycolipid or contain functional sulfhydryl
groups. The virion may have a specific protein projection that
is involved in the adsorption phenomenon, or a mosaic of several
capsid proteins may be involved. Presumably the initial
interaction is electrostatic and serves to orient the virus-cell
association for subsequent penetration or transport across the
cell membrane into the cytoplasm.

Although specific binding sites for viruses have been
demonstrated in cell membranes, the subsequent physiochemical
events concerned with transport across the cell membrane are
unknown. A major question is whether or not the transport of the
virus involves specific recognition of a virus protein by a host-
cell membrane macromolecule. If specific protein-protein
interactions are involved, can this be exploited as a target?
The virus after attachment may enter the cell by phagocytosis, by
fusion of the viral envelope with the plasma membrane, or by both
procedures. Other methods proposed are entry of the virus
through breaks in the plasma membrane or direct passage through
the membrane.

Various negatively charged polysaccharides such as heparin
prevent or decrease the rate of attachment of the virus to the
host cell, presumably by forming a complex with the virus (15).
Derivatives of _trans_-decalin as well as 3,4-dihydro-1-
iosquinoline acetamide-HCl prevent or delay the penetration of
virus into the cell. References are cited in Smith et al (11).

Before some viruses such as the myxovirus (influenza) or the
paramyxoviruses (measles virus), can penetrate into a host cell,
a virus protein must be activated by a host cell protease. Those
cells that do not have an activating protease are not susceptible
to infection. The specific proteolytic cleavage is required for
fusion of the viral and cell membranes, and hence the cellular
protease represents a target site. Thus a cellular protease
cleaves a large glycoprotein (F_0) of the paramyxovirus to form
the smaller virion glycoproteins (F_1) and (F_2) which are bonded
by a disulfide linkage. The myxovirus protein HA is similarly
cleaved to HA_1 and HA_2. The virions can also be activated by
incubation with trypsin. A new N-terminus on the F_1 polypeptide
is created, which has a highly hydrophobic amino acid sequence,
and this is believed to be involved in the interaction with a
receptor on the cell membrane (16-18).

Richardson et al. (18) has synthesized various oligopeptides resembling this hydrophobic region, and carbobenzoxy-D-Phe-L-Phe-Gly-D-Ala-D-Val-D-Ile-Gly and carbobenzoxy-D-phenylanine-L-phenylalanine-glycine, produced potent inhibition of the measles virus with an ED_{50} of 0.02 and 0.2 uM, respectively. Similarly the oligopeptide carbobenzoxy-glycine-L-leucine-L-phenylalanine-glycine produced good inhibition of the influenza A virus with an ED_{50} of 20 uM. These oligopeptides presumably compete with the N-terminus of the viral polypeptide and thereby prevent viral penetration. If the cell has no activating protease, then it is resistant to the virus infection. The consequence of inhibition of this cellular enzyme to the host in vivo remains to be determined.

Dubovi et al. (19) found several aromatic mono- and diamidines significantly retarded the penetration of the respiratory syncytial virus (a paramyxovirus) into the host cell. Bis(5-amidino-2-benzimidazolyl)methane inhibited or retarded viral penentration, but not the attachment of the respiratory syncytial virus to the host cell (19,20). Thus these compounds decrease the yield from multiple growth cycles (19). Two possible mechanisms were postulated by Dubovi et al. (19). One suggests a strong affinity for a specific binding site on the cell membrane in competition with the virus. The amidines have a strong cationic as well as a hydrophobic region which may afford such an interaction. The second hypothesis suggests these compounds function as potent protease inhibitors. Thus the protease target may be an activator of plasminogen which could be concerned with alteration of either the virus envelope or the host cell membrane.

Of interest is a recent report by Kantorovich Prokudina et al. (21) which described the ability of some protease inhibitors isolated from natural products (soy, potato, bean, honey, locust) to inhibit the replication of influenza A virus in cell culture. Whether a protease is the specific target or whether some other target is involved is yet to be clarified.

Penetration and/or Uncoating of Virus as Targets

Herpesvirus type 1 grown in the presence of 2-deoxyglucose was found by Spivack et al. (22) to have decreased infectivity as a result of a defect in its ability to penetrate into the cell or to uncoat. There was only a slight decrease relative to control virions for attachment to the host cells. Since the DNA from virions grown in the presence or absence of 2-deoxyglucose were equally infective, all steps subsequent to uncoating of the virus would appear to be unaffected. Although viral penetration or uncoating is prevented, the target site for deoxyglucose is that of glycoprotein synthesis, and this will be discussed in more detail later.

Amantadine is a drug approved for the prevention or early treatment of respiratory infections caused by the influenza A virus. This drug prevents penetration or uncoating and the target site for this compound has been postulated to be:

1. An interaction with membranes, thereby altering their surface charge.

2. An interaction with the M-protein which is coded by the virus and is required for penetration or uncoating of the virus.

3. An increase in the pH of lysosomes by amantadine, which is a primary amine, thereby preventing the lysosomal enzymes from uncoating the virus.

Intracellular Events as Targets

Once the virus has penetrated into the cell and its nucleic acid released from the virion, a vast array of events occur that are frequently unique to a specific virus. Our knowledge of these intracellular events is in the exponential phase of acquisition, and will provide future targets for antiviral drug development.

The released nucleic acid may either remain in the cytoplasm or will migrate into the nucleus of the cell, as with herpesvirus DNA. How is the nucleic acid protected from enzymic degradation after release? How is herpesvirus DNA transported into the nucleus? Will interference with stabilization of viral DNA transport constitute a viable target? If one can preferentially destroy the uncoated viral genome, then replication of the virus would obviously be prevented.

Another possible target could be the interaction of the viral but not the cellular genome with a tight binding or covalently bound substance, the consequence of which is interference or prevention of genome replication, or transcription, or both. The substance would be required to recognize only the viral genome. Can this be accomplished by the synthesis of sequence-specific non-polar oligonucleotides as described by Ts'O (23) for inhibition of neoplastic cells? He achieved a decrease in polarity of the oligonucleotides by conversion of the phosphodiester linkages between nucleosides to phosphotriesters, which contrary to general opinion are apparently quite stable. His laboratory (Miller et al. (24)) reported that tritium labeled oligodeoxyribonucleoside methyl phosphonates, with a chain length of up to nine nucleotidyl units, are taken up by intact mammalian cells in culture by passive diffusion. When [^{3}H] labeled d(Tp)$_8$ was incubated with mammalian cells, about 70 percent of the labeled thymidine was still associated with the intact oligomer 18 hours later. However, it is slowly degraded as evidenced by the formation of labeled dTTP and DNA-thymine. Colony formation was inhibited by this oligomer, but the molecular basis for the inhibition has not been ascertained as yet.

Bardos (see reference 25 for review) has synthesized partially thiolated polynucleotides as structural analogs of DNA and RNA polymerase templates for potential anticancer activity. Chandra and colleagues (26) have reported that poly C, containing 9% 5-mercaptocytidylate units had promising responses in the therapy of acute lymphocytic leukemia in children. Bardos and Ho (25) report that HSV-DNA polymerase was strongly inhibited by partially thiolated poly (U) with an I$_{50}$ of 15 uM. The target is postulated to be the inhibition of the dissociation of the enzyme-template complex.

Thus the potential for selectively targeting the viral DNA or RNA may indeed be a feasible approach. In fact Stebbing and Lindley (27) report that a mixture of single-stranded polynucleotides [(poly (I) and poly (5-hydroxymethyl-Cyd)] suppressed development of antiviral antibodies in mice infected with Semliki Forest virus. Although the basis for the antiviral effect is unknown, the inhibition was not by induction of interferon or by affecting the immune response.

Round and Stebbing (28) found the single-stranded polynucleotide copolymer of cytidine (9%) and 4-thiouridine (91%) not only inhibits the influenza virus transcriptase, but also had anti-influenza effects in experimental animals. The production of interferon or an antibody response was excluded, however stimulation of the cellular immune response is a possiblity under investigation. A more detailed account of antiviral effects of single stranded polynucleotide has been presented by Stebbing (29).

The synthesis of macromolecules required for eventual assembly into a virion may depend on the synthesis of viral specific enzymes, as well as unique structural proteins and nucleic acids. Interference with the synthesis of any of these virus-specific macromolecules would be a logical target for antiviral chemotherapy.

RNA-Synthesis as a Target

There are a number of sites that one could target to prevent the formation of RNA, or to increase its rate of degradation, or to produce non-functional RNA. Obviously, a major concern is selectivity of inhibition, since it would be of limited value to interfere simultaneously with RNA synthesis that is essential for the viability of an uninfected cell. However, in therapy of life-threatening viral infections, a certain amount of toxicity is acceptable. The risk-benefit ratio must always be considered in the use of any drug.

Can one interfere selectively with RNA biosynthesis? We indicated earlier that one might be able to synthesize a small oligonucleotide for selective hybridization with the viral template, and if the dissociation of the complex were low, then little or no viral m-RNA could be formed. At the present stage this approach is only hypothetical.

Another potential target may be related to the possible involvement of a viral specified protein in the control of gene expressions by splicing polycistronic primary transcripts, thereby allowing coding regions to be brought together into the same reading frame.

Another approach to decrease the availability of viral mRNA is to increase the rate of RNA degradation. This may be achieved with interferon or interferon inducers. One of the actions of interferon is degradation of m-RNA. Interferon refers to a family of proteins termed alpha (leucocyte), beta (fibroblast) and gamma (immune) and sub-groups also exist. Interferons have a broad spectrum of antiviral activity and are secreted by mammalian cells in response to viruses and a wide variety of substances both naturally occurring and synthetic. The secreted interferon binds to specific receptors on the cell surface and the events which follow are responsible for the creation of the antiviral state. The amino acid sequences of some interferons have been elucidated, and this should enable the preparation of synthetic interferons with high antiviral activity, but yet with little or no cytotoxicity.

One of the effects seen in extracts of interferon-treated cells is the activation of an enzyme which degrades m-RNA. This is brought about by the following series of events. A (2'-5')poly-A_N synthetase [ATP: (2'-5')oligo(A) adenyltransferase] is activated by a small amount of double-stranded RNA generated by the virus, and this converts ATP into a series of (2'-5')oligo adenylates with a triphosphate at the 5' end. This compound binds to a latent endoribonuclease which is thereby activated. The enzyme cleaves single-stranded RNA. The activation is reversible since on removal of (2'-5')poly A_N the enzyme reverts to the latent state. Degradation of viral m-RNA results in an inhibition of protein synthesis.

The 5'-dephosphorylated (core) 2'-5' oligo poly $(A)_3$ has been reported by Doetsch et al (30) and Henderson et al. (31) to replace human interferon in preventing the transformation by the Epstein Barr virus of human lymphocytes. These findings encourage the synthesis of analogs of 2'-5'poly(A). Analogs of (2'-5')poly (A) have been synthesized from 3'-deoxy ATP, rather than from ATP, and the 5'-triphosphate trimer of 3'-dATP, $(2'-5')ppp3'-dA(p3'dA)_2$, effectively inhibits translation, and is resistant to enzymic hydrolysis. Whether analogs of this type will prove to have clinical utility remains to be seen. Appropriate references include that by Ho (32), Dziewanowska and Pestka (33), Lengyl (34), DeClercq (35) and Pollard (36a).

Although the early transcripts upon infection with herpesvirus are produced from the input viral DNA, late RNA is transcribed from the newly formed DNA. Interference with the fidelity of the base sequence of the newly synthesized viral DNA will affect the formation of late RNA, and thus constitutes a viable target. Specificity is achieved by use of antiviral agents that are activated only in the virus-infected cell and are incorporated into the DNA or RNA.

Thus Otto and Prusoff (unpublished) found early transcription was not affected when HSV-1 was grown in the presence of 5-iodo-2'-deoxyuridine (IdUrd), 5-iodo-5'-amino-2',5'-dideoxyuridine (AIU, AIdUrd) or 5'-amino-5'-deoxythymidine (AdThd). Although total accumulation of [^{3}H]uridine into RNA in HSV-1 infected Vero cells was not significantly affected, the amount of HSV-1 specific poly A^+ RNA declined about 8 hours post infection, and became more pronounced at 18 to 24 hours after infection, relative to that observed in the infected cells in the absence of these compounds. These alterations in late HSV-1 specific RNA may be a consequence of the substitution of IdUrd, or AIdUrd, or AdThd for thymidine in the newly synthesized DNA, or an inhibition of polyadenylation by the nucleotide analogs. The consequence of producing such changes in RNA will be reflected in the proteins that are transcribed.

Methylation of RNA as a Target

Most viral and cellular mRNA contain a guanylate at the 5'-terminus which is methylated in the N-7 position by the enzyme guanine-7-methyltransferase. The methyl group is derived from S-adenosylmethionine which after donation of the methyl group forms S-adenosylhomocysteine which is then cleaved by S-adenosylhomocysteine hydrolase. Several antiviral compounds such as 9-β-D-arabinosyladenine (ara-A), (S)-9-(2,3-dihydroxypropyladenine, 3-deazaadenosine and the carbocyclic analog of 3'-deazaadenosine (36b) inhibit this hydrolytic enzyme. The consequence is an accumulation of S-adenosylhomocysteine which is a potent inhibitor of S-adenosylmethionine-dependent methylation reactions, such as of viral m-RNA (Fig. 1).

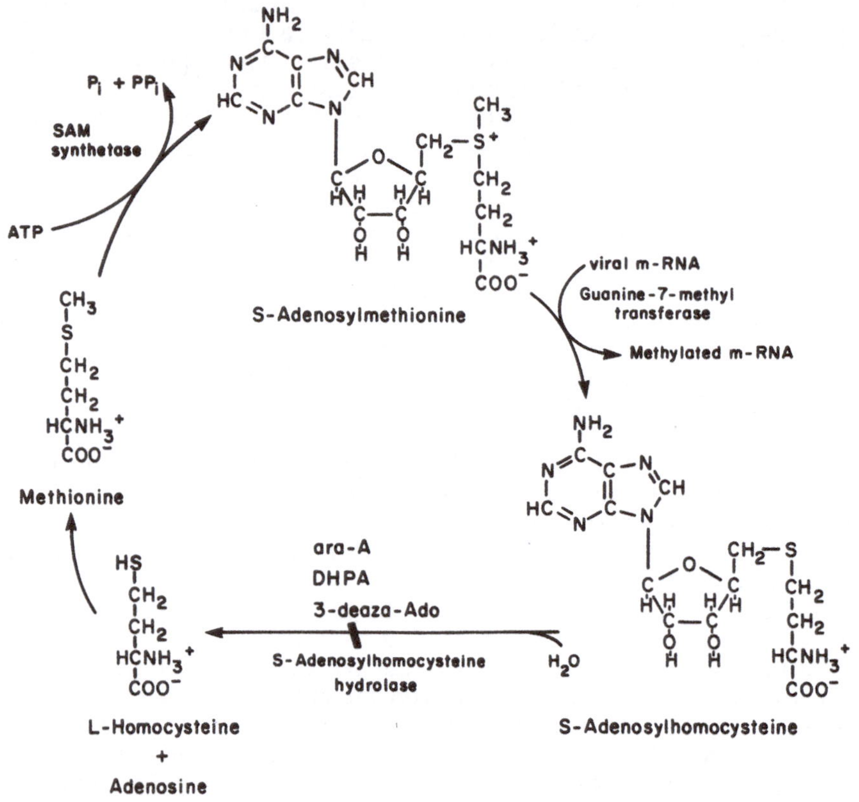

Fig. 1.   Inhibition of S-adenysylhomocysteine hydrolase.

1-β-D-Ribofuranosyl-1,2,4-triazole-3-carboxamide (ribavirin) is a triazole nucleoside, and X-ray crystallography revealed striking similarities to guanosine and inosine. It is of particular interest because it has a broad spectrum of antiviral activity against both RNA and DNA viruses. In addition to the monophosphate analog exerting an inhibition of IMP dehydrogenase, thereby decreasing the de novo formation of GTP and dGTP, the triphosphate analog also inhibits mRNA guanyltransferase. Thus the initial step in the capping of mRNA is inhibited, that is the transfer of GMP from GTP to the 5'-terminus of viral mRNA. The probability of this site of inhibition being a good target for antiviral chemotherapy has been a concern, since this reaction also appears to be essential for the capping of cellular mRNA also.

A recent report by Goswami et al (37) states that although the 2'-5' linked oligo A inhibits both viral and cellular mRNA (guanine-7-)-methyltransferase, the 3'-O-methylated analog, whether methylated in the 3'-terminal-OH, or methylated at all three 3'-OH groups, inhibited specifically the vaccinia viral mRNA with a K_i of 4 uM. These findings of specificity support the methylation reaction to be a viable target. The next step is to prepare derivatives that are permeable to the infected cell.

Transport of Nuclear RNA as a Target

Whereas mRNA transcripts formed in the cytoplasm have ready access to ribosomes, that formed from viral genomes present in the nucleus must first be transported across the nuclear membrane. Is there a difference in the transport of cellular- and viral mRNA that one can exploit as a target?

Protein Synthesis or Processing as a Target

Protein synthesis can occur in two stages. Early translation depends on mRNA transcribed form the input viral genome, whereas later translation is dependent upon progeny nucleic acid and hence can be affected by factors which do not impact on the early events.

A major target of interferon is an inhibition of protein synthesis. In addition to the activation of an endonuclease in extracts of interferon-treated cells which degrades mRNA and hence inhibits protein synthesis, there is an activation of a protein kinase eIF_2 which results also in an inhibition of protein synthesis. The activated protein kinase catalyzes the phosphorylation of the alpha subunit of the initiation factor eIF_2 thereby inhibiting the initiation of protein synthesis. Thus translation of viral mRNA in these cells would also be inhibited (38,39). The basis for specificity of viral versus cellular m-RNA is not clear.

Isatin-β-thiosemicarbazide inhibits late protein synthesis in poxvirus infected cells. Most late proteins are synthesized in the presences of IBT, but cleavage of the precursor for one of

the main virus structural polypeptides concerned with virus
maturation was inhibited (50). Specifically the integration of
polypeptides 4a and 4b into high density DNA-protein complex was
missing (51,52).

Carrasco and Smith (40) have found that the membranes of
virus-infected cells are altered both structurally and
functionally, and under these conditions allows the transport of
compounds into the cell that are impermeable to normal cells.
Such compounds should not exceed 1,500 daltons in molecular
weight and should be highly ionic to prevent uptake into normal
cells. Thus Carrasco (41) found that the GTP methylene analogue,
$GppCH_2p$, inhibited protein synthesis in the virus-infected cell
but not in normal cells. Membrane "leakiness" is produced by a
number of viruses, including: picornavirus, togavirus,
papovavirus, herpesvirus, rhabdovirus, and paramyxovirus.
Whereas most viruses produced membrane changes late in infection,
Carrasco and Esteben (42) found that the vaccinia virus produced
these changes during the first hour of viral adsorption.
However, Dawson et al. (43) were not able to find such an effect
with this protein inhibitor in encephalomyocarditis virus
infected cells or animals. Thus taking advantage of virus
induced membrane leakiness of the infected cell may be limited to
viruses which affect membrane permeability early in the infection
process.

Another potential target is selective binding of
macromolecules critically involved in protein synthesis. For
example, the studies of Honess and Roizman (44) indicate
formation and regulation of HSV α, β, and γ proteins in a
sequential and interdependent manner. Alpha-polypeptides are
required for synthesis of beta-polypeptides, which include
thymidine kinase and DNA-polymerase. Gamma-polypeptides in turn
shut off the synthesis of beta-proteins. Therefore, can we
develop a low molecular weight, stable polypeptide that can
specifically interact with the alpha-protein receptor, thereby
preventing essential viral protein synthesis? Gamma-proteins are
required to shut off synthesis of beta-proteins; hence a
substance that mimics gamma-proteins would also prevent formation
of the essential herpesvirus enzyme-DNA polymerase.

Glycoprotein Synthesis as a Target

RNA and DNA viruses, which have envelopes surrounding the
viral capsid, contain one or more glycoproteins as integral
membrane proteins with the carbohydrate-containing portion
exposed at the external surface. These glycoproteins may have 10
to 15 monosaccharides in either N- or O-glycosidic linkage to a
polypeptide asparagine, serine or threonine residue. Host cell

NATO ASI Series
Advanced Science Institutes Series

A series presenting the results of activities sponsored by the NATO Science Committee, which aims at the dissemination of advanced scientific and technological knowledge, with a view to strengthening links between scientific communities.

The series is published by an international board of publishers in conjunction with the NATO Scientific Affairs Division

A	Life Sciences	Plenum Publishing Corporation
B	Physics	New York and London
C	Mathematical and Physical Sciences	D. Reidel Publishing Company Dordrecht, Boston, and Lancaster
D	Behavioral and Social Sciences	Martinus Nijhoff Publishers
E	Engineering and Materials Sciences	The Hague, Boston, and Lancaster
F	Computer and Systems Sciences	Springer-Verlag
G	Ecological Sciences	Berlin, Heidelberg, New York, and Tokyo

Recent Volumes in this Series

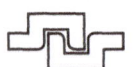

Series A: Life Sciences

glycosyl transferases are involved in the formation of these glycoproteins.

Several compounds are known which affect glycosylation reactions essential for the formation of viral glycoproteins, and the consequence is either an inhibition in the formation of virions or the formation of virions with decreased infectivity. Antiviral compounds which affect the synthesis of glycoproteins include 2-deoxy-D-glucose, D-glucosamine and tunicamycin.

Various sites of inhibition of glycosylation reactions have been reviewed by Schwarz and Datema (53) and by Schlotissek (54). Deoxyglucose is an analog of glucose and is utilized in substitution of glucose in the synthesis of guanosine diphosphate glucose and the formed GDP-deoxyglucose binds to the lipid carrier dolichol pyrophosphate. This event interferes with the synthesis of lipid-linked oligosaccharide intermediates that participate in the formation of N-glycosidically linked glycoproteins. These lipid linked intermediates, containing oligosaccharides of decreased molecular weight, are not transferred to proteins (50). In addition, UDP-deoxyglucose and deoxygluconic acid are also formed.

The virions produced in the presence of 2-deoxyglucose have a decreased infectivity and Spivack et al (55) have shown that the defect is not due to an inability to adsorb to the host cell, but rather involves the process of penetration or uncoating. Infectivity of virus grown in the presence of 2-deoxyglucose could be significantly enhanced after viral adsorption, by treating cell monolayers with polyethylene glycol to promote viral penetration. The viral DNA present in herpesvirus virions formed in the presence of an inhibitory concentration of 2-deoxyglucose is not defective, since it had equal ability as the DNA derived from control cells to transfect Vero cells (55).

Tunicamycin, a glucosamine-containing antibiotic, blocks the transfer of N-acetylglucosamine-1-phosphate from uridine diphosphate N-acetylglucosamine to dolichol monophosphate (56) and hence inhibits glycoprotein synthesis also.

Processing of Proteins as a Target

The processing of certain viral proteins by proteolytic cleavage is a critical event for many viruses (orthomyxovirus, paramyxovirus, picornavirus, togavirus). Thus for example poliovirus infected cells treated with protease inhibitors cannot process the high molecular weight polypeptides properly. Both cellular and viral encoded proteases are involved in this process. Thus inhibition of the viral protease would decrease

the rate of processing the polycistronic protein, and this would
constitute a viable target. Korant and Longberg-Holm (45) found
a protease coded by a picornavirus to be critical in the control
of replication of this virus. This enzyme is required for the
production of virus structural proteins, and regulates the viral
RNA replicase in the infected cell. This obviously would make a
good target (Fig. 2).

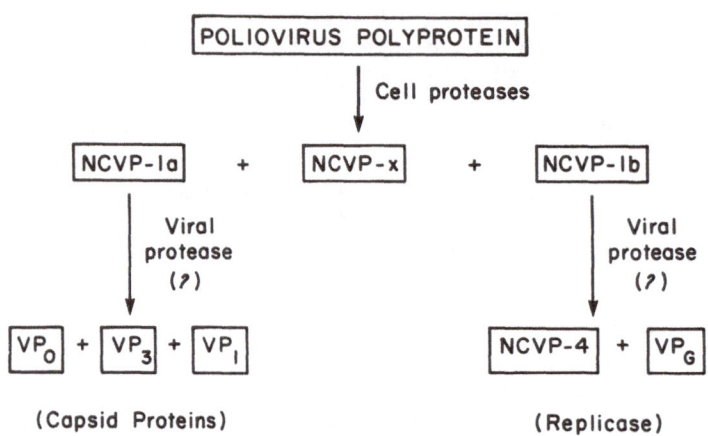

Fig. 2.   Proteolytic Processing of Poliovirus Protein

    Bis-benzimidazolamidine derivatives have been reported by
Tidwell et al (46) to inhibit argenine-specific esterproteases.
1,2-Bis(5-methoxy-2-benzimidazol-2-yl)-1,2-ethanediol has been
reported to have inhibitory activity against rhinovirus infection
both in vitro by Roderick et al (47), and in vivo by Shipkowitz
et al (48) but at toxic levels. However, inhibition of a
protease, as the basis for the antiviral activity was not
established. See reference by Eggers (49) for a review of this
area.

Enzymes as Targets

    A major target for development of antiviral agents is based
on the virus introducing into the infected cell enzymes that are
encoded by the viral genome (see review by Kit (57). These

unique enzymes may be associated with the virion or induced during the infective process. The viral encoded enzyme may catalyze reactions which are not only unique to the virus-infected cell, but also essential for viral replication. These enzymes are good targets for selective antiviral chemotherapy.

The RNA transcriptase of the influenza and parainfluenza viruses, the RNA replicase of the enteroviruses and rhinoviruses, and the reverse transcriptase of the RNA oncogenic viruses are examples of enzymes with unique activity. All negative-stranded viruses, such as the influenza virus, contain within the virion a RNA dependent RNA polymerase (Transcriptase) which is required for the synthesis of early viral m-RNA.

Retroviruses contain within the virion an RNA-dependent DNA polymerase which is required for the synthesis of minus-strand DNA from the RNA genome. The RNA of the DNA-RNA hybrid is then removed by a ribonuclease (RNAase H), and a plus strand of DNA is made forming a linear duplex DNA which migrates from the cytoplasm into the nucleus where it is converted into a circular form prior to integration into the cellular DNA.

In addition there are a number of viral encoded enzymes which catalyze reactions that normally occur in the uninfected cell. These enzymes may be sufficiently different from the comparable enzyme in the uninfected cell that selective targeting is possible. Differences between the viral and cellular enzymes may be found in migration during polyacrylamide gel electrophoresis, in isoelectric point, in immunological properties, in sensitivity to inhibition, in ability to utilize substrates, or in the binding affinity for the same substrate, etc.

Table 2 was prepared by Cheng et al (58) and indicates the various enzymes encoded by members of the herpesvirus family. These are all potential targets.

Table 2   Herpesvirus Enzymes*

	HSV-1	HSV-2	VZV	CMV	EBV
Thymidine Kinase	+	+	+	−	−
DNA Polymerase	+	+	+	+	+
DNase	+	+	+	+	+
Uracil-DNA-Glycosylase	+	+	?	?	?
dUTPase	+	+	?	?	?
Ribonucleotide Reductase	+	+	?	?	+

*With permission from the authors (58) and the publisher.

Although an increase in enzyme activity may occur during infection, one cannot assume that it is viral encoded. For example, marked increases in thymidine kinase activity are induced by SV40, polyoma and adenoviruses, but PAGE and isoelectric focusing analysis suggest these are cellular enzymes.

Nevertheless an increase in the activity of a particular enzyme that is restricted to the infected cell could be exploited as a target for preferential accumulation of an antiviral agent.

Thymidine kinase

Thymidine kinase specified by the herpesvirus genome is critical for the activation of a number of antiviral agents with diverse structures (Fig 3), which are not substrates for cellular thymidine kinase. The virus induced thymidine kinase differs from the comparable host cell enzyme in molecular weight, substrate specificity, electrophoretic mobility, isoelectric point and immunological properties (57).

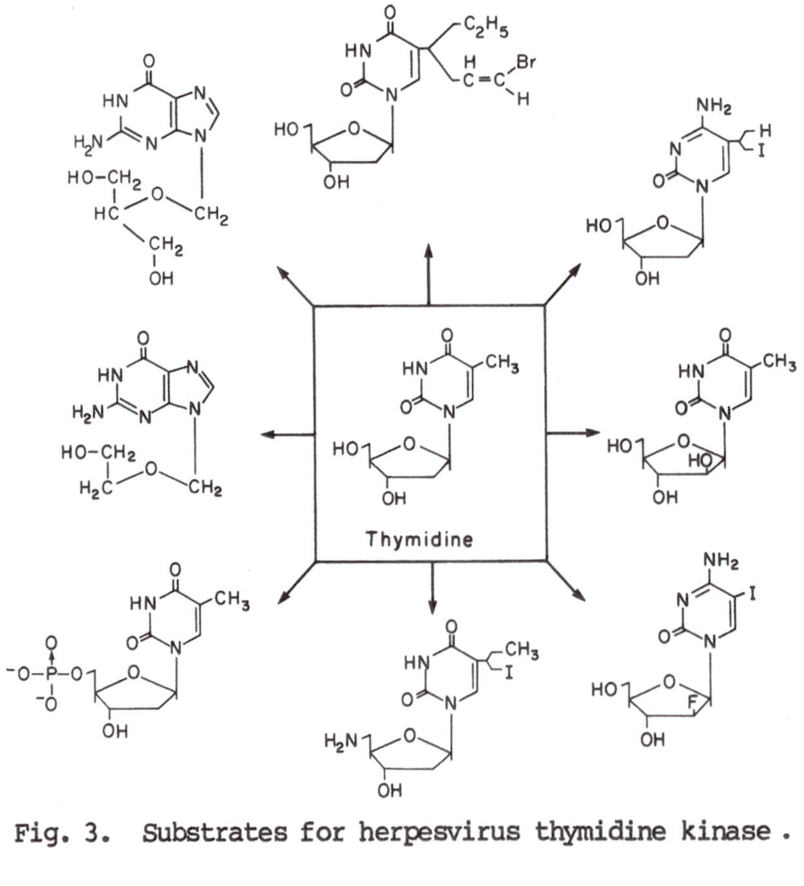

Fig. 3. Substrates for herpesvirus thymidine kinase .

The HSV-1 enzyme catalyzes the phosphorylation of not only thymidine, but also thymidine-5'-phosphate (59), deoxycytidine (60) as well a host of nucleoside analogs which as seen in Fig 3, include modifications such as (1) replacement of a pyrimidine by a purine as in acyclovir, (2) replacement of the methyl moiety of dThd by ethyl (ETdUrd) or propyl (PrdUrd), or 2-bromovinyl (BVdU), (3) replacement of the deoxyribose by an arabinose as in ara-T and FIAC, (4) replacement of deoxyribose by an acyclic component as in acyclovir, or (5) replacement of the 5'-OH by an amino group as in AIdUrd and AdThd. These unique biochemical properties make this enzyme a prime target for selective activation of antiviral agents in the HSV-1, HSV-2 or varicella-zoster virus infected cell.

Falke et al (61) have described the induction by HSV-1 of an enzyme which phosphorylates thymidine but utilizes AMP as the phosphate donor. This enzyme is termed adenylic acid: deoxythymidine 5'-phosphotransferase, and in addition there may be an ADP-dependent phosphotransferase (62). The relationship among these enzymes as well as their physiological role is yet to be elucidated. Their respective potential as targets for drug activation is not clear.

The basis for the therapeutic selectivity of IdUrd and F_3dThd in the treatment of herpetic keratitis is the marked increase in thymidine kinase activity in the viral infected cells. These analogs are phosphorylated by both cellular and viral thymidine kinases; however the viral enzyme has a 10-fold greater binding affinity for dThd, IdUrd and F_3dThd, and hence the viral infected cells trap more IdUrd or F_3dThd as the phosphorylated derivatives than the non-infected cells. These are further phosphorylated to the triphosphate derivatives and are incorporated into the DNA. There is a direct correlation between the incorporation of IdUrd into HSV-1 DNA and loss of infectivity.

Acyclovir (9-(2-hydroxyethoxymethyl)guanine) is used clinically for topical therapy of primary genital herpes and for non-life threatening mucocutaneous herpes, and intravenous therapy of severe initial genital herpes and initial or recurrent mucocutaneous herpes in immunocompromised children and adults. The molecular basis for its selective antiviral activity is the preferential phosphorylation by the viral encoded thymidine kinase, and the subsequent conversion by cellular enzymes to the triphosphate derivative (ACV-TP). The triphosphate analog may either preferentially inhibit the HSV-DNA polymerase, for which it has about a 100-fold greater binding affinity than for dGTP, or it may be incorporated into the viral DNA and function as a chain terminator. In addition the ACV-substituted DNA template may form a complex with the viral DNA polymerase thereby inactivating the enzyme. Thus the target for activation of ACV is the viral thymidine kinase, and the target for inhibition is termination of viral DNA biosynthesis by either inhibiting DNA polymerase, or terminal incorporation into DNA, or both.

There are a number of other compounds such as AdThd, AIdUrd, BrVdUrd, FIAC, 5-ETdUrd, 5-PrdUrd, ara-T, DHPG, which are uniquely or preferentially activated by phosphorylation in the HSV-infected cell. The proceedings of a recent symposium on Acyclovir has been published (63).

DNA Polymerase

All the members of the herpesvirus family induce a unique DNA polymerase which should be a unique target since this enzyme is an absolute requirement for replication of these viruses. Phosphonoformate (Foscarnet) has been recently reviewed by Oberg (64) and is in clinical trial against herpes infections. This agent is unique in that it does not have to be activated before attacking its target which is the herpesvirus DNA polymerase. There is a direct interaction with the pyrophosphate binding site on the herpesvirus DNA polymerase. This compound also inhibits the influenza virus RNA polymerase, reverse transcriptase, herpesvirus 3'-5'-exonuclease and the hepatitis B-DNA polymerase.

Ara-A is used clinically for topical therapy of herpetic keratitis and systemically for therapy of herpetic encephalitis. The prime target site for ara-A is more complex since we know a number of biochemical sites of inhibition (Table 3).

Table 3   Ara A-Targets

1. Inhibition of HSV-DNA polymerase
2. Incorporation into viral and cellular DNA
3. DNA elongation slowed
4. Inhibition of ribonucleoside diphosphate reductase
5. Inhibition of m-RNA polyadenylation
6. Inhibition of methylation of m-RNA
7. Inhibition of RNA-dependent RNA polymerase
8. Inhibition of terminal deoxynucleotidyl transferase

The basis for the selectivity of ara-A is not clear. There is no evidence for the viral induction of a new kinase for preferential or increased phosphorylation in the virus infected cell. The binding affinity of ara-ATP to the herpesvirus encoded DNA polymerase is greater than that to cellular DNA polymerase, and this is considered by some to be the prime target site. However it is incorporated into viral as well as cellular DNA, and the consequence of this is yet to be clarified.

Ribonucleotide Reductase

Ribonucleotide reductase is induced by HSV-1, HSV-2 and the Epstein Barr virus (65,66). The viral induced enzyme differs from the corresponding cellular enzyme in a lack of a Mg^{2+} requirement as well as being resistant to inhibition by dTTP and dATP. Whereas the cellular ribonucleotide reductase has an absolute requirement for ATP, the viral induced enzyme is inhibited by ATP (66). Evidence was obtained recently by Dutia (67) using temperature sensitive mutants that the HSV-1 induced ribonucleotide reductase is at least partially encoded by the herpesvirus.

Thiosemicarbazone derivatives have been known to have antiviral activity for over 30 years (68, 69). A recent study by Shipman et al (70) describes the antiviral activity of 2-acetylpyridine-thiosemicarbazone derivatives against HSV-1 and HSV-2. The biological activity of various other thiosemicarbazone derivatives (Purines, pyridine, isoquinoline, and isatin derivatives) have been evaluated, and references to these studies have been cited by Shipman et al (70).

Cavrini et al (71) synthesized thiosemicarbazones of hydroaramatic ketones and found that of 2,3-dihydro-4- (1H)-phenthrenone to be the most active against vaccinia and parainfluenza viruses. Brockman et al (72) found a good correlation between the inhibition by a series of heterocyclic thiosemicarbazones of ribonucleotide reductase and the formation of infectious herpesvirus virions.

Isatin-β-thiosemicarbazone (IBT) is a potent inhibitor of poxvirus, and although a thiosemicarbazone derivative, the target site is believed to be late protein synthesis, however the ultimate event responsible is not clear.

Protein Kinase

HSV-1 contains an envelope-associated cyclic AMP-independent protein kinase (73-75). Adenovirus, vaccinia viruses as well as many other enveloped viruses have been reported to have virion-associated protein kinases (references cited by Blue and Stobbs (75)). It is not clear whether the virus induced a host cell enzyme not normally produced in the uninfected cell, or whether the enzyme is a virus-induced alteration of a host cell protein kinase, or is a viral encoded enzyme.

An attractive role for this enzyme is an involvement in the
sequential synthesis of alpha, beta and gamma HSV-proteins, which
if confirmed would make this enzyme a good target for antiviral
chemotherapy.

Assembly and Release of Virion as a Target

The macromolecules synthesized during the infective process
must interact to form a mature virion. The molecular basis for
this assembly process is not completely understood, but should
provide a good target.

The proteins are synthesized in the cytoplasm and must be
transported into the nucleus since the formation of the
nucleocapsids of herpesviruses, adenoviruses and papovaviruses
takes place in the nucleus. The herpesvirus nucleocapside is
enveloped while passing through the nuclear membrane which has
been modified by incorporation of glycosylated herpesvirus-
specified proteins.

Other viruses may acquire their envelopes by passing through
(budding) the cytoplasmic or plasma membrane.

Those viruses that form their nucleocapsid in the cytoplasm
obtain their envelope by budding through a modified portion of
the cell membrane which has incorporated viral glycoproteins.
Thus the viral envelope contains host-cell lipid, but only viral-
specified proteins.

Perhaps once we acquire some understanding of the forces
directing these events we can exploit them as targets.

Conclusion

As indicated earlier serendipity was primarily responsible
for development of the antiviral drugs which we have today. The
targets for their antiviral activity were elucidated
subsequently. Now that we know a number of appropriate targets,
how can we exploit this knowledge for the development of
selective antiviral drugs or use the available drugs more
effectively?

With the exception of amantadine, the clinically approved
drugs are targeted against only certain HSV-1 and HSV-2
infections. We are in desperate need to find exploitable targets
for treatment or prevention of recurrent genital herpes. Is the
latent viral genome responsible for recurrent genital herpes a
viable target? What are the factors that control the rate of

recurrences? If a polypeptide, for example, were responsible for maintenance of the quiescent state by binding to the integrated DNA, then one could envision the administration of a tight binding polypeptide or oligonucleotide to a specific sequence of the herpesvirus DNA. Thus the question is "can we target the molecular basis which triggers the transformation of the quiescent state into a recurrence? We also must find targets to attack the rhinoviruses responsibile for respiratory infections and infections of the lower gastro-intestinal tract. There is also a great need to find exploitable targets in order to treat infections produced by the cytomegalovirus, Epstein-Barr virus, hepatitis B virus, and adenovirus. Other viral infections for which exploitable targets would be welcome include severe enteroviral infections caused by the coxsackie and echo viruses, and acute respiratory disease in small children and old people caused by adenovirus, parainfluenza virus and respiratory syncitial virus.

There is a need to find other target enzymes for activation of drugs other than by herpesvirus thymidine kinase, as well as for antiviral drugs that do not become incorporated in DNA as a primary or even as a secondary target.

As our molecular understanding of virus replication and virus-host interrelations increase, new targets for antiviral chemotherapy will no doubt evolve. In addition to new targets, we have problems as to how to best deliver the drug to the target sites. What is the potential of combination therapy to attack several target sites and should this be done simultaneously or sequentially. Thus it is very clear that although we have accomplished a great deal in the past few decades, we still have a long way to go before we can more effectively utilize not only the established targets, but also those that are today merely hypothetical.

Acknowledgement

The research emanating from our laboratory was supported by Grants CA-05262, T32-CA-09085 and F32-CA-06723 from the National Cancer Institute, United States Public Health Service.

REFERENCES

1. Black, F.L. (1979) Pharmac. Ther. 6 221-234.
2. Hillemann, M.R. reference cited by Maugh II, T.H. (1981) Science 214 1113.

3. Kleid, D.G., Yansura, D., Small, B., Dowbenko, D., Moore, D.M., Grubman, M.J., McKercher, P.D., Morgan, D.O., Robertson, B.H., and Bachrach, H.L. (1981) Science 214 1125-1129.

4. Bittle, J.L., Houghten, R.A., Alexander, H., Shinnick, T.M., Sutcliffe, J.G., Lerner, R., Rowlands, D.J. and Brown, F. (1982) Nature 298 30-33.

5. Arnon, R. (1980) Ann. Rev. Microbiol 34 593-618.

6. Baron, M.H. and Baltimore, D. (1982) J. Virology 43 969-978.

7. Lerner, R.A., Green, N., Alexander, H., Liu, F.-T., Sutcliffe, J.G. and Shinnick, T.M. (1981) Proc. Natl. Acad. Sci. 78 3403-3407.

8. Allen, W.P. and Rapp, F. (1982) J. Infect. Dis. 145 413-421.

9. Joklik, W.K. (1980) "Prinicples of Animal Virology" Appleton-Century-Crofts, New York.

10. Steele, F.M. and Black, F.L. (1967) J. Virol. 1 653-658.

11. Sands, J., Auperin, D. and Snipes (1979) Antimicrob. Ag. Chemother. 15 67-73.

11a. Smith, R.A., Sidwell, R.W. and Robins, R.K. (1980) Ann. Rev. Pharmacol. Toxicol. 20 259-284.

12. Yamamoto, N., Bister, K., zurHausen, H. (1979) Nature 278 553-554.

13. Zeng, Y., Zhou, H.M. and Xu, S.P. (1981) Intervirology 16 29-32.

14. Reinhardt, A., Sands, A. and Snipes, W. (1979) Antimicrob. Agents Chemother. 16 421-423.

15. Vaheri, A. (1964) Acta Pathol. Microbiol. Scand., Suppl. 171 1-98.

16. Choppin, P.W., Richardson, C.D., Merz, D.C., Hall, W.W. and Scheid, A. (1981) J. Infect. Dis. 143 352-363.

17. Choppin, P.W. Richardson, C.D., Merz, D.C. and Scheid, A. (1980) In Ciba Found. Symp. May 13-15, p. 252-269.

18. Richardson, C.D., Scheid, A. and Choppin, P.W., (1980) Virology 105 205-222.

19. Dubovi, E.J., Geratz, J.D., Shaver, S.R. and Tidwell (1981) Antimicrob. Agents and Chemother. 19 649-656.

20. Dubovi, E.J., Geratz, J.D. and Tidwell, R.R. (1980) Virology 103 502-505.

21. Kantorovich Prokudina, E.N., Semenova, H.P., Berezina, O.N. and Mosolov, V.V. (1982) Vopr. Virusol. 27 452-456.

22. Spivack, J.G., Prusoff, W.H. and Trtton, T. (1982) Virology 123 123-138.

23. Ts'o, P.O.P. (1983) In "The Development of Target-Oriented Anticancer Drugs" Eds. Y.C. Cheng, B. Goz and M. Minkoff, Raven Press. In Press.

24. Miller, P.S., McParland, K.B., Jayaramen, K. and Ts'o, P.O.P. (1981) Biochemistry 20 1874-1880.

25. Bardos, T.J. and Ho, Y.-K. (1982) In "New Approaches to the Design of Antineoplastic Agents" Eds. T.J. Bardos and T.I. Kalman. Elsevier Biomedical New York, Amsterdam, Oxford, pg. 315-332.

26. Chandra, P., Steel, L.K., Ebener, U., Woltersdorf, M., Laube, M., Kornhuber, B., Mildner, B. and Goetz, E. (1977) Pharmac. Ther. A. 1 231-287.

27. Stebbing, N. and Lindley, I.J.D. (1980) Arch. Virol 64 57-66.

28. Round, E.M. and Stebbing, N. (1981) Antiviral Res. 1 237-248.

29. Stebbing, N. (1979) Pharmacol. Ther. A. 6 291-332.

30. Doetsch, P., Suhadolnik, R.J., Sawada, Y., Mosca, J.D., Flick, M.B., Reichenbach, N.L., Dang, A.Q., Wu, J.M., Charubala, R., Pfleiderer, W. and Henderson, E.E. (1981) Proc. Natl. Acad. Sci. 78 6699-6703.

31. Henderson, E.E., Doetsch, P.W., Charubala, R., Pfleiderer, W. and Suhadolnik, R.J. (1982) Virology 122 198-201.

32. Ho, M. (1982) Pharmacol Reviews 34 119-129.

33. Dziewanowska, Z.E. and Pestka, S. (1982) Med. Res. Reviews 2 325-353.

34. Lengyl, P. (1982) Ann. Rev. Biochem. 51 251-282.

35. DeClercq, E. (1980) Antibiotics Chemother 27 251-287.

36a. Pollard, R.B. (1982) Drugs 23 37-55.

36b. DeClercq, E. and Montgomery, J.A. (1983) Antiviral Res. 3 17-24.

37. Goswami, B.G., Crea, R., VanBoom, J.H. and Sharma, O.K. (1982) J. Biol. Chem. 257 6867-6870.

38. Farrel, P.J., Sen, G.C., Dubois, M.-F., Ratner, L., Slattery, E. and Lengyl, P. (1978) Proc. Natl. Acad. Sci. 75 5893-5897.

39. Samuel, C.E. and Knutson, G.S. (1982) J. Biol. Chem. 257 11791-11795.

40. Carrasco, L. and Smith, A.G. (1980) Pharmacol. Therap. 9 311-355.

41. Carrasco, L. (1978) Nature 272 694-699.

42. Carrasco, L. and Esteben, M. (1982) Virology 117 62-69.

43. Dawson, K.M., Stewart, A. and Stebbing, N. (1979) J. Gen. Virol. 45 237-240.

44. Honess, R.W. and Roizman, B. (1974) J. Virol. 14 8-19.

45. Korant, B.D. and Longberg-Holm, K.K. (1981) Acta Biol. Med. Ger. 40 1481-1488.

46. Tidwell, R.R., Geratz, J.D., Dann, O., Volz, G., Zeh, D. and Loewe, H. (1978) J. Med. Chem. 21 613-623.

47. Roderick, W.R., Nordeen, C.W., Von Esch, A.M. and Appel, R.N. (1972) J. Med. Chem. 15 655-658.

48. Shipkowitz, N.L., Bower, R.R., Schleicher, J.B., Aquino, F., Appell, R.N. and Roderick, R.R. (1972) Appl. Microbiol. 23 117-122.

49. Eggers, H.J. (1982) In "Chemotherapy of Viral Infections", Eds. R.E. Came and L.A. Caliguiri, pp. 377-417, Springer Verlag Berlin, Heidelberg, New York.

50. Katz, E., Margalith, E., Winer, B. and Goldbaum, N. (1973) Antimicrob. Ag. and Chemother. 4 44-48.

51. Katz, E., Margalith, E., and Winer, B. (1978) J. Gen. Virol. 40 695-699.

52. Cooper, J.A., Moss, B. and Katz, E. (1979) Virology 96 381-392.

53. Schwarz, R.J. and Datema, R. (1980) Trends in Biol. Sci. 5 65-67.

54. Schlotissek, C. (1975) Curr Top. Microbiol. Immunol. 70 101-119.

55. Spivack, J.D., Prusoff, W.H. and Tritton, T.R. (1982) Virology 123 123-138.

56. Struck, D.K. and Lennarz, W.J. (1980), In "Biochemistry of Glycoproteins and Proteoglycans" Ed. W.J. Lennarz, Plenum, New York, p. 35.

57. Kit, S. (1979) Pharmacol. Ther. 4 501-585.

58. Cheng, Y.-C., Nakayama, K., Derse, D., Bastow, K., Ruth, J., Tan, R.S., Dutschman, G., Caradonna, S.J. and Grill, S. (1982) In "Herpesvirus, Clinical, Pharmacological and Basic Aspects" Eds. H. Shiota, Y.-C. Cheng and W.H. Prusoff, Excerpta Medica, Amsterdam-Oxford-Princeton, pg. 47-56.

59. Chen, M.S. and Prusoff, W.H. (1978) J. Biol. Chem. 23 1325-1327.

60. Jamieson, A.T., Gentry, G.A. and Subak-Sharpe, J.H. (1974) J. Gen. Virol. 24 465-480.

61. Falke, D., Nehbass, W., Brauer, D. and Mueller, W.E.G. (1981) J. Gen. Virol. 53 247-255.

62. Labenz, J., Friedrich, D., Muller, W.E.G. and Falke, D. (1983) Intervirol. 19 77-84.

63. "Proceedings of a Symposium on Acyclovir" Eds. D.H. King and G. Galasso (1982) American J. Medicine 73 No. 1A, pg. 1-392.

64. Oberg, B. (1983) Pharmac. Ther. 19 387-415.

65. Ponce deLeon, M.R., Eisenberg, R.J. and Cohen, G.H. (1977) J. Gen. Virol. 36 163-173.

66. Huszar, D. and Bacchetti, S. (1981) J. Gen. Virol. 37 580-588.

67. Dutia, B.M. (1983) J. Gen. Virol. 64 513-521.

68. Hamre, D., Bernstein, J. and Donovick, R. (1950) Proc. Soc. Exp. Biol. Med. 73 275-278.

69. Bauer, D.J. (1955) Brit. J. Exp. Pathol. 36 105-114.

70. Shipman, Jr., C., Smith, S.H., Drach, J.C. and Klayman, D.L. (1981) Antimicrob. Ag. and Chemother. 19 682-685.

71. Cavrini, V., Drusiani, A.M., Gatti, R., Giovanninetti, G., Peissi, L., Francki, L. and Nanetti, A. (1981) Arch. Pharm. (Weiheim) 314 372-376.

72. Brockman, R.W., Sidwell, R.W., Arnett, G. and Shaddix, S. (1970) Proc. Soc. Exp. Biol. Med. 133 609-614.

73. Lemaster, S. and Roizman, B. (1980) J. Virology 35 798-811.
74. Rubenstein, A.S., Gravell, M. and Darlington, R. (1972) Virology 50 287-290.
75. Blue, W.T. and Stobbs, D.G. (1981) J. Virology 38 383-388.

HERPESVIRUS TARGET CONSIDERATIONS FOR THE DESIGN OF ANTIVIRAL AGENTS

Fred Rapp and Brian Wigdahl

Department of Microbiology and Cancer Research Center
The Pennsylvania State University College of Medicine
Hershey, Pennsylvania, 17033 USA

INTRODUCTION

The herpesviruses, herpes simplex virus (HSV) type 1 (HSV-1), HSV type 2 (HSV-2), varicella-zoster virus (VZV), cytomegalovirus (CMV), and Epstein-Barr virus (EBV) can cause mild to severe diseases in humans.[1,2] In addition to causing clinically apparent disease due to lytic virus replication, this group of viruses has been shown to be harbored in the human population in persistent and latent forms.[3-6] Furthermore, several members have also been implicated in the etiology of certain human neoplasias.[7,8] The herpesviruses contain a double-stranded DNA genome enclosed in an icosahedral capsid structure surrounded by a lipid envelope, and replicate virus DNA in the cell nucleus. Although the herpesviruses have several biological characteristics in common, the individual viruses differ in multiple parameters. A partial list is shown in Table 1.

In general, the primary objective in the design of antiviral agents is to obtain compounds that effectively inhibit lytic virus replication without detriment to uninfected cells. This has, and should remain an important end in the design of chemotherapeutic agents for use in the treatment of herpesvirus infections. However, it has become increasingly evident that consideration should be given to therapy of latent herpesvirus infections as a means to control herpetic disease. In addition, because herpesviruses have been implicated in the causation of human cancers,[7,8] the potential role of latent virus infection in the oncogenic process and the effect of chemotherapy on this process must also be assessed.

Table 1. Variations in the Biological and
          Physical Properties of Herpesviruses

Genome structure
Virion structure
Functions encoded by genome
Control of genome expression
Host cell target for productive replication
Site of persistence or latency
Sensitivity to inhibition by antiviral
    agents

In the past decade, a multitude of information has been
collected on compounds with varying degrees of efficacy in
inhibiting herpesvirus replication in general and HSV in particular. In initial investigations, the most promising antiherpesviral agents were obtained by screening compounds originally
synthesized for treatment of human neoplasias. 5-Iodo-2'-
deoxyuridine (iodoxuridine, 5-iodo-dUrd) and 5-trifluoromethyl-
2'-deoxyuridine (trifluridine, 5-CF$_3$-dUrd), compounds used in
the past for the clinical management of HSV-induced ocular
keratitis[9,10] and still used for topical treatment of certain
ocular infections, are prime examples of compounds originally
synthesized as antineoplastic agents that possess antiviral
potency. Although 5-iodo-dUrd and 5-CF$_3$-dUrd are licensed for
use in the topical treatment of HSV-1-induced ocular keratitis
and have demonstrated varying degrees of antiviral effectiveness,
both compounds once phosphorylated to the nucleoside 5'-triphos-
phate are incorporated into the DNA of uninfected as well as
infected cells.[11,12] Therefore, little, if any, antiviral
specificity could be expected from these or other similar
derivatives unless they are phosphorylated specifically by the
HSV-induced deoxythymidine kinase present in the infected cell.
With present treatment regimens using 5-iodo-dUrd and 5-CF$_3$-dUrd
this does not appear to be the case. Therefore, these compounds
are fairly cytotoxic and their antiviral activity primarily
results from their inhibitory effect on host cell DNA synthesis.
Although 5-CF$_3$-dUrd is more efficacious for topical treatment
of HSV-1 keratitis than 5-iodo-dUrd[13,14] and less toxic during
ocular therapy than expected from in vitro cytotoxicity deter-
minations, the safety of this drug has been questioned.[15]

The basic underlying problem confronted during the development of antiviral agents or more specifically antiherpesvirus

agents is the strict dependence of the virus replicative processess on host cell metabolic pathways. However, in recent years as the herpesvirus replication cycles have been analyzed in increasing detail, it has become evident that there are both qualitative and quantitative differences between the metabolic state of uninfected and herpesvirus-infected cells. Perhaps the most extensively studied herpesvirus has been HSV. A variety of virus-coded HSV-specific functions have been identified and characterized[15,16] including thymidine (dThd)/deoxycytidine (dCyd) kinase, ribonucleotide diphosphate reductase, DNA polymerase, 3' to 5'-exodeoxyribonuclease, and deoxyuridine-5'-triphosphate pyrophosphatase. These individual virus-induced enzymes are either absent from the uninfected cell or are basically different from their host cell counterparts. The enzymes most exploited to date are the HSV-induced dThd/dCyd kinase and DNA polymerase. Due to the wide substrate specificity range of these two enzymes, nucleoside and nucleoside-5'-triphosphate analogs that are not recognized as substrates for the host cell dThd/dCyd kinases or DNA polymerases, respectively, would therefore be restricted to phosphorylation and DNA incorporation in the virus-infected cell. The final result, being inhibition of HSV replication without impeding uninfected cell metabolism.

The selective virus-specific approach to chemotherapy of HSV infections, exemplified by 9-(2-hydroxyethoxymethyl)-guanine (ACV, acyclovir), (E)-5-(2-bromovinyl)-2'-deoxyuridine (BVDU, bromovinyl deoxyuridine), and 1-(2-fluoro-2-deoxy-β-D-arabino-furanosyl)-5-iodocytosine (FIAC, fluoroiodoaracytosine), has reoriented antiviral chemotherapy and should lead to the development of even more potent and selective anti-HSV compounds. In addition, as more information is obtained on the virus-specific replicative events of other members of the herpesvirus group, a site-specific approach to the design of antiviral agents can be extended to these viruses.

Prior to a discussion of the potential targets for the design of specific herpesvirus antiviral agents it is essential to briefly review the virus-host cell interactions commonly associated with herpesvirus infections: (i) productive replication, (ii) persistence and latency, and (iii) virus-induced cell transformation. Since detailed reviews have been compiled on each of these three aspects, this discussion will only provide a framework to discuss targets already utilized, in addition to potential targets. Because information on each of the three cell-virus interactions are not equally well understood for HSV, VZV, CMV, and EBV, discussion will be limited to those systems most thoroughly documented.

PRODUCTIVE REPLICATION

The most thoroughly documented cell-virus interaction associated with herpesvirus infection is the productive or lytic virus replication cycle. In general, this process involves adsorption of the herpesvirus to host cell, followed by penetration, uncoating, early transcription and translation of virus-specific genes and mRNAs, DNA replication, late virus-specific transcription and translation, assembly and eggression of mature progeny virus with subsequent death of host cell. The availability of numerous permissive host cell systems, the relatively short virus replication cycle, and an accurate method to quantitate infectious virus production has promoted extensive investigations of the HSV productive replicative cycle. Recent advances in biochemical and molecular biological technologies have extended these studies into detailed analyses of HSV DNA, RNA, and protein structure and functions. The utilization of HSV as a model system for the regulation of herpesvirus productive replication has aided in understanding the replicative cycles of other members of the herpesvirus group.

The genomes of human herpesviruses range in size from approximately 80×10^6 to 150×10^6 daltons. The linear double-stranded DNA genomes of HSV-1 and HSV-2 have approximate size of 100×10^6 daltons[17] and average base compositions of 67% and 69% guanine plus cytosine,[17] respectively. Several characteristic structural features are common to the genomes of most herpesviruses. These include: (i) terminal repetitions that allow the genome to circularize, (ii) discontinuities in the sugar-phosphate backbone of each strand of the duplex DNA, and (iii) internal repeated sequences.[18] Under relatively stringent hybridization conditions HSV-1 and HSV-2 DNAs share 47 to 50% of their sequences with good base pair matching,[19] whereas the homology of HSV with other human herpesviruses amounts to only a few percent of their sequences with relatively poor matching of base pairs. Specifically, the HSV genomes consist of two covalently linked components, designated L and S, that comprise 82% and 18% of the virus DNA respectively.[20] Each component consists of predominantly unique sequences (U_L or U_S) bracketed by inverted repeats.[21] Another property common to many of the herpesvirus genomes is isomerization of genome subcomponents. The HSV L and S components can invert relative to each other.[22,23] Therefore, DNA extracted from virions consists of four populations of DNA molecules differing only in the relative orientation of the two components. [22,23] The relationship between HSV genome subcomponent isomerization and virus DNA replication has been discussed previously in great detail.[18]

Approximately 50 HSV-infected cell polypeptides (ICPs) have been identified to date, ranging in size from 12,000 to 220,000 in molecular weight.[24,25] They have been classified into at least three groups, designated α, β, and γ or immediate-early, early, and late, respectively.[26] Their synthesis is coordinately regulated and ordered in a sequential cascade fashion. The α polypeptide synthesis reaches maximal rates 2-4 hr postinfection, and to date, four α or immediate-early genes 4,0,22, and 27 have been identified.[27] α Polypeptides are subsequently phosphorylated and transported into the infected cell nucleus.[28] Functional α polypeptides are required for activation of transcription of β genes, ultimately resulting in the synthesis of β polypeptides. β Polypeptides are synthesized maximally from 5-7 hr postinfection and include enzymes involved in virus DNA synthesis, such as dThd/dCyd kinase and DNA polymerase. In turn, one or more functional β polypeptides are required for the cessation of synthesis of α polypeptides and activation of the expression of γ genes. The synthesis of γ polypeptides occurs between 12 and 17 hr after infection and includes mostly virion structural polypeptides.[26]

In addition to synthesis of virus-specific components involved directly or indirectly in the synthesis of progeny virus, HSV shuts off host cell DNA[29] and protein synthesis[30] during the course of infection and reduces host cell RNA synthesis.[31] The inhibition of host cell protein synthesis apparently occurs in two discrete steps. The first step, possibly occurring prior to transcription of the parental genome, is most probably mediated by a nonstructural protein.[32] The second step requires the synthesis of virus proteins and coincides approximately with the synthesis of the β polypeptides.[26,33] With the onset of virus DNA and late protein synthesis, assembly of infectious progeny virus particles is initiated. In contrast to HSV, CMV does not code for its own dThd/dCyd kinase activity although it does code for its own DNA polymerase.[34] However, CMV does stimulate cellular DNA[35] and RNA[36] and protein[37] synthesis as well as cellular RNA polymerase,[38] dThd kinase,[39,40] and ornithine decarboxylase.[41] In addition to containing DNA and numerous structural polypeptides (15 to 33 in number), the completed HSV virion contains several additional components common to most other human herpesviruses: (i) glycosylated proteins, (ii) lipids, (iii) polyamines, and (iv) enzymatic activities.

LATENCY

In contrast to the productive replicative cycle, which ultimately results in cell death, herpesviruses can also be maintained in a noninfectious or latent form in the nervous system and other tissues of asymptomatic individuals. Latent

VZV most probably resides in the dorsal root ganglia.[41] The
primary infection, varicella or chickenpox is usually a benign
infection occurring in childhood. In some individuals, the
virus reactivates resulting in zoster or shingles. Although
VZV infection is usually a self-limited disease, it can be
life-threatening in neonates and immunocompromised persons,
such as children with leukemia.[42] CMV can remain latent, most
probably in lymphocytes,[43] with reactivation occurring predominantly during immunosuppressive therapy, as is the case in
renal transplant patients, certain cancers, and during immmuno-
suppressive cancer therapy.[43] However, the predominance of
information on characterization of latent herpesvirus infection
of the lymphoid or nervous system relates to EBV and HSV.

It has been thoroughly documented that latent EBV infections of the lymphoid system preferentially involve B-lympho-
cytes.[44] Perhaps the most informative results on the mechanism
of EBV latency and reactivation have been obtained from analyses
of several EBV genome-containing lymphoblastoid cell lines.
Experimental results obtained with Raji cells,[45] a lymphoblastoid
cell line derived from a Burkitt's tumor, indicate that this
cell line contains approximately 50 genome equivalents of
virus DNA, with most of the virus DNA present as closed,
circular molecules and a minor portion possibly covalently
associated with host cell DNA.[46,47] The retention of the
complete EBV genome in at least a subfraction of the Raji cell
population is suggested by the induction of infectious virus
in a small percentage of the cells after treatment with 5-bromo-
deoxyuridine.[48] Subsequent studies with EBV-containing lymphoblastoid cell lines derived from normal individuals, infectious
mononucleosis patients, or tumors suggest that the virus DNA
was present in a predominantly nonintegrated form, with a
minor integrated species suggested.[46,47] This indicates that
both DNA forms may be essential for the establishment of the
latent state as well as for its maintenance. However, an
originally non-EBV-containing B-cell line (Ramos) infected in
vitro with EBV, was shown to carry one integrated EBV genome
equivalent per cell.[49] In addition, human cord blood lymphocytes infected in vitro and then serially propagated carry
10-15 nonintegrated EBV genome equivalents of DNA.[50]

The expression of the EBV genome during in vitro virus
latency in several lymphoblastoid cell lines has also been
investigated. From 10-30% of the virus DNA was transcribed,
with about 5% of this represented in the polysomal fraction.[51]
In the Raji and Namalwa cell lines, the polyadenylated and
polyribosomal fractions were enriched for the same class of
EBV RNA and were coded for by about 5% of the EBV genome.[52]
This result contrasted with that of productively infected cells,
where all the transcripts were polyadenylated and appeared in

the polyribosomal fraction, with 45% of the genome being transcribed.[52] It appears that at least two of the translation
products produced during virus latency are Epstein-Barr nuclear
antigen (EBNA) and lymphocyte-detected membrane antigen.
However, the role these proteins play in the establishment or
maintenance of latency is not known.

Results with somatic cell hybrids suggest that reactivation involves both positive and negative control mechanisms.
A positive control mechanism was indicated by the fact that
fusion of a nonproducer B-cell line with a producer line
resulted in a producer heterokaryon. However, fusion of a
generally nonproducer cell line (Raji) with the Daudi or
Jijoye cell lines (lines in which some cells replicate virus
DNA) resulted in a block in virus DNA synthesis,[45] suggesting
a functional negative control mechanism. Establishment and
maintenance of the latent state may also be modulated by host
cell factors, as suggested by synchronization of latent virus
DNA replication with that of the host cell[53] and by the fact
that different sources of lymphocytes, when transformed by the
same substrain of virus, express virus-specific genetic information to different extents.[45]

HSV is also responsible for recurrent infections in man,
most often as self-limited, periodic, cutaneous, oral or
genital mucosal infections. However, the virus can cause
serious generalized infections in newborns or meningitis,
myelitis or encephalitis in adults.[54] Although HSV is capable
of lytic replication in a wide variety of cell types in vitro
and in vivo, the sensory and autonomic ganglia of animals and
man are the sites where the virus persists in a latent form.[3-6]
In experimental animals inoculated at external sites, the
virus spreads within axons to the dorsal or autonomic ganglion.[3-6]
After productive infection subsides, the chronic or latent
stage is established with infectious HSV demonstrated only by
explantation of ganglia or by cocultivation with suitable
indicator cells.[3-6] As shown by infectious center analysis,
the fraction of ganglion cells containing virus decreases from
1% during the acute phase to 0.1% during the latent phase in a
mouse model system.[55] Two basic theories regarding the state
of the HSV genome during virus latency, static or dynamic,[6]
have been proposed. During static latency the virus genome is
thought to be maintained in a nonexpressive form, while in
dynamic latency the virus genome is expressed at a low level
continuously[6,56] with infectious virus produced at levels that
are below detection limits.

Recently, with the advent of more sensitive techniques,
several investigators have examined acute and latent HSV
infections in vivo to determine the level of virus-specific

DNA and RNA present in cells. During the acute phase, HSV was
recovered from mouse ganglionic homogenates, and virus DNA was
detected at 1.2-2.0 genome equivalents per cell and at 0.11
genome equivalents per cell after establishment of the latent
phase using ^{125}I-labeled virus DNA as probe.[57] Virus-specific
RNA was detected at 0.1-0.2 copies per cell during the acute
phase. However, during the latent phase less than 0.002
copies per cell were detected in the ganglia.[57] Galloway and
coworkers[58] have examined the amount of RNA present in paravertebral ganglia removed from humans at autopsy by DNA-RNA in
situ hybridization using [^{3}H]HSV DNA probe. Their results
suggest that 0.4-0.8% of the neurons present in the ganglia
expressed HSV RNA. These results suggest that many latently
infected ganglion cells contain HSV DNA with only a few of
these cells expressing virus genes, or that only some neurons
are preferentially infected. In both instances, it would
appear that the latent state involves a virus-host interaction
in which complete expression of the HSV genome is repressed by
some undefined mechanism. However, it is possible that the
HSV RNA detected in the ganglia subsequent to death resulted
from some biological consequence of death rather than latency-
specific HSV transcription. More recently, Galloway and
coworkers[59] have examined human paravertebral ganglia at
autopsy for virus-specific sequences by in situ hybridization,
using as probes HSV-2 restriction endonuclease DNA fragments
purified from agarose gels after electrophoretic separation
and recombinant plasmid DNAs containing HSV-2 DNA inserts.
Virus-specific RNA from the left-hand 30% of the virus genome
was detected in all HSV-2-positive samples. RNA homologous to
other DNA sequences from the L component were present at a
reduced frequency. No virus-specific RNA from the S component
was detected. These results suggest that differential transcription of the HSV genome occurs in latently infected human
ganglion cells.

Studies in rabbit and guinea pig model systems by Tenser
and coworkers[60] indicate that HSV thymidine kinase (TK) expression may be necessary to establish latent ganglionic infection.
Additional studies by Fong and Scriba[61] using [^{125}I]deoxycytidine
as a specific probe detected TK activity during latent infection.
However, it is possible that TK is required for establishment
of latency but not for maintenance of the latent state.

Indirect immunofluorescence studies by Green et al.[62]
using monospecific antiserum to VP175 have suggested the
presence of HSV VP175 or ICP-4 in trigeminal ganglia of latently
infected rabbits. VP175 is a nonstructural polypeptide of the
α group and is synthesized at maximal rates 3-4 hr postinfection.
The protein has DNA-binding properties[63] and it has been

suggested that VP175 may be required continuously to maintain
early protein synthesis and may act to inhibit immediate early
or α protein synthesis.[64] In light of the putative regulatory
properties of VP175 during productive virus replication,
speculation on the functional role of this protein in the
establishment and/or maintenance of HSV neuronal latency may
be warranted.

Stevens and coworkers have utilized HSV temperature-sensitive
(ts) mutants[65,66] that are restricted at 38°C (normal mouse
temperature) in cell culture systems to screen for their
capacity to establish latent infections in the central and
peripheral nervous systems in mice. Several points can be
made from these studies. (i) The mutants can be catagorically
divided into two groups that differ in their ability to establish
latent infections. (ii) Both DNA[+] and DNA[-] mutants can establish
latent infections, suggesting that virus DNA replication is
not required for establishment of latent infection. (iii)
HSV-specified proteins in addition to the immediate early (α)
polypeptides are required to establish latent virus infections.
(iv) Genes involved in establishment and maintenance of latency
are not clustered on the genetic or physical map.

Sokawa and coworkers[67] have shown that the activity of
interferon (type I) and that of an interferon-induced enzyme
that catalyzes the formation of an unusual oligoadenylate with
2'-5' linkage from adenosine triphosphate in the presence of
double-stranded RNA (2',5'-oligoadenylate synthetase) parallel
virus synthesis during the acute phase of HSV infection in
mouse trigeminal ganglia subsequent to cheek skin inoculation.
The interferon and 2',5'-oligoadenylate synthetase appeared
shortly after initiation of virus replication and their activity
declined with the disappearance of virus from tissue. In
contrast, antibody against HSV appeared in the sera at a time
when the amounts of virus and the synthetase were declining
and the interferon activity had already diminished. Although
unconfirmed, this result may indicate that interferon and the
immune system play a role in the establishment of virus latency.
In addition, it has been demonstrated that cell-mediated
immunity plays a role in the resistance of mice to infection
with HSV.[68] However, passively transferred serum anti-HSV
antibody was also effective in preventing the spread of HSV.[69]
Additional information is required to define the role of the
immune system in the modulation of latency.

Perhaps central to the issue of HSV neuronal latency is
the form of the virus genome during latent infection. Several
forms of the virus genome during latency have been proposed
(Table 2).

Table 2.   State of Latent Virus Genome

Complete or incomplete virion or nucleocapsid structure

Nonintegrated genome
    Nuclear or cytoplasmic
    Linear
    Concatameric
    Circular
    Nucleosomal
    Methylated
    Independent L and S genome components

Integrated genome
    Single site
    Multiple sites
    Random or specific sites
    Independent L and S genome components

These include complete or incomplete virion structures, with
uncoating and virus gene expression inhibited by some unknown
mechanism; free nucleic acid located in either the nucleus or
cytoplasm, with virus genome present as a linear or plasmid DNA
form or as a concatameric DNA molecule; virus genome integrated
in a specific site or multiple sites in chromosomes of latently
infected neurons; or a combination of these alternatives.
Recent studies by Rock and Fraser[70] have demonstrated by DNA
blot hybridization that most, if not all, of the HSV-1 genome
was present in central nervous tisse of latently infected mice.
In addition, although the HSV-1 junction DNA fragment could be
detected, authentic HSV-1 terminal DNA fragments could not be
detected in total mouse brain or pooled trigeminal ganglia DNA,
suggesting a nonintegrated plasmid form of DNA.   In contrast,
Fraser and coworkers[71] have detected HSV-1 terminal and junction
DNA fragments of authentic size in human brain DNA suggesting
that the HSV-1 genome was in a nonintegrated linear form in at
least a portion of the specimens examined.

     In addition to direct studies on human tissue and experi-
mental animal models, numerous investigators have developed in
vitro cell culture systems that have enabled studies of the
maintenance and expression of virus genetic information in
persistent or latent HSV infections.[72-74]   In vitro systems to
study HSV latency often eliminate or minimize important modulating factors that control virus replication, such as serum
antibody,[69] cell-mediated immunity,[68] interferon,[75] and hormones.[76]   However, their innate simplicity provides a means to

analyze the intracellular molecular mechanism of HSV latency
and reactivation. These studies coupled with the results
obtained with animal model systems may provide insights into
HSV latency in humans.

In vitro latency models have utilized primary rodent neuronal
cells, transformed neuronal cell lines, semipermissive cells of
non-neuronal origin, and permissive cells of non-neuronal origin.
Essential to the use of permissive cells for replication of HSV,
regardless of origin, is the necessity to block the cytocidal
potential of HSV by treatment with HSV-specific antisera, chemical
inhibitors, ultraviolet (UV) irradiation of virus, change in
cultivation temperature, or a combination of these treatments.

We have previously described infections of human embryo
lung fibroblast (HEL-F) cells with HSV-1 or HSV-2 in which the
virus genome was maintained in a repressed form by increased
temperature subsequent to partially blocking virus replication
by treatment with antiviral compounds.[77-79] Recent studies
have used human leukocyte interferon (IFN-α) in combination
with a relatively noncytotoxic compound BVDU[80,81] and increased
temperature (40.5°C) to establish and maintain the HSV-1 genome
in a repressed form after high multiplicity infection of HEL-F
cells with HSV-1.[82]

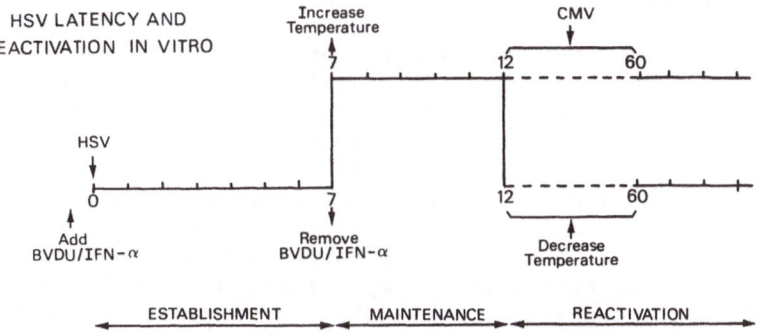

Figure 1. HSV-1 latency and reactivation in vitro

HSV-1 was activated by either reducing temperature or superinfecting with human CMV (HCMV).

The validity of the HEL-F cell-virus interaction has been
analyzed in a cell of neuronal origin.[83] Treatment of HSV-1-
infected rat fetal neurons isolated from the dorsal root ganglia
with BVDU alone or in combination with IFN-α resulted in repression of HSV-1 replication. Analogous to repressed HSV-1 infections of HEL-F cells, the HSV-1 genome could be maintained in a

repressed form in rat fetal neurons by increasing the incubation
temperature to 40.5°C after inhibitors were removed. In addition,
virus could be predictably reactivated from neurons by reducing
the incubation temperature. Current studies are directed at
determining the state the virus genome and its expression during
establishment and maintenance of latency and during reactivation. In addition, these in vitro HSV latency models will
provide a method to analyze chemotherapy of latently infected
cells.

TRANSFORMATION AND HUMAN NEOPLASIA

Herpesviruses, ubiquitous human pathogens, are suspected
carcinogenic agents and have been associated with several human
tumors by epidemiologic studies. In addition, probable cause
is suspected because these viruses remain in association with
their host from an early age providing great opportunity for
virus genes and gene products to interact with host cells over
extensive time periods prior to tumor development. To date,
EBV, HSV-1, HSV-2 and HCMV have been associated with human
neoplasia while a link between VZV and human cancer has not
been established. However, recent reports suggest that VZV may
morphologically transform cells in culture.[84,85]

Significant evidence links EBV infection to Burkitt's
lymphoma (BL) and nasopharyngeal carcinoma (NPC). Seroepide-
miological studies have revealed the prevalence of high antibody titers to EBV-specified antigens in patients with these
tumors.[86] Nucleic acid hybridization analyses have demonstrated
virus DNA in a majority of tumor biopsies.[87] In addition,
EBV-specified nuclear antigens (EBNA) can be detected in the
cells from these biopsies. When tumor cells are removed and
grown in culture, 5-20% of the cells produce virus particles.
Various EBV isolates can transform normal human lymphocytes
into "immortal" or transformed lymphoblastoid lines.[86] Inoculation of purified EBV into owl monkeys or cottontop marmosets
induces lymphoproliferative disease resembling malignant lymphoma.[88] The reisolation of EBV from tumors induced in experimental animals has fulfilled Koch's postulates to the extent it
can be fulfilled for a human oncogenic virus. EBV infection is
also implicated in undifferentiated NPC, a squamous cell carcinoma
with lymphocyte infiltration. Familial clustering of NPC
suggests that a genetic predisposition and EBV infection act
together to induce this malignancy.[8] Virus obtained from NPC
cells can transform normal lymphocytes into lymphoblastoid
cells.[8] Although no in vitro cell culture system propagates
EBV to high titer, the EBV genome has recently been cloned into
plasmids permitting the generation of large quantities of EBV
DNA for molecular analyses.[89]

The association of HSV with the human genital tract is well documented.[8] Perhaps the greatest interest lies in the etiologic relationship of HSV-2 with cervical carcinoma. Antibodies against HSV-2 are found more frequently and at higher titers in women with cervical carcinoma, dysplasia, and carcinoma in situ than in healthy women or women with other malignancies.[8] Because HSV-2 is transmitted sexually, it is important to note that the same women who epidemiologically are at high risk for cervical cancer because of early sexual exposure and poor genital hygiene, are more likely to acquire HSV-2 infections. There have been reports that sera from cervical cancer patients react with a soluble antigen prepared from HSV-2-infected cells and with antigens isolated from HSV-transformed cells.[90] In addition, HSV-specific antigens have been detected in vulvar cancer and cervical cancer cells.[90]

A number of investigators reported that HSV-specific RNA was localized in cells from cervical carcinoma.[91] In vitro cell culture studies have demonstrated that HSV-2 inactivated by photodynamic treatment[92] or UV light[93] can transform a variety of different rodent cells. These tranformed cells when inoculated into syngeneic newborn hamsters, produced tumors capable of metastasizing to the lungs and other organs.[94]

Recent efforts have focused on direct identification of the region or regions of the HSV genome responsible for induction and/or maintenance of morphological transformation. Camacho and Spear[95] have reported the transformation of hamster embryo fibroblast cells with the XbaI-F DNA fragment which maps between positions 0.3 and 0.45 on the HSV-1 genome. In addition, Reyes and coworkers[96] have reported morphological transformation of BALB/c/3T3 cells with the HSV-1 DNA fragment mapping between positions 0.311 and 0.415 on the virus genome.

Two separate regions of the HSV-2 genome have been implicated in morphological transformation. Jariwalla and coworkers[97] reported morphological transformation of Syrian hamster embryo cells with a 16.5×10^6 dalton fragment of HSV-2 mapping between positions 0.43 and 0.58 on the virus genome. In contrast, Reyes and coworkers[96] reported morphological transformation of BALB/c/3T3 cells with the 4.6×10^6 dalton BglII-N DNA fragment of HSV-2 strain 333 which maps between 0.58 and 0.63 on the virus genome. Subsequent studies by Galloway and McDougall[98] have extended these findings by transforming primary rat embryo cells and NIH/3T3 cells with a recombinant DNA containing the BglII-N DNA fragment cloned into pBR322.

The acute clinical manifestations of congenital CMV infections have been studied extensively; however, the role of CMV in the etiology of human cancers is far less defined than that

of EBV or HSV. CMV has been detected in patients with many
types of neoplasias. However, as may also be true with EBV and
HSV-2, it is possible that the virus may be reactivated from a
latent state as a result of host immunological deficiencies.
CMV has been isolated from the urine of leukemia, Hodgkin's
disease, and lymphosarcoma patients, and from cell cultures
derived from biopsies of cervical carcinoma. In addition,
lymphoblastoid cell lines have been established from patients
with CMV mononucleosis.[8] More recently, CMV has been associated in human prostatic cancer and Kaposi sarcoma.[99]

CMV can only be propagated in human cells; however, the
virus can abortively infect cells derived from nonhuman sources,
e.g., hamster cells. Although CMV has not been shown to be
oncogenic after inoculation into laboratory animals, the virus
can malignantly transform cells in culture.[100] Hamster embryo
fibroblast cells can be transformed by UV-inactivated and non-
inactivated CMV.[101] Human CMV-transformed hamster cells after
continued passage in vitro produce fibrosarcomas when inoculated
into newborn hamsters. DNA-DNA renaturation experiments demonstrated 0.4 genome equivalents of CMV per early passage CMV-transformed human cell.[8] CMV antigens have been detected in cervical
cancer cells and CMV DNA has been detected in 4 of 7 tumors of
the colon.[8]

CHEMOTHERAPEUTICALLY ADVANTAGEOUS VIRUS TARGETS

The complexities of virus latency and oncogenic transformation associated with most human herpesviruses complicate
the already difficult task of virus-specific chemotherapy.
Nevertheless, significant advances have been made in the management of herpesvirus infections. After an era of antiherpetic
compounds with little, if any, antiviral selectivity, numerous
compounds have been isolated and/or synthesized which have one
or a few specific sites of action localized in the virus-infected
cell, affording a favorable chemotherapeutic index. The following
discussion of chemotherapeutically advantageous virus-specific
targets will center on the HSV replicative cycle, diverging to
other members of the herpesvirus group at applicable points.
The productive replicative cycle of herpesviruses can be divided
into several phases, according to events that (i) occur prior
to expression of the virus genome, including virus adsorption,
penetration, and uncoating; (ii) are associated with expression
of the immediate-early (α) or early (β) genes; (iii) are associated
with synthesis of virus DNA; (iv) are associated with expression
of late (γ) genes; and (v) occur during assembly and egression
of infectious virus (Table 3).

Table 3.   Selected Herpesvirus Chemotherapeutic
           Targets and Agents

General Target	Selected Agents
Adsorption, penetration, uncoating	Vaccines, heparin, arildone
Immediate early and early virus gene expression	BVDU, ACV, FIAC, PFA, ribavirin, interferon, isoprinosine
Virus DNA synthesis	BVDU, ACV, FIAC, PFA
Late virus gene expression	Ribavirin, interferon, isoprinosine
Virus assembly and egression	2-deoxy-D-glucose, tunicamycin

In recent years numerous experimental studies have approached
the issue of development, efficacy, and safety of vaccines for
control of herpesvirus infections. In general, immunization
with live attenuated and inactivated virus vaccines as well as
administration of hyperimmune serum have been a standard clinical
practice for preventing virus infections. In fact, for animal
herpesvirus infections, extraordinary success in the prevention
of Marek's disease has been achieved by immunization.[102]
However, due to the lack of a complete understanding of the
pathogenesis of herpesvirus infections, specifically the possible
role of herpesviruses in certain human neoplasias, administration
of live virus vaccines and inactivated preparations containing
virus DNA has been approached with caution. Perhaps molecular
biological technology will enable a more complete understanding
of the structure and function of the herpesvirus genomes and
emergence of molecularly engineered virus vaccines.

A large number of studies have stressed the role of the
immune system in the pathogenesis of herpesvirus infections.
Experiments performed in animal model systems have pointed to
the role of cell-mediated immunity and the role of macrophages.[103]
Antibody given parenterally or acquired in utero in newborn
animals may also protect against acute HSV disease.[104] Antibody
also reportedly influences the control of latent HSV infection,
via immunological surface modulation.[69] Furthermore, binding
of immunoglobulin to Fc receptors present on the surface of

HSV-infected cells and the resultant inhibition of infectious
virus release may be relevant to immune modulation of HSV
infection.[105] Immunological control of herpesvirus infections
may, therefore, result in limitation of acute disease as well
as modulation of reactivation.

The use of chemotherapeutic agents to prevent herpesvirus
adsorption, penetration, or uncoating is, in theory, an attractive possibility. Inhibition of these early virus replicative
events before involvement of a large number of cellular processes
in the synthesis of progeny virus would result in a most favorable
chemotherapeutic index. However, to date, very few compounds
act at these points in the herpesvirus replicative cycle. A
negatively charged mucopolysaccharide (heparin) prevents virion
adsorption to the host cell by forming a heparin-virus complex,
presumably through electrostatic interactions; however, the
reaction is reversible and the virus is not inactivated.[106,107]
Arildone, 4-[6-(2-chloro-4-methoxy) phenoxy]-hexylheptan-3,5-dione,
an aryloxy alkyl diketone and a member of a relatively new and
novel group of antiviral compounds, inhibits rhinovirus and
poliovirus replication by interacting with the protein capsid
of the virus and thereby blocking the uncoating process.[108,109]
This group of compounds represented by arildone and its pyrazole
derivative, 4-[6-(2-chloro-4-methoxy)phenoxy]hexyl-3,5-diethyl-
1H-pyrazole methanesulphonate has been somewhat effective in
the topical management of HSV infections in mice and guinea
pigs.[110] Current studies are directed at delineating the
mechanism of action of HSV inhibition. A focal point for these
studies will most likely be the HSV nucleocapsid uncoating
process.

The second phase of the HSV replicative cycle (expression
of immediate-early (α) and early (β) genes) is associated with the
induction of two virus-specific enzymes, dThd/dCyd kinase[111,112]
and DNA polymerase.[113] To a large extent, the antiviral specificities of ACV, BVDU, and FIAC depend on the phosphorylation to
the nucleoside-5'-monophosphate by the virus-encoded dThd/dCyd
kinase, thereby restricting the production of phosphorylated
derivatives to virus-infected cells.[114,116] Specifically, the
K_I for BVDU with respect to HSV-1- HSV-2- and VZV-induced dThd
kinase are 0.24, 4.24, and 0.07 μM, respectively, compared with
greater than 150 μM for the K_I for cytosolic dThd kinase.[114]
Thus, BVDU has a much higher affinity for the virus-encoded
dThd/dCyd kinase than for the cytosol dThd kinase, the predominant
species of cellular dThd kinase. In addition, BVDU has a
significantly higher affinity for the HSV-1 dThd/dCyd kinase
than for the HSV-2 dThd/dCyd kinase, which appears to correlate
with its differential inhibitory activity towards these two
viruses. The minimal inhibitory concentration of BVDU for
HSV-1 ranges from 0.007 to 0.01 μg/ml,[81] whereas that for HSV-2

is 100 to 200 times greater (1.0 to 2.0 µg/ml). Also relevant
to the differential inhibition of HSV-1 and HSV-2 is the role
of the additional phosphorylating activity associated with the
virus-encoded dThd/dCyd kinase enzyme[117] that converts BVDU to
its nucleoside-5'-diphosphate form; the HSV-2 dThd/dCyd kinase
has not been shown to contain this enzymatic activity.[118,119]
In addition, there appears to be a direct correlation in the
different susceptibilities of HSV-1 and HSV-2 strains to BVDU
and their ability to induce a deoxypyrimidine triphosphatase
(dPyTPase) activity. HSV-1 strains, which are highly sensitive
to BVDU, induce dPyTPase activity, whereas HSV-2 strains,
which are comparatively resistant to BVDU, fail to induce the
enzyme.[120] ACV (an acyclic analog of guanosine) like BVDU and
several other selective antiherpetic compounds must be activated
by a virus-specified kinase.[116,121] Analysis of the dTHd/dCyd
kinase-inducing capabilities of several human and animal
herpesviruses that are susceptible to ACV suggests that the
virus-encoded dCyd kinase rather than dThd kinase phosphorylates
ACV. HSV-1, HSV-2, and VZV induce a kinase activity that
contains the dThd and dCyd kinase activities within the same
molecule; other herpesviruses, (pseudorabies virus, infectious
bovine rhinotracheitis virus, and herpesvirus saimiri) encode
only a dThd kinase activity and are not inhibited by ACV.
However, these viruses are inhibited by BVDU, suggesting that
the antiviral activity of BVDU depends on the phosphorylation
of dThd kinase, whereas the virus-encoded dCyd kinase moiety
is responsible for phosphorylation of ACV. In contrast to HSV
and VZV, CMV is relatively resistant to inhibition by BVDU;
however, this is not unexpected because CMV does not encode a
dThd kinase activity.[39,40] However, ACV has shown slight to
moderate antiviral activity against CMV, suggesting an additional
mechanism of action. Related to this mechanism may be the
fact that ACV (200 µM) treatment of CMV-infected cells resulted
in transient inhibition of the synthesis of two CMV late
polypeptides (67 and 150 kilodaltons),[122] without arresting
virus DNA synthesis, or possibly the induction of an alternate
cellular phosphorylating activity.[123] Although no information
on the latter possibility has been obtained, CMV does stimulate
a multitude of cellular processes after infection, including
DNA,[35] RNA,[36] and protein[37] synthesis as well as several
enzymes including dThd kinase.[39-41]

After initial phosphorylation or activation of BVDU and
ACV by virus-induced enzymatic activities, further phosphorylation by cellular kinases results in production of nucleoside
5'-triphosphates. Subsequent to activation of ACV or BVDU,
the target for BVDU and ACV is most likely virus DNA synthesis
or the virus DNA. BVDU-5'-triphosphate inhibits HSV-1-induced
DNA polymerase to a greater extent than the cellular α, β, and
γ DNA polymerases.[124] The fact that BVDU-5'-triphosphate inhibits

the HSV-2-induced DNA polymerase to the same extent as the
HSV-1 polymerase[125] is consistent with the hypothesis that the
differential susceptibility of HSV-1 and HSV-2 to BVDU resides
at a level other than the virus-induced DNA polymerase. Using
[32]P-radiolabeled orthophosphate to monitor DNA synthesis it has
been demonstrated that BVDU inhibited DNA synthesis in HSV-1-
infected cells at a 3,000-fold lower concentration than DNA
synthesis in uninfected cells.[125] BVDU, ACV, and FIAC all
demonstrated highly selective inhibitory effects on HSV-1 DNA
synthesis compared with cellular DNA synthesis in uninfected
cells. Both ACV and BVDU 5'-triphosphate can substitute for
dThd-5'-triphosphate as an alternate substrate and can be
incorporated into DNA; this incorporation is restricted primarily
to the virus-infected cell.[126,127] However, unlike ACV, BVDU
can be incorporated internally into DNA. The incorporation of
ACV into DNA leads to a premature termination of DNA synthesis
since ACV does not offer the 3'-hydroxyl group required for
further chain elongation.[126] The internal incorporation of
BVDU into DNA has little, if any, effect on the primer-template
activity of the resultant DNA.[128] If HSV-1-infected cells are
treated with BVDU (10 µg/ml), it is incorporated into both
virus and cell DNA,[127] wherease BVDU treatment at lower concentrations (0.03-0.1 µg/ml) results in incorporation into virus
DNA only.[129] The extent of the incorporation is directly
correlated with the reduction in virus yield, and since the
BVDU-substituted HSV-1 DNA has an increased susceptibility to
single-strand breakage,[129] the incorporation of BVDU into virus
DNA may well be central to the antiviral activity of the nucleoside analog. Although BVDU-5'-triphosphate is recognized by
DNA polymerase as an alternate substrate and is incorporated
into the interior of the DNA molecule, one would expect (E)-5-
(2-bromovinyl)-2-(β-D-arabinofuranosyl)uracil (BVaraU) to be
incorporated at the 3'-end of the replicating DNA, thereby
acting as a chain terminator. However, BVaraU is not an efficient
substrate for HSV-1 DNA polymerase.[128,130]

     Genetic analysis has demonstrated that both virus-induced
dThd/dCyd kinase and DNA polymerase play important roles in the
selective antiherpetic activity of ACV and BVDU. Two distinct
classes of ACV- and BVDU-resistant mutants with mutations in
the genes coding for dThd/dCyd kinase and DNA polymerase have
been isolated. In addition, ACV- and BVDU-resistant HSV mutants
have been isolated which are cross-resistant to BVDU and ACV,
indicating that the loci conferring drug resistance may be
overlapping.[131]

     In contrast to ACV and BVDU, phosphonoformic acid (PFA)
does not require prior activation by host cell or virus-specific
enzymatic activities, but acts directly by inhibiting the
preferentially virus-induced DNA polymerase by complexing with

the enzyme at the pyrophosphate binding site.[132,133] This interaction results either in dead end inhibition by decreasing the concentration of the active form of the enzyme available for the polymerization step or in alternative product inhibition, in which the terminal nucleoside monophosphate is covalently bound to phosphonoacetic acid and released from the primer DNA chain.[132] As expected, PFA-resistant HSV mutants have been isolated with mutations mapping within the DNA polymerase gene.[134]

Ribavirin is active against a large number of RNA and DNA viruses including several herpesviruses. The mechanism of action of this compound is uncertain and may depend on the virus. However, it suggests additional virus targets. The antiviral activity of the compound was originally attributed to its inhibitory effect on the synthesis of guanosine monophosphate (GMP), which is eventually converted to guanosine triphosphate and deoxyguanosine triphosphate for subsequent incorporation into RNA and DNA. Ribavirin-5'-monophosphate inhibits inosine monophosphate dehydrogenase, which converts inosine monophosphate to xanthine monophosphate the immediate precursor of GMP.[135] The resultant depletion of GDP inhibits cellular DNA and RNA synthesis, which may impair cellular metabolic functions in the synthesis of progeny virus. Additional data suggest that ribavirin-5'-triphosphate preferentially inhibits HSV DNA polymerase.[136] In addition, in ribavirin-treated, vaccinia virus-infected cells, virus DNA and polypeptides are synthesized but little, if any, progeny virus was formed, as suggested by the failure of the virus DNA to acquire resistance to deoxyribo- nuclease. These data suggest that ribavirin inhibited the coating of virus DNA by virus-coded polypeptides.

Several antiviral agents inhibit the translation of viral mRNA into protein. Although this group of chemotherapeutic agents is pleotropic in their mechanism of action, it has been theorized that their virus selectivity would be derived from structural differences between host cell and virus-specific mRNAs. Interferons, a group of host cell-coded glycoproteins that are produced in response to virus infection or some nonvirus inducers, inhibit both RNA and DNA viruses including herpes- viruses. Interferons are associated with a wide range of biological activities and extensive discussion is beyond the scope of this review. However, pertinent to the subject of virus chemotherapy is the inhibitory effect of interferon on host cell protein synthesis. The binding of interferon at the host cell surface results in the synthesis or stimulation of host cell proteins which effect selective inhibition of the translation of virus mRNA, thereby inhibiting virus replica- tion. To this end, interferon stimulates the action of at least three enzymatic pathways that can inhibit mRNA trans-

lation. (i) Eukaryotic initiation factor 2 is inactivated by
an interferon-mediated, ribosome-associated kinase phosphorylation,
resulting in inhibition of Met-tRNA$_f$ binding to the 40S ribosomal
subunit, thereby inhibiting the formation of protein synthesis-
initiating complex.[137] (ii) Interferon induces an enzymatic system
that enhances the degradation of RNA by induction of a (2'-5')
(A)$_n$ synthetase which converts ATP into oligonucleotides; these
(2'-5') adenylates then activate a latent endoribonuclease that
degrades mRNA and, therefore, inhibits protein synthesis.[138]
(iii) Interferon increases the activity of a cellular phosphodiesterase that degrades the amino acid receptor site of transfer
RNA resulting in a transfer RNA-reversible inhibition of protein
synthesis.[139] However, the basis for the selectivity of interferon is not well understood nor is the mechanism of inhibition
of herpesvirus replication.

Isoprinosine, a paracetamidobenzoic acid salt of inosine
dimethylamino-isopropanol (1:3 mole ratio) also inhibits both
RNA and DNA viruses, including herpesviruses. The exact mechanism
of action of isoprinosine is uncertain; however, several lines
of experimental evidence suggest a unique selective inhibition
of translation of virus mRNAs. These include the fact that
isoprinosine (i) increases the rate of synthesis of polyribosomal
RNA and protein in uninfected cells, whereas in infected cells
isoprinosine reduces the rate of polyribosomal RNA synthesis
and the expression, and presumed synthesis of virus-specific
protein; (ii) enhances the in vitro translation of host mRNA
while reducing the translation of exogenous mRNA; and (iii)
initiates an alteration in host ribosomal structure.[140,141]
The authors postulated that isoprinosine selectively enhances
the synthesis and transport of host cell mRNA, while on the
basis of quantity, is preferentially translated by the limited
host ribosomes. In addition, the ribosomes are altered in some
manner such that virus mRNA is translated less efficiently than
host mRNA.[140,141] These theories have yet to be confirmed.

The last process in the herpesvirus replication cycle is
expression of late virus-specific genes, assembly of infectious
virions, and egression from host cell. Inhibitors of protein
glycosylation, 2-deoxy-D-glucose, D-glucosamine and tunicamycin
inhibit the replication of herpesviruses. The antiviral activity
has been attributed to a deficiency in the glycosylation of
virus glycoproteins resulting in loss of proper function. In
the case of tunicamycin, the compound prevents the formation of
N-acetylglucosamine-dolichylphosphate, an intermediate in the
synthesis of the core oligosaccharides of glycoproteins.
Subsequent to alteration in glycosylation, virus glycoproteins
may be hamperd in their transfer from the rough endoplasmic
reticulum to the plasma membrane where they are incorporated
into the virion structure during the egression process. Altered

glycoproteins, therefore, would interfere with the final stages of virion assembly. Other mechanisms have been proposed for inhibitors of glycosylation. Tunicamycin and 2-deoxy-D-glucose may inhibit the glycosylation of host cell glycoproteins essential for virus DNA synthesis.[142]

CONCLUSIONS AND PROJECTIONS

The herpesviruses are similar in several biological and physical properties. However, each individual virus presents a unique therapeutic problem for the development of effective treatment. The past decade has seen the rational development of therapeutic agents that specifically inhibit selected virus replicative events with only minimal impairment of host cell function. Compounds such as BVDU and ACV, although highly selective antiherpetic compounds, should be used as developmental models for synthesis of additional, more selective agents. Compounds that are effective against HSV have, in general, been effective in treatment of VZV, with lesser antiviral activity against EBV. However, CMV has posed chemotherapeutic difficulties due to the lack of induction of a virus-specific dThd kinase activity. Agents directed at the CMV-coded DNA polymerase or possibly a combination chemotherapy approach may prove to be useful. As immediate-early replicative events of the herpes-viruses are analyzed by molecular biological techniques, new avenues of virus-specific chemotherapy that block virus replica-tion at very early points in the life cycle of the virus can be developed, thereby minimizing effects on host cell metabolism even further. Due to the ability to establish latent infections and the potential oncogenicity of herpesviruses an early block would be useful.

ACKNOWLEDGMENTS

We thank Luzi Pfenninger, Helen Traglia and Judy Raker for technical assistance on research described and supported by Grants CA 18450, CA 27503, CA 34479 and CA 09124 awarded by the National Cancer Institute to F.R. We are also grateful to Melissa Reese for editorial assistance and Carol A. Buck and Elaine K. Neidigh for typing support.

REFERENCES

1. A. J. Nahmias, W. R. Dowdle, and R. F. Schinazi, "The Human Herpesviruses: An Interdisciplinary Perspective," Elsevier, New York (1981).

2. B. Roizman, "The Herpesviruses," Vol. 2, Plenum Press, New York (1983).

3. J. R. Baringer, Herpes simplex virus infections of the nervous system, in: "Handbook of Clinical Neurology: Infections of the Nervous System," Vol. 34, Part II," P. J. Vinken and G. W. Bruyn, eds., North-Holland, New York (1978).

4. J. G. Stevens, Herpetic latency and reactivation, in: "Oncogenic Herpesviruses," Vol. II, F. Rapp, ed., CRC Press, Boca Raton, Florida (1980).

5. R. J. Klein, The pathogenesis of acute, latent, and recurrent herpes simplex virus infections, Arch. Virol. 72:143 (1982).

6. P. Wildy, H. J. Field, and A. A. Nash, Classical herpes latency revisited, in: "Virus Persistence Symposium," B. W. J. Mahy, A. C. Minson and G. K. Dardy, eds., Cambridge University Press, Cambridge (1982).

7. D. M. Knipe, Cell growth transformation by herpes simplex virus, Prog. Med. Virol. 28:114 (1982).

8. F. Rapp, Viral carcinogenesis, in: "Aspects of Cell Regulation, International Review of Cytology," J. F. Danielli, ed., Academic Press, Inc., New York (1983).

9. H. E. Kaufman and C. Heidelberger, Therapeutic antiviral action of 5-trifluoromethyl-2'-deoxyuridine in herpes simplex keratitis, Science 145:585 (1964).

10. H. E. Kaufman, Clinical cure of herpes simplex keratitis by 5-iodo-2'-deoxyuridine, Proc. Soc. Exp. Biol. Med. 109:251 (1962).

11. W. H. Prusoff and B. Goz, Potential mechanisms of action of antiviral agents, Fed. Proc. 32:1679 (1973).

12. W. E. G. Müller, Mechanisms of action and pharmacology: chemical agents, in: "Antiviral Agents and Viral Diseases of Man," G. J. Galasso, T. C. Merigan and R. A., Buchanan, eds., Raven Press, New York (1979).

13. W. H. Prusoff and D. C. Ward, Commentary: nucleoside analogs with antiviral potency, Biochem. Pharmacol. 25:1233 (1976).

14. D. J. Bauer, "The Specific Treatment of Virus Diseases," University Park Press, Baltimore (1977).

15. B. Roizman and D. Furlong, The replication of herpesviruses, in: "Comprehensive Virology," Vol. 3, H. Fraenkel-Conrat and R. R. Wagner, eds., Plenum Press, New York (1974).

16. P. Spear and B. Roizman, Herpes simplex viruses, in: "DNA Tumor Viruses," J. Tooze, ed., Cold Spring Harbor Laboratory, Cold Spring Harbor, New York (1980).

17. E. D. Kieff, S. L. Bachenheimer, and B. Roizman, Size, composition and structure of the DNA of subtypes 1 and 2 herpes simplex virus, J. Virol. 8:125 (1971).

18. B. Roizman, The structure and isomerization of herpes
    simplex virus genomes, Cell 16:481 (1979).
19. E. D. Kieff, B. Hoyer, S. L. Bachenheimer, and B. Roizman,
    Genetic relatedness of type 1 and type 2 herpes simplex
    viruses, J. Virol. 9:738 (1972).
20. P. Sheldrick and N. Berthelot, Inverted repetitions in the
    chromosome of herpes simplex virus, Cold Spring Harbor
    Symp. Quant. Biol. 39:667 (1975).
21. S. Wadsworth, R. J. Jacob, and B. Roizman, Anatomy of
    herpes virus DNA. II. Size, composition and arrangement
    of inverted terminal repetitions, J. Virol. 15:1487 (1975).
22. G. S. Hayward, R. J. Jacob, S. C. Wadsworth, and B. Roizman,
    Anatomy of herpes simplex virus DNA: evidence for four
    populations of molecules that differ in the relative
    orientations of their long and short segments, Proc. Natl.
    Acad. Sci. USA 72:4243 (1975).
23. H. Delius and J. B. Clements, A partial denaturation map
    of herpes simplex virus type 1 DNA: evidence for inversions
    of the unique DNA regions, J. Gen. Virol. 33:125 (1976).
24. R. W. Honess and B. Roizman, Proteins specified by herpes
    simplex virus. XI. Identification and relative molar
    rates of synthesis of structural and non-structural herpes-
    virus polypeptides in the infected cell, J. Virol. 12:1346
    (1973).
25. K. L. Powell and R. J. Courtney, Polypeptides synthesized
    in herpes simplex virus type 2-infected HEp-2 cells,
    Virology 66:217 (1975).
26. R. W. Honess and B. Roizman, Regulation of herpesvirus
    macromolecular synthesis. I. Cascade regulation of the
    synthesis of three groups of viral proteins, J. Virol.
    14:8 (1974).
27. S. Mackem and B. Roizman, Regulation of herpesvirus macro-
    molecular synthesis: Transcription-initiation sites and
    domains of α genes, Proc. Natl. Acad. Sci. USA 77:7122
    (1980).
28. K. W. Wilcox, A. Kohn, E. Sklyanskay, and B. Roizman,
    Herpes simplex virus phosphoproteins. I. Phosphate
    cycles on and off some viral polypeptides and can alter
    their affinity for DNA, J. Virol. 33:167 (1980).
29. B. Roizman and P. R. Roane, The multiplication of herpes
    simplex virus. II. The relation between protein syn-
    thesis and the duplication of viral DNA in infected HEp-2
    cells, Virology 22:262 (1964).
30. R. J. Sydiskis and B. Roizman, The disaggregation of host
    polyribosomes in productive and abortive infection with
    herpes simplex virus, Virology 32:678 (1967).
31. E. K. Wagner and B. Roizman, RNA synthesis in cells infected
    with herpes simplex virus. I. The patterns of RNA synthesis
    in productively infected cells. J. Virol. 4:36 (1969).

32. M. L. Fenwick and M. J. Walker, Suppression of the syn-
    thesis of cellular macromolecules by herpes simplex virus,
    J. Gen. Virol. 41:37 (1978).

33. B. Roizman, M. Kozak, R. W. Honess, and G. Hayward, Regula-
    tion of herpes virus macromolecular synthesis: evidence
    for multilevel regulation of herpes simplex 1 RNA and
    protein synthesis, Cold Spring Harbor Symp. Quant. Biol.
    39:687 (1975).

34. E. S. Huang, Human cytomegalovirus. III. Virus-induced
    DNA polymerase, J. Virol. 16:298 (1975).

35. S. C. St. Jeor, T. B. Albrecht, F. D. Funk, and F. Rapp,
    Stimulation of cellular DNA synthesis by human cytomeg-
    alovirus, J. Virol. 13:353 (1974).

36. S. Tanaka, T. Furukawa, and S. A. Plotkin, Human cyto-
    megalovirus stimulates host cell RNA synthesis, J. Virol.
    15:297 (1975).

37. M. F. Stinski, Synthesis of protein and glycoproteins in
    cells infected with human cytomegalovirus, J. Virol.
    23:751 (1977).

38. K. Hirai and Y. Watanabe, Induction of α-type DNA poly-
    merases in human cytomegalovirus-infected WI-38 cells,
    Biochim. Biophys. Acta 447:328 (1976).

39. V. Zavada, V. Erban, D. Resacova, and V. Vonka, Thymidine
    kinase in cytomegalovirus infected cells, Arch. Virol.
    52:333 (1976).

40. R. L. Miller, J. P. Iltis, and F. Rapp, Differential
    effect of arabinofuranosylthymine on the replication of
    human herpesviruses, J. Virol. 23:679 (1977).

41. H. C. Isom, Stimulation of ornithine decarboxylase by
    human cytomegalovirus, J. Gen. Virol. 42:265 (1979).

42. J. A. Zaia, Clinical spectrum of varicella-zoster virus
    infection, in: "The Human Herpesviruses: An Interdisci-
    plinary Perspective," A. J. Nahmias, W. R. Dowdle, and R.
    F. Schinazi, eds., Elsevier, New York (1981).

43. T. H. Weller, Clinical spectrum of cytomegalovirus infection,
    in: "The Human Herpesviruses: An Interdisciplinary
    Perspective," A. J. Nahmias, W. R. Dowdle and R. F. Schinazi,
    eds., Elsevier, New York (1981).

44. J. S. Pagano and J. G. Nedrud, Latency of the Epstein-Barr
    virus and cytomegalovirus, in: "The Human Herpesviruses:
    An Interdisciplinary Perspective," A. J. Nahmias, W. R.
    Dowdle and R. F. Schinazi, eds., Elsevier, New York (1981).

45. G. Klein, EBV-persistence in human lymphoid and carcinoma
    cells, in: "Persistant Viruses," J. G. Stevens, G. J.
    Todaro, and C. F. Fox, eds., Academic Press, New York
    (1978).

46. M. Nonoyama and J. S. Pagano, Separation of Epstein-Barr
    virus DNA from large chromosomal DNA in non-virus pro-
    ducing cells, Nature (New Biol.) 238:169 (1972).

47. A. Adams, T. Lindahl, and G. Klein, Linear association between cellular DNA and Epstein-Barr virus DNA in a human lymphoblastoid cell line, Proc. Natl. Acad. Sci. USA 70:2888 (1973).

48. P. Gerber, Activation of Epstein-Barr virus by 5-bromo-deoxyuridine in virus-free human cells, Proc. Natl. Acad. Sci. USA 69:83 (1972).

49. A. Adams, G. Bjursell, C. Kaschka-Dierich, and T. Lindahl, Circular Epstein-Barr virus genomes of reduced size in a human lymphoid cell line of infectious mononucleosis origin, J. Virol. 22:373 (1977).

50. M. Anderson and T. Lindahl, Epstein-Barr virus DNA in human lymphoid cell lines: in vitro conversion, Virology 73:96 (1976).

51. S. D. Hayward and E. D. Kieff, Epstein-Barr virus-specific RNA. I. Analysis of viral RNA in cellular extracts and in the polyribosomal fraction of permissive and nonpermissive lymphoblastoid cell lines, J. Virol. 18:518 (1976).

52. T. Orellana and E. Kieff, Epstein-Barr virus-specific RNA. II. Analysis of polyadenylated viral RNA in restringent, abortive, and productive infections, J. Virol. 22:321 (1977).

53. B. Hampar, A. Tanaka, M. Nonoyama, and J. G. Derge, Replication of the resident repressed Epstein-Barr virus genome during early S phase (S-1 period) of nonproducer Raji cells, Proc. Natl. Acad. Sci. USA 71:631 (1974).

54. A. J. Nahmias, J. Dannenbarger, C. Wickliffe, and J. Muther, Clinical aspects of infection with herpes simplex viruses 1 and 2, in: "The Human Herpesviruses: An Interdisciplinary Perspective," A. J. Nahmias, W. R. Dowdle and R. F. Schinazi, eds., Elsevier, New York (1981).

55. M. A. Walz, H. Yamamoto, and A. L. Notkins, Immunological response restricts number of cells in sensory ganglia infected with herpes simplex virus, Nature 264:554 (1976).

56. J. G. Stevens, Latent herpes simplex virus and the nervous system, Curr. Top. Microbiol. Immunol. 70:31 (1975).

57. A. Puga, J. D. Rosenthal, H. Openshaw, and A. L. Notkins, Herpes simplex virus DNA and mRNA sequences in acutely and chronically infected trigeminal ganglia of mice, Virology 89:102 (1978).

58. D. A. Galloway, C. Fenoglio, M. Shevchuck, and J. K. McDougall, Detection of herpes simplex RNA in human sensory ganglia, Virology 95:265 (1979).

59. D. A. Galloway, C. M. Fenoglio, and J. K. McDougall, Limited transcription of the herpes simplex virus genome when latent in human sensory ganglia, J. Virol. 41:686 (1982).

60. R. B. Tenser, R. L. Miller, and F. Rapp, Trigeminal ganglion infection by thymidine kinase-negative mutants of herpes simplex virus, Science 205:915 (1979).

61. B. Fong and M. Scriba, Use of [^{125}I] deoxycytidine to detect herpes simplex virus-specific thymidine kinase in tissues of latently infected guinea pigs, J. Virol. 34:644 (1980).

62. M. T. Green, R. J. Courtney, and E. C. Dunkel, Detection of an immediate early herpes simplex virus type 1 polypeptide in trigeminal ganglia from latently infected animals, Infect. Immun. 34:987 (1981).

63. K. L. Powell and D. J. M. Purifoy, DNA-binding proteins of cells infected by herpes simplex virus types 1 and 2, Intervirology 7:225 (1976).

64. R. A. F. Dixon and P. A. Schaffer, Fine-structure mapping and functional analysis of temperature-sensitive mutants in the gene encoding the herpes simplex virus type 1 immediate early protein VP 175, J. Virol. 36:189 (1980).

65. K. W. Lofgren, J. G. Stevens, H. S. Marsden, and J. H. Subak-Sharpe, Temperature-sensitive mutants of herpes simplex virus differ in the capacity to establish latent infection in mice, Virology 76:440 (1977).

66. J. C. Gerdes, H. S. Marsden, M. L. Cook, and J. G. Stevens, Acute infection of differentiated neuroblastoma cells by latency-positive and latency-negative herpes simplex virus ts mutants, Virology 94:430 (1979).

67. Y. Sokawa, T. Ando, and Y. Ishihara, Induction of 2',5'-oligoadenylate synthetase and interferon in mouse trigeminal ganglia infected with herpes simplex virus, Infect. Immun. 28:719 (1980).

68. D. L. Lodmell, A. Niwa, K. Hayashi, and A. L. Notkins, Prevention of cell to cell spread of herpes simplex virus by leukocytes, J. Exp. Med. 137:706 (1973).

69. J. G. Stevens and M. L. Cook, Maintenance of latent herpetic infection: an apparent role for antiviral IgG, J. Immunol. 113:1685 (1974).

70. D. L. Rock and N. W. Fraser, Detection of HSV-1 genome in central nervous system of latently infected mice, Nature (London) 30:523 (1983).

71. N. W. Fraser, W. C. Lawrence, Z. Wroblewska, D. H. Gilden, and H. Koprowski, Herpes simplex type 1 DNA in human brain tissue, Proc. Natl. Acad. Sci. USA 78:6461 (1981).

72. M. Levine, A. L. Goldin, and J. C. Glorioso, Persistence of herpes simplex virus genes in cells of neuronal origin, J. Virol. 35:203 (1980).

73. H. Youssoufian, S. M. Hammer, M. S. Hirsch, and C. Mulder, Methylation of the viral genome in an in vitro model of herpes simplex virus latency, Proc. Natl. Acad. Sci. USA 79:2207 (1982).

74. R. W. Price, R. Rubenstein, and A. Khan, Herpes simplex virus infection of isolated autonomic neurons in culture: viral replication and spread in a neuronal network, Arch. Virol. 71:127 (1982).

75. L. W. Catalano and S. Baron, Protection against herpes-virus and encephalomyocarditis virus encephalitis with a double-stranded RNA inducer of interferon, Proc. Soc. Exp. Biol. Med. 133:684 (1970).
76. W. A. Blyth, T. A. Hill, H. J. Field, and D. A. Harbour, Reactivation of herpes simplex virus infection by ultra-violet light and possible involvement of prostaglandins, J. Gen. Virol. 33:547 (1976).
77. A. M. Colberg-Poley, H. C. Isom, and F. Rapp, Reactivation of herpes simplex virus type 2 from a quiescent state by human cytomegalovirus, Proc. Natl. Acad. Sci. USA 76:5948 (1979).
78. B. L. Wigdahl, H. C. Isom, and F. Rapp, Repression and activation of the genome of herpes simplex viruses in human cells, Proc. Natl. Acad. Sci. USA 78:6522 (1981).
79. B. L. Wigdahl, H. C., Isom, E. De Clercq, and F. Rapp, Activation of herpes simplex virus (HSV) type 1 genome by temperature-sensitive mutants of HSV-type 2, Virology 116:468 (1982).
80. E. De Clercq, J. Descamps, P. De Somer, P. J. Barr, A. S. Jones, and R. T. Walker, (E)-5-(2-bromovinyl)-2'-deoxy-uridine: a potent and selective anti-herpes agent, Proc. Natl. Acad. Sci. USA 76:2947 (1979).
81. E. De Clercq, J. Descamps, G. Verhelst, R. T. Walker, A. S. Jones, P. F. Torrence, and D. Shugar, Comparative efficacy of antiherpes drugs against different strains of herpes simplex virus, J. Infect. Dis. 141: 563 (1980).
82. B. L. Wigdahl, A. C. Scheck, E. De Clercq, and F. Rapp, High efficiency latency and reactivation of herpes simplex virus in human cells, Science 217:1145 (1982).
83. B. L. Wigdahl, R. J. Ziegler, M. Sneve, and F. Rapp, Herpes simplex virus latency and reactivation in isolated rat sensory neurons, Virology 127:159 (1983).
84. L. D. Gelb, J. J. Huang, and W. J. Wellinghoff, Vari-cella-zoster virus transformation of hamster embryo cells, J. Gen. Virol. 51:171 (1980).
85. K. Yamanishi, Y. Matsunaga, Y., T. Ogino, P. Lopetegui, Biochemical transformation of mouse cells by varicella-zoster virus, J. Gen. Virol. 56:421 (1981).
86. G. Miller, Oncogenesis by Epstein-Barr virus, in: "The Human Herpesvirus: An Interdisciplinary Perspective," A. J. Nahmias, W. R. Dowdle and R. F. Schinazi, eds., Elsevier, New York (1981).
87. H. zur Hausen, H. Schulte-Holthausen, G. Klein, W. Henle, G. Henle, P. Clifford, and L. Santesson, EBV DNA in biopsies of Burkitt tumours and anaplastic carcinomas of the naso-pharynx, Nature (London) 228:1056 (1970).
88. M. A. Epstein, R. D. Hunt, and H. Rabin, Pilot experiments with EB virus in owl monkeys (Aotus trivirgatus). I.

Reticuloproliferative disease in an inoculated animal, Int. J. Cancer 12:309 (1973).

89. T. Dambaugh, C. Beisel, M. Hummel, W. King, S. Fennewald, A. Cheung, M. Heller, N. Raab-Traub, and E. Kieff, Epstein-Barr virus (B95-8) DNA. VII. Molecular cloning and detailed mapping, Proc. Natl. Acad. Sci. USA 77: 2999 (1980).

90. G. R. Pearson and L. Aurelian, Immunology of herpesvirus-associated cancers, in: "The Human Herpesviruses: An Interdisciplinary Perspective," A. J. Nahmias, W. R. Dowdle and R. F. Schinazi, eds., Elsevier, New York (1981).

91. N. J. Maitland, J. H. Kinross, A. Busuttel, S. M. Ludgate, G. E. Smart, and K. W. Jones, The detection of DNA tumour virus-specific RNA sequences in abnormal human cervical biopsies by in situ hybridization, J. Gen. Virol. 55:123 (1981).

92. F. Rapp, J. H. Li, and M. Jerkofsky, Transformation of mammalian cells by DNA-containing viruses following photo-dynamic inactivation, Virology 55:339 (1973).

93. R. Duff and F. Rapp, Properties of hamster embryo fibro-blasts transformed in vitro after exposure to ultraviolet-irradiated herpes simplex virus type 2, J. Virol. 8:469 (1971).

94. R. Duff and F. Rapp, Oncogenic transformation of hamster embryo cells after exposure to inactivated herpes simplex virus type I, J. Virol. 12:209 (1973).

95. A. Camacho and P. G. Spear, Transformation of hamster embryo fibroblasts by a specific fragment of the herpes simplex virus genome, Cell 15:993 (1978).

96. G. R. Reyes, R. LaFemina, S. D. Hayward, and G. S. Hayward, Morphological transformation by DNA fragments of human herpesviruses: evidence for two distinct transforming regions in HSV types 1 and 2 and lack of correlation with biochemical transfer of the thymidine kinase gene, Cold Spring Harbor Symp. Quant. Biol. 44:629 (1980).

97. R. J. Jariwalla, L. Aurelian, P. O. P. Ts'O, Tumorigenic transformation induced by a specific fragment of DNA from herpes simplex virus type 2, Proc. Natl. Acad. Sci. USA 77:2279 (1980).

98. D. A. Galloway and J. K. McDougall, Transformation of rodent cells by a cloned DNA fragment of herpes simplex virus type 2, J. Virol. 38:749 (1981).

99. G. Giraldo, E. Beth, and E.-S. Huang, Kaposi's sarcoma and its relationship to cytomegalovirus (CMV). III. CMV DNA and CMV early antigens in Kaposi's sarcoma, Int. J. Cancer 26:23 (1980).

100. L. Geder, R. Lausch, F. O'Neill, and F. Rapp, Oncogenic transformation of human embryo lung cells by human cytomega-lovirus, Science 192:1134 (1976).

101. T. Albrecht and F. Rapp, Malignant transformation of hamster embryo fibroblasts following exposure to ultra-virolet-irradiated human cytomegalovirus, Virology 55: 53 (1973).

102. A. E. Churchill, L. N. Payne and R. C. Chubb, Immunization against Marek's disease using a live attenuated virus, Nature (London) 221:744 (1969).

103. B. Rager-Zisman and A. C. Allison, Mechanism of immuno-logic resistance to herpes simplex virus 1 (HSV-1) in-fection, J. Immunol. 116:35 (1976).

104. S. Baron, M. G. Worthington, J. Williams and J. W. Gaines, Postexposure serum prophylaxis of neonatal herpes simplex virus infection of mice, Nature (London) 261:505 (1976).

105. J. Costa, A. S. Rabson, C. Yee, and T. S. Tralka, Immuno-globulin binding to herpesvirus-induced Fc receptors inhibits virus growth, Nature (London) 269: 251 (1977).

106. A. Vaheri and K. Cantell, The effect of heparin on herpes simplex virus, Virology 21:661 (1963).

107. A. J. Nahmias and S. Kibrick, Inhibitory effect of heparin on herpes simplex virus, J. Bacteriol. 87:1060 (1964).

108. J. J. McSharry, L. A. Caliguiri, and H. J. Eggers, Inhibition of uncoating of poliovirus by arildone, a new antiviral drug, Virology 97:307 (1979).

109. L. A. Caliguiri, J. J. McSharry, and G. W. Lawrence, Effect of arildone on modifications of poliovirus in vitro, Virology 105:86 (1980).

110. F. Pancic, B. A. Steinberg, G. D. Diana, P. M. Carabateas, W. G. Gorman, and P. E. Came, Antiviral activity of win 41258-3, a pyrazole compound, against herpes simplex virus in mouse genital infection and in guinea pig skin infection, Antimicrob. Agents Chemother. 19:470 (1981).

111. S. Kit, D. R. Dubbs, and M. Anken, Altered physical properties of thymidine kinase after infection of mouse fibroblast cells with herpes simplex virus, J. Virol. 1:238 (1967).

112. A. T. Jamison, G. A. Gentry, and J. Subak-Sharpe, In-duction of both thymidine and deoxycytidine kinase ac-tivity by herpes viruses, J. Gen. Virol. 24:465 (1974).

113. H. M. Keir, J. Subak-Sharpe, W. I. H. Shedden, D. H. Watson, and P. Wildy, Immunological evidence for a specific DNA polymerase produced after infection by herpes simplex virus, Virology 30:154 (1966).

114. Y. -C. Cheng, G. Dutschman, E. De Clercq, A. S. Jones, S. G. Rahim, G. Verhelst, and R. T. Walker, Differential affinities of 5-(2-halogenovinyl)-2'-deoxyuridines for deoxythymidine kinases of various origins, Mol. Pharmacol. 20:230 (1981).

115. K. A. Watanabe, U. Reichman, K. Hirota, C. Lopez, and J. J. Fox, Nucleosides. 110. Synthesis and antiherpes virus activity of some 2'-fluoro-2'-deoxyarabinofuranosylpyrimidine nucleosides, J. Med. Chem. 22:21 (1979).

116. G. B. Elion, P. A. Furman, J. A. Fyfe, P. De Miranda, L.
     Beauchamp, and H. J. Schaeffer, Selectivity of action of
     an antiherpetic agent, 9-(2-hydroxyethoxymethyl) guanine,
     Proc. Natl. Acad. Sci. USA 74:5716 (1977).
117. M. S. Chen and W. H. Prusoff, Association of thymidylate
     kinase activity with pyrimidine deoxyribonucleoside kinase
     induced by herpes simplex virus, J. Biol. Chem. 253:1325
     (1978).
118. J. Descamps and E. De Clercq, Specific phosphorylation of
     E-5-(2-iodovinyl)-2'-deoxyuridine by herpes simplex virus-
     infected cells, J. Biol. Chem. 256:5973 (1981).
119. J. A. Fyfe, Differential phosphorylation of (E)-5-(2-bromo-
     vinyl)-2'-deoxyuridine monophosphate by thymidylate kinases
     from herpes simplex viruses type 1 and 2 and varicella-zoster
     virus, Mol. Pharmacol. 21:432 (1982).
120. F. Wohlrab, D. R. Mayo, G. D. Hsiung, and B. Francke,
     Correlation of the expression of deoxypyrimidine tri-
     phosphatase with sensitivity to E-5-bromovinyl-2'-deoxy-
     uridine in clinical isolates of herpes simplex virus type
     1 and type 2, in: "International Workshop on Herpesviruses,"
     A. S. Kaplan, M. La Placa, F. Rapp and B. Roizman, eds.,
     Esculapio Publishing Co., Bologna (1981).
121. J. A. Fyfe, P. M. Keller, P. A. Furman, R. L. Miller, and
     G. B. Elion, Thymidine kinase from herpes simplex virus
     phosphorylates the new antiviral compound, 9-(2-hydroxyethoxy-
     methyl) guanine, J. Biol. Chem. 253:8721 (1978).
122. E.-C. Mar, P. C. Patel, and E.-S. Huang, Effect of 9-(2-hy-
     droxyethoxymethyl) guanine on viral-specific polypeptide
     synthesis in human cytomegalovirus-infected cells, Am. J. Med.
     73:82 (1982).
123. G. B. Elion, Mechanism of action and selectivity of acyclovir,
     Am. J. Med. 73:7 (1982).
124. H. S. Allaudeen, J. W. Kozarich, J. R. Bertino, and E. De
     Clercq, On the mechanism of selective inhibition of herpes-
     virus replication by (E)-5-(2-bromovinyl)-2'-deoxyuridine,
     Proc. Natl. Acad. Sci USA 78:2698 (1981).
125. E. De Clercq, BVDU, (E)-5-(2-bromovinyl)-2'-deoxyuridine,
     in: "Antiviral Drugs and Interferon: The Molecular Basis
     of Their Activity," Y. Becker, ed., Martinus Nijhoff, The
     Hague (1982).
126. P. A. Furman, P. V. McGuirt, P. M. Keller, J. A. Fyfe, and
     G. B. Elion, Inhibition by acyclovir of cell growth and
     DNA synthesis of cells biochemically transformed with
     herpesvirus genetic information, Virology 102:420 (1980).
127. H. S. Allaudeen, M. S. Chen, J. J. Lee, E. De Clercq, and
     W. H. Prusoff, Incorporation of E-5-(2-halovinyl)-2'-
     deoxyuridines into deoxyribonucleic acids of herpes simplex
     virus type 1-infected cells, J. Biol. Chem. 257:603 (1982).
128. J. L. Ruth and Y.-C. Cheng, Nucleoside analogues with
     clinical potential in antivirus chemotherapy. The effect

of several thymidine and 2'-deoxycytidine analogue 5'-triphosphates on purified human (α,β) and herpes simplex virus (types 1, 2) DNA polymerases, Mol. Pharmacol. 20:415 (1981).

129. W. R. Mancini, E. De Clercq, and W. H. Prusoff, The relationship between incorporation of $\underline{E}$-5-(2-bromovinyl)-2'-deoxyuridine into herpes simplex virus type 1 DNA with virus infectivity and DNA integrity, J. Biol. Chem. 258:792 (1983).

130. J. Descamps, R. K. Sehgal, E. De Clercq, and H. S. Allaudeen, Inhibitory effect of E-5-(2-bromovinyl)-1-β-D-arabinofuranosyluracil on herpes simplex virus replication and DNA synthesis, J. Virol 43:332 (1982).

131. E. De Clercq, Selective antiherpes agents, Trends Pharmacol. Sci. 3:492 (1982).

132. S. S. Leinbach, J. M. Reno, L. F. Lee, A. F. Isbell, and J. A. Boezi, Mechanism of phosphonoacetate inhibition of herpesvirus-induced DNA polymerase, Biochemistry 15:426 (1976).

133. J. A. Boezi, The antiherpesvirus action of phosphonoacetic acid, Pharmacol. Ther. 4:231 (1979).

134. D. M. Coen, P. A. Schaffer, P. A. Furman, P. M. Keller, and M. H. St. Clair, Biochemical and genetic analysis of acyclovir-resistant mutants of herpes simplex virus type 1, Am. J. Med. 73:351 (1982).

135. D. G. Streeter, J. T. Witkowski, G. P. Khare, R. W. Sidwell, R. J. Bauer, R. K. Robbins, and L. N. Simon, Mechanism of action of 1-β-D-ribofuranosyl-1,2,4,triazole-3-carboxamide (vivazole). A new broad spectrum antiviral agent, Proc. Natl. Acad. Sci. USA 70:1174 (1973).

136. B. Oberg and E. Helgstrand, Selective inhibition of viral polymerases by ribavirin triphosphate, in: "Current Chemotherapy," Vol. 1, W. Siegenthales and R. Lüthy, eds., American Society for Microbiology, Washington, D. C. (1978).

137. C. E. Samuel, Mechanism of interferon action: phosphorylation of protein synthesis initiation factor eIF-2 in interferon-treated human cells by a ribosome-associated kinase possessing site specificity similar to hemin-regulated rabbit reticulocyte kinase, Proc. Natl. Acad. Sci. USA 76:600 (1979).

138. I. M. Kerr and R. E. Brown, pppA2'p5'A2'p5'A: An inhibitor of protein synthesis synthesized with an enzyme fraction from interferon-treated cells, Proc. Natl. Acad. Sci. USA 75:256 (1978).

139. A. Schmidt, Y. Chernajovsky, L. Shulman, P. Federman, H. Berissi, and M. Revel, An interferon-induced phosphodiesterase degrading (2',5') oligoisoadenylate and C-C-A terminus of t-RNA, Proc. Natl. Acad. Sci. USA 76:4788 (1979).

140. P. Gordon and E. R. Brown, The antiviral activity of
     isoprinosine, Can. J. Microbiol. 18:1463 (1972).
141. P. Gordon, B. Ronsen, and E. R. Brown, Anti-herpesvirus
     action of isoprinosine, Antimicrob. Agents Chemother.
     5:153 (1974).
142. K. D. Radsak and D. Weber, Effect of 2-deoxy-D-glucose on
     cytomegalovirus-induced DNA synthesis in human fibroblasts,
     J. Gen. Virol. 57:33 (1981).

PICORNAVIRUSES AND TOGAVIRUSES: TARGETS FOR DESIGN OF ANTIVIRALS

Bruce D. Korant, Karl Lonberg-Holm and Paolo LaColla*

Central Research & Development Department, E. I. du Pont de
Nemours & Company, Wilmington, Delaware 19898 USA; and
Department of Microbiology, Virology Institute, University
of Cagliari, Sardinia, Italy*

MEDICAL SIGNIFICANCE

The picornaviruses and togaviruses are important pathogens in
man and some other animal species. Human infections caused by
picornaviruses are found in all geographical regions and climates.
The prototype of the group are the polioviruses, causitive agents of
poliomyelitis, a devastating disease of the central nervous system.
Thirty or more years ago poliovirus epidemics caused paralytic
disease in thousands of children and young adults, especially among
the middle and upper class of Western countries. The terrifying
nature of these epidemics helped launch high visibility,
public-funded animal virology research in the U.S. and western
Europe. Today the fruits of that support are visible; the
picornaviruses are among the best understood of all viruses at the
biochemical level. Togaviruses are usually arthropod borne but they
are not limited to tropical regions. Their infections are often
subclinical but sometimes lead to encephalitides and hemorrhagic
fevers which rank among the most lethal of all virus diseases.

The important groups of picornaviruses and togaviruses are
listed, together with the infections attributed to them in (1).

The diseases caused by these viruses range from life-threatening
to the mild and self-limiting (such as common colds). The
rhinoviruses alone cause about one-half billion colds every year in
the U.S. Except for the poliovirus subgroup, and a few of the

togaviruses, there are no practical vaccines, and approved clinical
antivirals are non-existent. The relatively extensive biochemical
studies of these two virus groups and their importance as pathogens
makes them particularly attractive candidates for antiviral design.

General Properties of Picornaviruses and Togaviruses

The viruses of these two groups are relatively small and simple
when compared to other animal viruses. A rather complete picture of
their structure, molecular biology and replication has been
constructed over the past twenty years. Recently the biochemistry
of the two groups has been reviewed in detail (2,3). Table 1
provides their salient features.

Table 1. Generalized Physical Properties of Picorna- and Togaviruses

Characteristic	Picornavirus	Togavirus
A. Virions	Particle wt. 6.5×10^6, diam. 30nm, do not contain carbohydrate or lipid, assembly in cytoplasm	Particle wt. $> 50 \times 10^6$, diam. 45-70nm, contain 50% protein, 10% carbohydrate, 30% lipid, assembly by budding
B. Nucleic acid	SS genome RNA of 7,500 nucleotides, mol. wt. 2.5×10^6 (35S), positive strand, with 3' poly A, capping 5' protein; mRNA similar, but no capping protein	SS genome RNA of 13,000 nucleotides, mol. wt. 4×10^6 (42-49S) positive strand, with 3' poly A, capping 5' nucleotide, genomic mRNA for non-structural proteins, mRNA (26S) for structural proteins
C. Proteins	4 structural plus RNA capping peptide, several non-structural, extensive protein cleavages	3 or 4 structural, three non-structural, extensive protein cleavages (alphas), glycosylation, covalent lipid
D. Viral-coded enzymes	RNA polymerase(s), protease	RNA polymerase(s), transcriptase, protease (?)

EARLY VIRUS-CELL INTERACTIONS

The early nonreplicative events in virus life cycles seem to have evolved under two major constraints. First, the need to take advantage of normal cellular processes in order to enter the cell and uncoat the genome. Second, the need to escape neutralizing antibodies prior to entry and hence the need to maximize the individuality of the surface structure of each serotype of virus. Broadly specific antiviral agents which interfere with normal cellular processes are likely to be toxic and agents which bind to and modify virion structures are likely to be too specific for utility. However, investigation of the mechanisms of action of antiviral agents in the laboratory can contribute significantly to understanding the early events in virus infection of host cells.

Virion Structure and the Identity of the Virus Attachment Protein

The togaviruses contain one nucleocapsid protein and usually two integral membrane glycoproteins, E1 and E2, which as heterodimers form the morphological "spikes" (4). The glycoproteins E1 and E2 also contain covalently bound fatty acids (5). A third envelope protein E3, may also be found in some togaviruses. Long flexible spikes such as these may have evolved for attachment to minimize the decrease in freedom of motion (entropy) for the virus during the initial steps of binding (6). Identification of which virion envelope protein recognizes the cellular receptor unit has been complicated by the fact that there are two mechanisms for attachment. Attachment to host cells at neutral pH differs from nonspecific attachment to host cells, erythrocytes, and liposomes at acid pH (7,8,9). The larger glycoprotein E1 probably plays the major role in both specific and nonspecific attachment based upon the effects of specific antibodies (10,11,12).

Picornaviruses are small nonenveloped viruses which contain four nonglycosylated capsid proteins; VP1, VP2, VP3, and VP4. Their capsid is constructed with sixty copies of each (13). The identity of the capsid attachment protein has been debated (14,15). VP1 is the leading candidate in most cases, and a basic portion of aphthovirus VP1 has been suggested as a site for receptor recognition (16). Since all of the picornavirus capsid proteins are in intimate contact, it is possible that the recognition of cellular receptors will not be attributable to a single polypeptide chain, in analogy with the heavy and light chain contribution to the antigen binding site of the immunoglobulins.

The Conformation of the Picornavirus Capsid

The picornavirus capsid is metastable and can undergo a profound conformational alteration which is required for progression of cell-associated virus along the pathway leading to infection.

Paradoxically, this conformation change can lead to irreversible
loss of infectivity when it occurs prematurely. The product of the
conformation change with enteroviruses or rhinoviruses is the
"A-particle" which shows increased sensitivity to proteases,
decreased stability to dissociating detergents, increased negative
charge, and which also sediments 10-20% more slowly than native
virions (15,17). Formation of these can be catalyzed by the host
cell (below) or by mild denaturation with heat, urea, and in some
cases acid pH. A-particles have lost native antigenic determinants
and the ability to attach to the specific cellular receptors. They
have also lost VP4. However, VP4 is not exposed on the surface of
native virions and probably does not confer the native properties to
the virus (15,17).

Cardioviruses undergo a conformational alteration which does not
lead to A-particles. Instead they "fall apart" into 14S fragments
containing five copies each of the three major polypeptides (13).
Native virions are sensitive to both pH and chloride ion
concentration, for example ME virus is unstable at pH5-6 in 0.1$\underline{M}$
chloride but stable in the absence of chloride. Aphthoviruses also
"fall apart" to 12S fragments and virion stability is dependent upon
both pH and chloride ion concentration (18,19).

The native conformation of some enteroviruses may also be
altered under mild conditions. For example Coxsackievirus A13 (20)
and poliovirus type 2 strain 712Ch2ab (21) are both unstable in
dilute neutral buffer at 37°. A-particles are formed and this may
be partly dependent upon the chloride ion concentration (21). At
the present time there is an obvious need for a systematic
investigation of the roles of different ions in the stability of
nonenveloped virus capsids.

Exposure of capsid proteins of a number of native picornavirus
virions and of altered subviral particles has been compared by
measuring their reactivities with various chemical labeling agents
(reviewed in (15)). The results generally indicate that VP1 is the
most exposed capsid protein in native virions and that the
conformational alteration to either A-particles or empty capsids
coincides with major alteration in the relative exposure of the
other polypeptides. With several picornaviruses, isolated VP1
(22,23) or a fragment of VP1 (24,25,26) elicit weak neutralizing
antibodies, and for poliovirus some, but not all neutralizing
monoclonal antibodies can recognize isolated VP1 (27,28,29). The
molecular locus of poliovirus neutralization has been mapped at
about 1/3 the distance from the N-terminal end of VP1 (30).
However, neutralization of poliovirus does not block attachment to
receptors, at least when using potent polyclonal antibody (31).

Isoelectric focusing can detect alteration in capsids. When
rhinovirus type 2 virions or "native" empty capsids are exposed to

pH5, they are irreversibly converted to A-particles or conformationally altered empty capsids and these have isoelectric points decreased by 2 pH units and do not attach to cells (32).

Mandel found that poliovirus can undergo a reversible conformational alteration with a 3 pH unit reduction in isoelectric point (33). Unlike rhinovirus, poliovirus virions can be intercoverted between the acidic and neutral forms by simply adjusting the pH of the mixture before the beginning of the electrophoretic separation. However, heated particles which have been irreversibly altered also have an acidic isoelectric point. Mengovirus, a cardiovirus, also has two "native" and fully infectious isoelectric forms but the acid form seems to have slightly greater hemagglutinating activity (34). Neutralization of poliovirus with polyclonal antibody also coincides with an "all or none" conformational transformation which "locks" the native virions in an acidic isoelectric form (35).

Oxidizing agents and UV light (36) can also catalyze formation of poliovirus A-particles. Poliovirus is unusually sensitive to inactivation by lactoperoxidase mediated iodination (37) and we have confirmed this (Table 2). We also found that incubation of lactoperoxidase-treated virions at 37° for one hour caused a further 98% inactivation of the small residual infectivity, i.e., that surviving virus was unstable (data not shown). When similar experiments were run with trace amounts of ^{125}I-iodide, we found that on the average 34% of the iodide became incorporated into the virions, but of this only 0.5% was recovered covalently bound to the capsid proteins following disruption with sodium dodecyl sulfate. The iodinated particles have sedimentation rates identical with native virions but they have about 50% greater electrophoretic mobility then native virions, presumably due to an increased net negative charge. They are unstable at 37° and are converted to particles which contain RNA, lack VP4, and which resemble A-particles in sedimentation (Figure 1).

Table 2. Lactoperosidase Inactivation of Poliovirus 2

	NaI	Lactoperoxidase	H_2O_2	$log_{10}\left(\dfrac{pfu\ treated}{pfu\ untreated}\right)$
1	--	--	200 nmol	0.1
2	--	25 µg	200 nmol	0.0
3	10 nmol	25 µg	--	0.1
4	10 nmol	25 µg	200 nmol	-2.9

Purified virus (712-Ch2ab) and reagents were mixed in phosphate buffer, pH 7.2, in a total volume of 1.0ml and incubated 10 minutes at 25°. Samples were then diluted directly in serum-containing medium for measurement of infectivity.

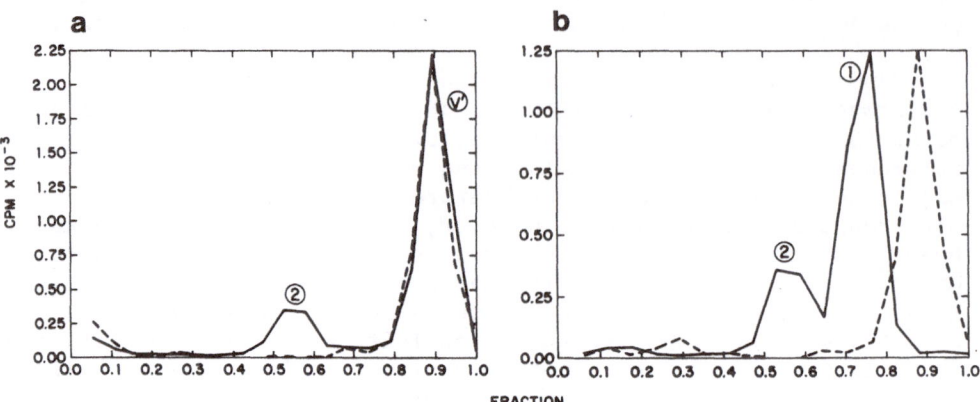

Figure 1. Sucrose gradient sedimentation of ^{14}C- poliovirus particles after iodination with lactoperoxidase and sodium iodide. Highly purified amino acid-labeled poliovirus 2 (approximately 100 µg) was mixed with 40 nmol NaI, 800 nmol H_2O_2, and 100 µg lactoperoxidase in 4 ml 0.05M phosphate buffer pH 7.2. The mixture was incubated 10 minutes at 25° and then 2.5 mM glutathione was added to stop the reaction. Aliquots were sedimented into 10-25% sucrose gradients before (A) or after (B) incubation at 37° for 2 hours. ^{32}P-labeled untreated virions were added as internal markers, sedimentation is from left to right. ^{125}I cpm x 10^{-3} is shown by the continuous line.

Our results are consistent with the following interpretation. Poliovirus reacts with iodine in the presence of the peroxidase system in two ways. A small amount of iodine is added to tyrosine exposed at the capsid surface. A larger fraction reacts with sulfhydryl groups in hydrophobic "folds" in the capsid. The products are unstable sulfenyl iodides which hydrolyze upon exposure to water; this exposure is slow at low temperature but rapid at 37°C. A similar hypothesis was used to explain the unstable product of iodination of tobacco mosaic virus with ^{125}I-I$_3$ (38). We calculated that approximately 20 atoms of ^{125}I were incorporated unstably into each polio virion. These particles, despite their increased negative charge could attach to host cells at 25° (data not shown). They had a greatly reduced infectivity, probably because they underwent "eclipse" at a premature stage of infection. At 37° the unstable virions (V', Figure 1) are rapidly converted to "A-particles" (Component 1, Figure 1B) and we separately confirmed that these did not attach to cellular receptors.

A second product of the iodination reaction (component 2) appears to consist of empty capsids.

It may be asked if the normal process of cell-mediated eclipse of P2 involves peroxidase. We do not believe this to be so, on the basis of other evidence. Peroxidase inhibitors (1 mM thiourea, 0.1 mM propyl thiouracil, or 0.1 mM ergothionine) failed to slow cell-mediated eclipse of unmodified virus. Also, the thiopyrimidine S-7, a potent inhibitor of cell-mediated eclipse (see below) does not prevent inactivation of P2 in the lactoperoxidase system.

The features of the poliovirus capsid which lead to its sensitivity to lactoperoxidase catalyzed iodination are not shared by human rhinovirus 2. This suggests that enzymatic iodination is not a universally applicable virucidal process. It would of be of interest to determine if inactivation of enteroviruses by iodine, bromine, chlorine, and ozone induce changes similar to those observed with the iodide-lactoperoxidase system we have described with poliovirus.

Processes other than conformational change can lead to inactivation of picornaviruses. Under exceptional circumstances RNA can be removed from native HRV2 viruses without changes in the capsid (32). RNA can also be damaged in situ. For example, in moderately alkaline buffers ammonia cleaves RNA within the poliovirus capsid but does not damage isolated RNA (39). A structure activity relationship among low molecular weight amines (40) has been found and ethyl amine is most active. The mechanism seems to be formally analogous to the action of "lysosomotropic bases" on whole cells; the uncharged lipophylic amine penetrates the virus capsid or intracellular endosomal vesicle and effectively raises the pH; this will be discussed below.

Attachment to Cellular Receptors

Virions usually recognize and attach to specific cellular receptors by multiple noncovalent bonds. It has often been possible to demonstrate partial saturation of specific cellular receptors with excess togavirus particles (7,41). With isolated octameric spike protein micelles from Semliki Forest virus it is possible to obtain true saturation, but care most be taken to avoid nonspecific attachment, at low pH (7). There often seems to be 5×10^4 to 5×10^5 specific sites per cell for togaviruses and hence more than 5×10^5 receptor units per host cell. Specific attachment of togaviruses is dependent upon ionic strength (42), pH (7,43,44), and may also be dependent upon temperature (45). The receptors may be inhibited or blocked by proteases and lectins (41) and are probably cell surface glycoproteins (46). Some work has suggested that HLA or H2 antigens are togavirus receptors (46a) although this has also been disputed (75). Much work remains before receptor "families" can be established for representative togaviruses.

Attachment of picornaviruses to specific cellular receptors has been extensively reviewed (14,17,47). The rates of attachment vary widely and can be sensitive to temperature, pH, and ionic composition (12,48,49) as well as the nature of the host cell (48,50,51,52). Since virions must diffuse to the receptor site, the maximum attachment rate is diffusion dependent and for picornaviruses and suspensions of cells, this can be calculated to be $1-2 \times 10^{-7}$ cm^3 per min (49). This rate is slow enough to be measured with some manual dexterity and has been observed for the rapid attachment of the cardiovirus EMC to HeLa cells (48). It can be calculated that if 1 in 10^4 collisions results in permanent attachment, the rate will be half maximal (17) and in most cases attachment is considerably slower than half maximal. Thus most viruses must "bump" up against the cell a great number of times to become permanently attached. In order to appreciably decrease the apparent rate of attachment, an agent or set of conditions must be chosen to reduce the efficiency of the collisions by a very large factor.

There are often about 10^4 specific picornavirus receptor sites per cell which suggests that there may be about 10^5 receptor units per cell, assuming that one site can contain about 12 subunits (17). Thus the receptor subunits are typically abundant surface molecules.

Cardioviruses, equine rhinovirus, and a bovine enterovirous attach to sialic acid residues on cell surface glycoproteins. Two different viruses may recognize the same glycoprotein (for example EMC and influenza virus both attach to glycophorin) but they recognize different residues on the same molecule and these residues differ in sensitivity to enzymatic inactivation (53). However, it seem likely from binding competition studies that even closely related cardiovirus strains can recognize entirely different receptor molecules on selected host cells (50). These differences may be related to virulence; it has long been postulated that receptor specificity mediates the tropism of picornavirus infection (51,52,54).

Binding competition usually clearly establishes "receptor families" among those picornaviruses which recognize non-sialic acid determinants. We have found four "families" of viruses which share saturable receptors on HeLa cells: Two families of human rhinoviruses, the poliovirus family and the Coxsackievirus B family (55). Interestingly, Coxsackievirus A21 shares receptors with one of the human rhinovirus families and also shares a simliar pattern of pathogenicity (upper respiratory infection). Also, a group of adenoviruses share the HeLa cell receptors recognized by the B group Coxsackieviruses. Thus it can be predicted that the purified receptor for the adenoviruses should be identical to the purified Coxsackievirus B receptor. Despite progress in purification of these receptors (47,56) it is not yet possible to test this.

Purification of virus receptors has proven exceedingly difficult, despite use of affinity chromatography, but use of monoclonal anti-receptor antibodies may speed up progress. Although antiviral agents which inactivate or compete with cellular receptors may seem attractive, it seems a priori unlikely that they will be practical. This is because receptors probably play essential roles for normal cells and because a significant decrease in the rate of virus attachment to cells requires an almost complete block in the proportion of successful virus-cell collisions. Some natural products are known to compete with or block cellular receptors. For example, concancavalin A (57) agglutinates togaviruses and also reduces attachment of picornaviruses. Slowing of virus attachment may also have undesirable consequences, some strains of virus with reduced avidity for cellular receptors have increased pathogenicity, probably because they disseminated more widely in infected animals (44,48,58).

Penetration and Uncoating of Togaviruses

The past four years has seen a dramatic increase in our understanding of the mechanism of entry of enveloped viruses into host cells, and this has been aided by use of chemicals which block early events of infection in tissue culture (59-67). Helenius and coworkers (68) found that Semliki Forest virus(SFV), is first bound to multivalent sites distributed on plasma membrane of host cells. SFV then migrates to clathrin coated pits and enters the cell, by adsorptive endocytosis, into vesicles which are non-coated or partly coated. These vesicles or "endosomes" (69,70) are formed by a process which occurs at temperatures above 10° and which is only partly inhibited by metabolic poisons and cytochalasin B (60,63) Endosomes are converted slowly to lysosomal vesicles by fusion with lysosomes at temperatures above 20°. Prior to this and a few minutes after formation, the pH of the endosome is lowered to about 5 (71,72,73) by means of an ATP-driven proton pump (74).

The pH within the endosome promotes fusion of the togavirus envelope with the membrane of the vesicle, therby permitting release of the core into the cytoplasm. This model also resolves the long standing debate (17,75) over fusion versus viropexis mechanisms for penetration of enveloped viruses. Entry of cell-associated SFV into the endosome can be monitored by loss of susceptibility to elution by proteases, while fusion and removal of the envelope may be monitored by susceptibility of the virion RNA to ribonuclease (60,63).

The acidic environment of the endosome promotes fusion of enveloped viruses by inducing a conformational alteration in their spike proteins, as has been detailed by biophysical studies of the myxovirus hemagglutinin (76). Fusion of three families of enveloped viruses (toga-,myxo- and rhabdoviruses) with cell plasma membranes (65,77) erythrocytes (78) or directly with cholesterol-containing liposomes (8,68) can be induced by simply adjusting these systems to

pH below 6. The mechanism for this may be related to the non-specific
attachement of the spikes to erythocytes and other cells at low pH, as
already discussed. Fusion of togaviruses to host cells at low pH
prior to entry into the endosome can lead to productive infection
(68,77).

Elucidation of the strategy for togavirus penetration and
uncoating has relied heavily on chemical inhibitors which prevents
acidification of endosomes. These inhibitors are lysosomotropic
amines, ionophores, and metabolic poisons. Lysosomotropic amines
(68,70,79) penetrate the cell in the uncharged basic form, and then
neutralize compartmentalized acids. They include ammonia,
methylamine, tributyl amine, and more complicated molecules such as
chloroquine and amantadine (Figure 2a and 2b). Ionophores such as
monensin or FCCP which are carriers of selected cations or protons,
also reduce pH gradients within cellular compartments (73,80).
Metabolic poisons which greatly reduce intracellular ATP prevent the
ATP dependent proton pump from maintaining the acid pH of endosomes
and/or lysosomes (72). Poisoning of both glycolysis and oxidative
phosphorylation seems to be required.

There is a good correlation between ability to raise the endosomal
pH and the inhibition of infection by SFV (68,73). Levels of some
lysosomotropic amines and other agents which can bring the pH of
fibroblast endosomes from 5 to about 6,2 are: methylamine(10^{-2}M),
ammonia(5×10^{-3}M), chloroquine (1.4×10^{-4}M), FCCP(10^{-5}M), and
monensin(6×10^{-6}M). Many small amines such as procaine may also be
lysosomotropic (81) and older reports of antiviral activity by local
anesthetics (82) may have been a result of this action.

The molecular details of the conformational changes in the
togavirus spike protein which permit fusion with lipid bilayers below
pH 6 are not yet known. Neither is it known why cholesterol is a
necessary ingredient of liposomes (8,9) for fusion with virions. A
specific structure appears to be exposed only below pH 6. In the
para- or orthomyxoviruses the amino acid sequence of one relevant
portion of the fusion protein or hemagglutinin spike is known, and
peptide analogs have been found to inhibit fusion (83).

Penetration and Uncoating of Picornaviruses

It seems reasonable to discuss a model for early events of
infection which compromises the differences between the "viropexis"
and "direct fusion" theories for uncoating and penetration (Figure 3).
In this hypothetical model virions (a) recruit receptor units on the
plasma membrane (b) and then enter an endosome (c). The endosome is
acidified and the virion is conformationaly altered, in some cases as
a direct consequence of the pH(d). As is often the case (84) the
receptor units are also recycled to the plasma membrane since they are
no longer able to be attached to the conformatonally altered surface

of the virion. If virions are conformationally altered prior to entry
into the endosome, they may be "sloughed" from the cell surface. The
conformationally altered particles have acquired properties which
permit intercalation into the endosome membrane (e) and they may
either penetrate into the cytoplasm (f) or the RNA alone may
penetrate. The model is less likely to be relevant to picornaviruses
such as the cardioviruses and apthoviruses which do not form
A-particles than it is to the enteroviruses and rhinoviruses.

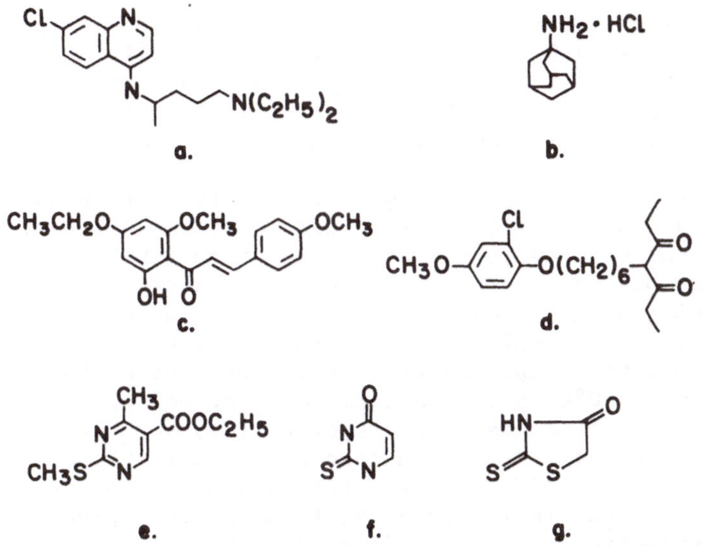

Figure 2. Agents which block the uncoating steps of togaviruses
and picornaviruses. (a) chloroquine, (b) amantadine
hydrochloride, (c) Ro 09-0410, (d) arildone, (e) S-7, (f)
2-thiouracil, (g) rhodanine.

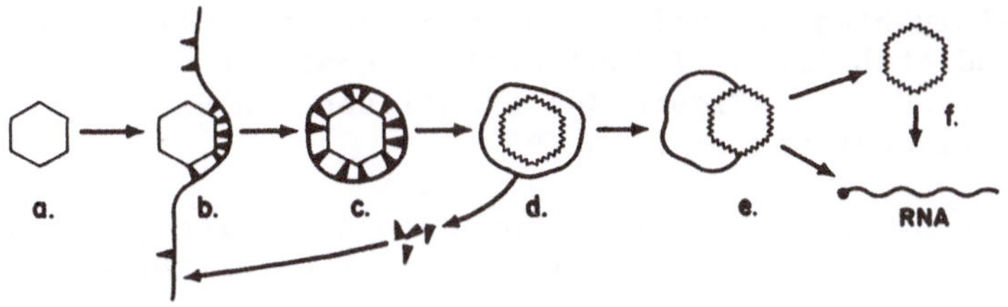

Figure 3. Hypothetical model for early steps in picornavirus
infection.

Although direct evidence for the model of Figure 3 is lacking, a number of observations are fully compatible with it. For example, rhinoviruses rapidly form A-particles at the pH of the endosome. Rhinovirus and poliovirus A-particles attach firmly to liposomes (21). Metabolic poisons appear to reduce the number of receptors for human rhinovirus 2 (HRV2) on host cells after reducing cellular ATP by about 90% (85). It is known that acidification of endosomes (an ATP-dependent process) is required for recycling of many plasma membrane receptors which otherwise rapidly accumulate in the endosome (84,86,87). (This explantion for the effects of poisons on HRV2 attachment could be tested by examining the effects of lysosomotropic amines which should act in the same way.) Also, metabolic poisons were found to reduce the rate of eclipse of HRV2 and poliovirus 2 by 93% and 82% (85). Thus the living cell has been shown to contribute to the eclipse process and it seems likely that isolated and purified receptors alone will be inefficient inducers of the conformational alteration which leads to eclipse.

A number of antiviral agents have been found to stabilize native virions in vitro and also to block cell-mediated eclipse. A simple example is sodium dodecyl sulfate (SDS) which inhibits HRV2 plaque formation at 0.1 mM levels. This antiviral activity was also detected using human rhinovirus 1A, but not a number of other rhinoviruses or with poliovirus 2. Antiviral activity paralled the ability of SDS to inhibit in vitro acid mediated eclipse, heat inactivation at 50°C, and also cell-mediated eclipse (88). A similar mode of action has been reported (89) for the compound Ro 09-0410 (Figure 2c). At 0.1 μM concentration it binds to HRV2, prevents infection and stabilizes the virions against heat inactivation. The binding is slow, temperature dependent and requires extraction with chloroform for reversal.

Arildone (Figure 2d) blocks cell-mediated uncoating of poliovirus at 0.3 μM levels and also protects the virus capsid from inactivation by heat and alkali (90,91). At substantially higher concentrations antiviral activity is detected for a Coxsackievirus, a togavirus, a rhabdovirus, and herpes viruses, but the mode of action at these higher levels appears to involve inhibition of RNA synthesis (92,93). The thiopyrimidine S-7 (Figure 2e) specifically and reversibly binds to polioviruses, prevents cell-mediated eclipse, and stabilizes virions against heat inactivation (94). Similarly, rhodanine (Figure 3g) stabilizes echovirus 12 virions and prevents cell-mediated eclipse (95).

Other agents have also been found to stabilize picornaviruses against eclipse through covalent modification of the capsid. For example, 1 mM 2-thiouracil (Figure 2f) reduces the specific infectivity of poliovirus type 1 by 90% (96). The residual infectivity of the treated virus is then stabilized against heat inactivation at 50°. Incubation of poliovirus with L-cystine prior to addition of 2-thiouracil prevented the reaction with thiouracil but also stabilized virus against heat inactivation (96). Thiouracil and cystine appear to react covalently with a SH group which is relatively sequestered in the poliovirus capsid, but not all strains of poliovirus appear to be susceptible (97). Inorganic polysulfides can also react in place of cystine and ^{35}S-Na$_2$S$_4$ was shown to be incorporated into virions (98), and then be removed again by prolonged incubation with cysteine or reduced glutathione (98,99). The modified and stabilized virus can still react with neutralizing antibodies (99) and hence retains native antigenicity. It is of curious interest that polioviruses treated with neutralizing antibody appear to be stabilized against cell-mediated uncoating (31).

A different mode of action may be involved in the relatively nonspecific abilities of glutathione and dithio-threitol at high concentrations to stabilize picornavirus virions against cell-mediated eclipse and heat inactivation (100,101). In contrast to the picornaviruses, SH reducing agents such as dithiothreitol inactivate both alphaviruses and flaviviruses (102).

In summary, it is possible to block picornavirus infection with agents which prevent conformational alteration of the capsid. Those agents which are effective at low concentration are usually very specific for a few viruses and do not have the broad spectrum of activity needed in the clinic. They are, however, useful laboratory tools and might also conceivably be useful as adjuvants during immunization since they should preserve the native (neutralizable) antigenicity of a viral immunogen. Much work needs to be done to understand how those agents interact with virions. For example, do uncharged agents (such as Ro 09-0410, arildone, S-7, thiouracil, or rhodanine) cause a conformational alteration in native virions, or do they modify the interconversions between acidic and neutral forms of virus detected by isoelectric focusing? Also, what, if any, is the relationship between the actions of those agents and the effects of neutralizing antibody on capsid conformation?

Other steps in penetration also need closer investigation. Do A-particles intercalate directly into lipid bilayers and if so, do they change the ionic permeability? If penetration of membranes occurs via exposure of "fusion" sites on the capsid protein, what are these sites and can ligands be designed to prevent or block their exposure?

VIRAL RNA SYNTHESIS

Perhaps the most studied event in the replication of picorna- and togaviruses is the synthesis of viral RNA. Yet the crucial events and detailed mechanisms are not well understood. In large

part this process is characterized by what chemists would term
"heterogeneous catalysis", with several viral proteins acting in
concert with host proteins in ultrastructural assemblages or
"bodies" in a multi-step process which amplifies the viral genome
10^5-fold in a few hours.

The kinetics of viral RNA synthesis are readily assayed;
addition of actinomycin selectively blocks transcription of cellular
RNA. Enucleated cells also support viral RNA synthesis normally.
The genome RNA is infectious, and incoming coat protein is not
required. Infecting RNA serves as messenger for production of a
RNA-dependent RNA polymerase, which catalyzes synthesis of a minus
strand. For picornaviruses the same or a different polymerase then
produces from the minus template many copies of plus strand. As
infection proceeds, proportionately more of the replicating plus
strand RNA is withdrawn into mature virions. The togaviruses are
similar, except that a subgenomic mRNA is transcribed from the 5'
third of the minus strand, and is used to produce structural
proteins. Viral RNA synthesis has been reviewed (Perez-Bercoff in
2, and Kennedy in 3). Greater than 90% of viral RNA made is of the
genomic type. Complementary or minus strands are not free in the
cell, but are associated with plus strands and polymerases. About
six nascent plus strands are associated with any one minus strand at
the mid-cycle of infection.

Full-length, double-stranded RNA is probably not a bona fide
intermediate in viral RNA synthesis. However, electron
photomicrographs show circular double-stranded structures in
replicating RNA preparations (103). Most would agree that viral RNA
is base-paired to a complementary strand for at least part of its
length during replication or transcription. This introduces the
possibility of using intercalating agents as antiviral drugs.
Ideally, the agent would be designed to intercalate into double-
stranded RNA, but not double-stranded DNA.

The lack of rigorous data on structural differences between ds
DNA and ds RNA, hinders the rational modification of well-known DNA
intercalators, such as mitomycin or acridine for this purpose.
Since picorna- and togaviruses produce their RNA in the cytoplasm,
another approach might be to modify a known DNA intercalating agent
so that it could enter the cytolasm, but not the nucleus. An
intercalating drug specific for replicating RNA would be expected to
interfere with normal polymerase action, and reduce the quantity of
viral RNA. It might also cause mis-reading by a polymerase passing
intercalated sites, introducing lethal mutations into the progeny.

Are the Ends the Means?

It is now becoming feasible to design synthetic oligonucleotides
which competitively bind to the viral polymerase or transcriptase,

or to sequences of the genome required for polymerase binding, and thus prohibit initiation of genome or mRNA synthesis. Sequence analyses which may provide a blue-print for such inhibitory oligonucleotides are now available in great detail for several picornaviruses and alpha togaviruses.

The 5' or 3' end sequences which would naively be the most crucial to initiate complementary strand synthesis. There is supporting evidence for this from analysis of defective (DI) togavirus genomes which effectively interfere with RNA synthesis of the parent virus. Kennedy (in 3) reported that minimal-size DI togaviruses contain both 5' and 3' end sequences of the viral genome, but lack virtually all (90%) of the internal sequence. As few as 750 nucleotides (not counting poly A) are sufficient.

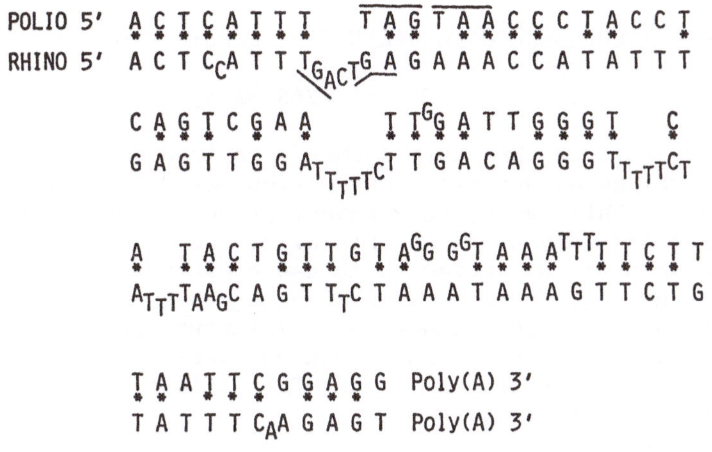

```
POLIO 5'  A C T C A T T T   T A G T A A C C C T A C C T
          *   * * * * *     * * * *   * * * * *
RHINO 5'  A C T C_C A T T T_G_ACT_G A G A A A C C A T A T T T

          C A G T C G A A     T T^G G A T T G G G T   C
          * * * * * * *       * * *   * * * *         *
          G A G T T G G A_T_TTT_C^T T G A C A G G G T_TTT_C^T

          A   T A C T G T T G T A^GG G^GT A A A^TTT T C T T
          * * * * * * * * * * *   * * *   *   * * * * *
          A_TTT_T_A^A^gC A G T T_T^C T A A A T A A A G T T C T G

          T A A T T C G G A G G  Poly(A) 3'
          * * * * *   * * * *
          T A T T T C_AA G A G T  Poly(A) 3'
```

Figure 4. Comparison of rhinovirus type 2 and poliovirus 3' non-coding regions.

The 3' sequences of a poliovirus and a rhinovirus are shown in figure 4. The comparison illustrates a number of interesting features. First, there are homologies between the two sequences. Since neither codes for protein, homology suggests conservation of structural features in the RNA. These features may be needed for binding of polymerase or structural proteins during virus assembly. Another feature seen frequently at the 3' ends of viral (and some cellular mRNAs) is a low G-C content, indicating a relatively

unpaired, open structure. Present also in some picornaviruses is
the sequence -AAUAAA-, which is thought to provide transcription
termination/poly A addition signals in mRNAs (rev. by Fellner in 2).
However poliovirus, FMDV and alpha togaviruses lack this sequence.
The alphaviruses do contain a highly conserved 3' sequence of 20
bases: 5'-AUUUUUUUUUAACAUUUC-poly A-3' (104,105,106). Defective
interfering particles also possess this sequence, and most
interestingly, a Sindbis virus mutant resistant to DI virus has two
base changes within the twenty nucleotide sequence (107). There are
also a number of possible palindromic, or mirror-image, sequences at
the 3' ends of the togavirus RNAs, and one palindrome has been
reported in foot-and-mouth disease virus (108). Reiterated
sequences are also present.

     Internal sequences are important in regulation of togavirus
transcription. A detailed study was made of the sequences which
initiate 26S RNA synthesis in several alphaviruses (109,110) and
this clearly revealed a 21 base sequence homology:

5'——————————ACCUCUACGGCGGUCCUAAAU (49S genome RNA)——————————3'

                    5'-AU (26S mRNA)——————————3'

A cautionary note is that this sequence is in a coding portion of a
non-structural gene, and its conservation may reflect translational
requirements. This aside, the transcriptase, which synthesizes coat
protein mRNA probably binds at this site, and it may be possible to
synthesize an oligonucleotide which can compete. It may be inferred
from the large differences between the sequence above and the 3'
terminal sequence of the genome that different specifities are
involved in the enzyme(s) which produces minus strand and mRNA. A
third sequence occurs at the 3' end of the minus strand for the
alphavirus, Sindbis (111):

     3'-GUAUCCGCCGCAUCAUGUGUG ——————————————————————————5'

This implies that a third enzyme may be necessary for plus strand
synthesis. It is interesting that there are three non-structural
alphavirus proteins present in the RNA replication complex (see
below).

     What likelihood is there that a conventional oligonucleotide can
penetrate intact cells to reach a polymerase binding site? This is
a difficult question although some positive evidence has been
reported (for reviews see 112,113). Even if uptake occurred, could
an oligonucleotide compete effectively with extended binding regions
of the much larger viral genome and its polymerase? Although these
are difficult concepts, good model systems are available with toga-
and picornaviruses for testing it. In addition, alterations in
permeability of virus-infected cells (176) may permit use of
compounds which are impermeable in healthy cells.

Viral RNA Polymerases

The RNA polymerases of toga- and picornaviruses are associated with replicating viral RNA structures in cytolasmic vesicles. Conflicting data place different viral polypeptides in these structures. All the data taken together do not exclude any of the viral proteins from playing some role in initiation, elongation or release of viral RNA from replication complexes. The situation is less confusing when enzymatic properties and polypeptide compositions have been determined for purified polymerases. Non-structural polypeptides of molecular weights 86, 72 and 70×10^3 were found in the isolated Semliki Forest virus polymerase (114). More recent studies detected three unique, stable polypeptides of mol. wts. 89, 76, and 60×10^3 (Schlesinger and Kaariainen in 3). If we accept that there are three distinct non-structural polypeptides coded by the alphaviruses, it is tempting to assign to them the three functions required for synthesis of the 3 viral RNAs discussed above.

Some progress has also been made with the picornavirus polymerase. The enzyme prepared from poliovirus-infected cells copies poliovirus genome RNA into a product containing all of the complementary sequences (115,116). Priming is accomplished either by oligouridylate or a host factor, probably a protein. The enzyme can also be forced to produce negative strand copies of other poly-adenylated mRNAs, including plant viruses and cellular mRNA (115). The possible role of a cellular protein is the initiation reaction may be a problem unless specific inhibitors can be developed.

The most ubiquitous viral protein in the picornavirus polymerase preparations is variously called p56, p63, NCVP4, or E (117, Diskin in 2). Fortunately, these all turn out to be homologues of the poliovirus sequence comprised by amino acid 1728 through 2207 (118). Antibodies against this protein (119,120) or synthetic peptides corresponding to it (121) inhibit polymerase activity and the protein probably has a function in synthesis of viral RNA. In our opinion, precursors to this protein also play a role in control of initiation of viral RNA synthesis. The replicase of poliovirus is proteolytically unstable (122) but NCVP4 is not. This implies that a larger, unstable polypeptide, possibly NCVP1b (mol. wt. 85,000) is the functional initiator of viral RNA synthesis, and it is also a possible target for designed antivirals.

Structure of NCVP1b

Having nominated NCVP1b as a picornavirus polymerase, we are fortunate in having its entire amino acid sequence, derived from the genome sequence (118). The protein is a superb candidate for a multi-functional enzyme. Near its amino-terminus (amino acid 1456)

is the sequence for VPg, the genome-linked peptide.
Carboxyl-terminal to VPg is a proteolytic enzyme which ensures
correct cleavage of the preceding and following protein domains
(123), and then the coding sequence for NCVP4 follows. Antibodies
to both VPg and NCVP4 prevent proper polymerase function, and thus
the extreme ends of NCVP1b are both essential for the polymerase.
Unfortunately, at the present time, no detailed analysis of the
secondary and tertiary features of NCVP1b or NCVP4 have been
presented. We are not aware of sequences in the polio polymerase
reminscent of other polymerases or other enzymes. The amino acid
content is not remarkable, nor are there obvious domains responsible
for nucleotide binding. The functional anatomy of this protein must
be elucidated before rational design of inhibitors is possible.

    The reactions involving VPg are of interest as targets (rev. in
124). The peptide becomes covalently attached to viral genome RNA
and minus strands through a tyrosine hydroxyl at position 3 (see
below).

    GAYTGLPNKKPNVPTIRTAKVQ (Polio VPg)
       |
       O
       |
    U-viral RNA

The peptide is rich in basic residues, and several prolines provide
strong secondary folding. Infecting viral RNA has its VPg removed
by a cellular enzyme, but attachment of RNA to VPg is carried out in
an unknown manner. These steps are undoubtedly important and
interference with them would impair replication. Proteolytic
processing of VPg will be treated in a separate section.

    Synthetic inhibitors of the polymerase of picornaviruses include
a wide variety of structures (125), for example certain crown
ether-containing compounds (126).

An 18, Crown 6 Ether Antiviral

Such compounds bind to NCVP1b and its products, and are potent inhibitors of viral RNA synthesis at submicromolar levels. Resistance is rare, but occurs at a frequency of 10^{-5} to 10^{-6}. A general property of 18,6 crown ethers is their chelation of metal ions. The polymerase of polio contains zinc (116,120) and possibly other metals, and selective chelating agents may be inhibitors. Active crown ethers may both bind to an amino acid sequence of the polymerase, and also chelate the metal coenzyme. Many 15,5 crown ethers were found inactive as inhibitors, and they are not effective metal chelators. The most active compounds in the series are not cytotoxic at levels 50-fold above the antiviral ED90.

Guanidine ($CN_3H_6^+$)

This classic inhibitor of the RNA polymerase of certain picornaviruses is still a valid object for study after more than twenty years of research. Inhibition is seen only at relatively high levels ($\sim$1mM), but these levels are not cytotoxic. Viral RNA synthesis is rapidly affected by the compound, and other effects on viral protein synthesis and assembly seem to be secondary (reviewed in 127). Various reports have assigned the site of action of guanidine to the polio polymerase (127), structural proteins of poliovirus (128,129) and a non-structural polypeptide of FMDV of unknown function (130). Whatever the site of action of guanidine, changes in viral sensitivity to it are frequent (10^{-3}) and dramatic. Thus the site is quite different from that bound by crown ethers.

TRANSLATION AND CO- OR POST-TRANSLATIONAL MODIFICATIONS OF VIRAL PROTEINS

Viral protein synthesis and processing in cells infected by picornaviruses and alpha togaviruses has been extensively studied (rev. 2,3). A favorable feature for these studies is the drastic inhibition of cellular protein synthesis (and protein synthesis of most other coinfecting animal viruses) by the small positive-stranded RNA viruses. This effect permits radioisotope labeling solely of viral proteins shortly after infection with moderate to high multiplicities. The viral proteins are in the main found as insoluble aggregates in lysed cells; however, addition of ionic detergents and reducing agents, followed by heating, solubilizes the aggregates, and permits their separation in polyacrylamide gels according to size. Alternatively they may be resolved by isoelectric focusing in pH gradients containing 8 molar urea.

The shut-off of host protein synthesis by picornaviruses has been of long-standing interest as a dramatic example of translational control. Following infection there is usually, but not always, viral-induced inactivation of factors required in

translation of cellular proteins, particularly the binding of
cellular mRNA to the initiation complex. The mechanism is of
fundamental interest. However, at present, no clear foundation for
design of viral inhibitors based on the reactions is available.

The initiation of viral translation may be a process which could
be specifically inhibited, but little is known about the molecular
details. Virus-specific stem and loop structures have been
postulated in poliovirus mRNA sequences upstream of the
ribosome-binding sites and prior to translational start signals
(118,131). Such a model needs further verification and to our
knowledge, there have been no reports of synthetic virus-specific
inhibitors which act by this mode (125). Picorna- and togaviruses
may be blocked at this stage by endogenous agents induced by
interferon; this aspect of interferon action needs much further
study to be definitive.

The mRNAs of picorna- and togaviruses appeared at one time to be
unique in the sense that they coded for several proteins, but used
only a single translation start signal. Their translation product
is a polyprotein containing several domains, each with a distinct
function. Gene expression is controlled by elaborate post-
translational proteolytic events, which generate viral structural
proteins, polymerases, etc. (reviewed in 2,3). This surprising
result gave immediate promise of an opportunity for intervention.
However, it has become apparent that this type of regulation is not
limited to picorna- or togaviruses. In fact, most animal viruses
use some degree of proteolytic processing for structural proteins
(122). And it has recently become more clear that there are also
polyprotein precursors for some classes of cellular proteins. The
neuropeptide hormones provide several very clear examples of large
mRNAs with single initiation sites, producing a number of different
protein/peptide products by cleavage (132).

A different strategy, possibly unique, occurs in the production
of flavivirus proteins. The viral mRNA is translated to yield
several protein products, by virtue of several postulated internal
translation initiation sites (by Westaway, in 3). This has not been
clearly described before and the viral-specific internal initiation
sites, once characterized, should yield significant data about
virus-specific translation initiation signals and co-factors (133).
Such a strategy is, of course, well-known with the small
bacteriphages, as well as poly-cistronic bacterial messages with
internal start and stop signals.

Proteolytic Processing of Viral Polypeptides

While many animal viruses (and others infecting insects, plants
and bacteriophages) use proteolytic processing during virion
assembly, the picorna- and alpha togaviruses are the most

outstanding examples, because they use cleavages to form all viral
polypeptides.

In picornaviruses, the cleavages of the structural precursors do
not cause the products to separate. To the contrary; there is
alteration in conformation, antigenic structure and isoelectric
point and the cleavage products assemble into larger and more
complex structures, culminating in the viral capsids (rev. in 134).
When the RNA associates with the capsids it is locked in by a final
protein cleavage which also locks the virion into a specific native
configuration which recognizes cellular receptors. Proteolytic
cleavages regulate RNA synthesis as well, by activating and
deactivating the viral polymerase, and also by generating the
genome-linked peptide. For a review of protein cleavage pathways of
picornaviruses, see Rueckert in (2).

The alphaviruses behave similarly, except that the
non-structural and structural polypeptides are produced from
different mRNAs. The Semliki Forest virus uses at least four
protein cleavages (104). The core protein is cleaved from the
nascent chain as soon as its translation is completed, this is
mediated by the core itself or a protease associated with the viral
polysomes. By comparison, the cleavages of envelope proteins 2 and
3 occurs about thirty minutes after synthesis, possibly at the
plasma membrane. Cleavage of a 6K polypeptide from the amino
terminus of envelope protein 1 occurs during intracellular
transmembrane movement, possibly in a manner analogous to trimming
of signal sequences from secreted cellular protein hormones. For a
full view of translation and proteolytic processing of alphavirus
polypeptides, see Schlesinger and Kaariainen in (3).

Although the phenomenon of protein cleavages in virus-infected
cells is throughly documented, the actual function or role of the
cleavages is not clear. For example, the extensive proteolysis of
the picornavirus structural precursor does not cause dissociation of
the products, but rather alters their folding. Why this cannot be
accomplished without cleavage remains to be understood. It seems
likely that cleavage is used for modulation of protein function in a
system where transcriptional and translational controls are not
favored.

Cleaved Sites in Viral Precursor Proteins

In the past several years the amino acid sequences of the
cleavage recognition sites have become known for several picorna-
and togaviruses. This data has in part been provided by end-group
analyses of viral proteins (135,136), but more information has been
deduced from sequencing of viral genomes, mRNA, or DNA complementary
to the viral RNA. The latter technique has permitted analysis of
large portions of togaviruses (104,105), picornaviruses,

(137,138,139) and the entire sequence of a poliovirus (118).

Comparisons of the cleaved sites in precursor proteins of two picornaviruses and two alphaviruses are shown in Table 3. The picornavirus sequences have several rather distinct classes of cleaved sites. First are the nascent cleavages sites, processed almost instantly during translation. They are reminiscent of chymotryptic cleavage sites; the new carboxyl termini being donated by aromatic (tyrosine) or hydrophobic residues. Next are the rapidly processed sites, such as those in structural precursors (half-life of 5 min. in infected cells). For the picornaviruses these contain glutamine-glycine or glutamic acid-X residues, often surrounded by hydrophobic (leucine, isoleucine, valine) sequences. Sites which are more slowly cleaved also contain glutamine or glutamic acid, but are in more neutral or hydrophilic regions of the precursors, which may alter their affinity for the participating proteases. Lastly, there are distinctly specific sites which are processed during combination of the viral RNA and the coat proteins, the so-called maturation cleavage.

There is greater diversity at cleavage sites of alphavirus structural proteins, although alanine residues tend to be present. While there is moderate to high conservation of the individual sites between two alphaviruses, different sites show substantial chemical differences, suggesting participation of several proteases. All cleavages can be considered chymotryptic-like, except for the di-basic site cleaved between envelope proteins 2 and 3. The latter trypsin-like cleavage is commonly found in the intracellular proteolysis generating cellular secretory proteins (140). The occurrrence of highly conserved peptide regions at processing sites of viral precursor proteins may offer an avenue for design of selective inhibitors (see below).

Origin of the Processing Proteases

At the outset, it should be emphasized that the important cleavages of picorna- and togavirus precursor proteins in infected cells do not resemble the normal degradations of intracellular proteins. As a rule, the ends of proteolytic cleavages of viral proteins are at highly site-specific regions (see Table 3), and no small soluble peptides or amino acids are released. Thus, the cleavages are not part of the well-known degradative pathways present in all cells.

From a number of lines of evidence, it appears that specific cleavages of nascent viral polypeptides are carried out by cellular proteases associated with the translational apparatus. These may be the membrane-bound proteases which process signal sequences from nascent cellular secretory proteins. Another possibility is a ribosomal protease of animal cells (141). The cleavage specificity

Table 3. Viral Protein Cleavage Sites.

Picornaviruses

| Poliovirus (nascent) | leu thr thr tyr/gly phe gly his |
| | asp ala met tyr/gly thr asp gly |

Poliovirus (intermediate)	pro arg leu gln/gly leu pro val*
	ala leu ala gln/gly leu gly gln*
	ala leu phe gln/gly pro leu gln*
	ala met gln gln/gly ile thr asn
	val ile lys gln/gly asp ser trp
	ala gly his gln/gly ala tyr thr

| Poliovirus (maturation) | pro met leu asn/ser pro asn ile |

FMDV (structural)	pro ser lys gln/gly ile phe pro
	pro arg thr gln/thr thr ser thr
	lys gln leu leu/asn phe asp leu

| FMDV (non-structural) | glu/gly leu ile val |

| FMDV (maturation) | ala leu leu ala/asp lys lys thr |

Alphaviruses

| Core/E3 (nascent) | ser glu glu trp/ser ala pro leu | SFV |
| | thr • • •/ • • • • | Sindbis |

| 6K/E1 (nascent) | ala pro val ala/cys ile leu ile | SFV |
| | lys val asp •/tyr gln his ala | Sindbis |

| E3/E2 | arg his arg arg/ser val ser gln | SFV |
| | • ser lys •/ • • ile asp | Sindbis |

| E2/6K | arg ala his ala/ala ser val ala | SFV |
| | ser • asn •/glu thr phe thr | Sindbis |

* – rapid
• – identical

is reminiscent of chymotrypsin, a serine-active site protease. Both
the nascent cleavage reactions and the ribosomal protease are
blocked by diisopropylphosphofluoridate (DFP) a classic inhibitor of
serine proteases (141,142). In our view, if as seems likely, the
nascent cleavages, are carried out by a cellular protease of unknown
host function, it is not a suitable target for antiviral agents.

A body of evidence from several laboratories established that
the intermediate protein cleavages of picornavirus precursors, some
of which yield the viral structural proteins, are carried out by a
virus-coded protease. The enzyme activity is not detected in
extracts of uninfected cells, but is found in lysates of infected
cells in quantities which increase in time after infection and with
the amount of infecting virus (143). Cell-free protein synthesizing
systems, programmed with picornaviral RNA, produce a specific
proteolytic activity which also processes the capsid polypeptides
(144). In both infected cells and cell-free extracts it was shown
that the protease is not efficient at cleaving proteins of
heterologous viruses or cellular proteins (145,146). Fewer studies
have been done with the alphaviruses, but there is some evidence for
an auto catalytic cleavage of the core polypeptides (147,148). This
latter finding needs additional confirmation.

The picornavirus protease is apparently a member of the class of
proteases with cysteine at the active site, based on inhibitor
sensitivity (143,144). Other members are the well-known plant thiol
proteases,; papain, ficin, etc. Some lysosomal proteases of
mammalian cells (cathepsin B) are also of this class. In general,
cysteine proteases are usually not highly specific with regard to
substrates cleaved, and thus the specifity determinants of the viral
enzymes are of innate biochemical interest. Glutamine or glutamic
acid cleaving proteases are described in bacteria, but have not been
well-characterized.

The identity of the picornavirus protease has been a matter of
some controversy. An initial report claimed that a structural
polypeptide of EMC virus possessed protease activity (149).
However, a subsequence study (144) indicated that a non-structural
translation product was required. It was then proposed that a
polioviral non-structural polypeptide of approximate molecular
weight 40,000, probably NCVPX, was the polioviral processing
protease (143). This result was then challenged by three groups,
based on independent purification methods or antibody studies
(150,151,152). A way to reconcile all the data would be to nominate
a 22,000 molecular weight polypeptide (designated 7C in poliovirus
and p22 in EMC) as the minimum-size version of the enzyme. This
would explain its misassignment as a structural protein, since
structural polypeptide gamma of EMC nearly co-migrates with p22.
The protease sequence is also present in larger precursors,
including a 40,000 molecular wieght form, designated p6a of

poliovirus (118). This larger form of 7E has protease activity, as do poliovirus NCVP2 and polypeptide D of EMC (153 and our unpublished results). Ultimately, proof of coding of the protease by the picornavirus genome and its identity will come from a combination of genetic analysis of protease mutants and recombinant DNA studies which demonstrate that a particular sequence of viral cDNA expresses the enzyme in a heterologous host.

Inhibitors of Viral Protein Cleavages

This topic has been reviewed recently (154). The approach seems to be promising for design of antiviral agents for those animal viruses which produce their own specific proteolytic enzymes, that is picornaviruses, retroviruses and adenoviruses, and possibly alphaviruses, myxo- and paramyxoviruses, poxviruses, and herpesviruses.

The mode of action of cleavage inhibitors can be classified in either of two general categories; alteration of the substrate to prevent its proper cleavage, or inhibitors of the protease catalytic activity. The first category contains certain protein modifying reagents, amino acid analogs, high or low temperature and some metal ions, zinc being the most often reported. Inspection of this category reveals that none of these treatments is likely to lead to specific alteration of viral proteins. For example, analogs will be incorporated into cellular as well as viral proteins. High temperature effects, as well as being interesting, are possibly of clinical significance, since with some viruses, 38°-39° is sufficient to alter normal processing. This is a temperature which is often encountered during fever episodes accompanying clinical viral infections. One of the more useful effects is that of zinc ions. Selected members of the picornavirus group, particularly human rhinoviruses, are very susceptible to zinc ions added to tissue culture medium. A very rapid inhibition of rhinoviral protein cleavage, particularly of the structural protein protein precursors occurs in 0.1 mM zinc medium (155). The effect is reversed by replacing the culture fluid with medium lacking zinc. The zinc appears to alter precursor structure by binding to cysteine on other residues in the substrate rather than a direct action on the viral protease (156). A potential problem is utilizing zinc is the frequent selection (10^{-4} to 10^{-5}) of resistant mutants, which map in the structural proteins. Nevertheless, animal and human studies have indicated in vivo efficacy of zinc ions versus rhinoviruses (G. Eby, personal communication).

The second approach is targeted at the proteolytic enzymes. The picornavirus protease is very sensitive to inhibitors of thiol proteases, eg mercurials, N-ethyl maleimide, and idoacetate. We have begun to design inhibitors which mimic the structure of the cleavage sites, based in part on their peptide sequences. For

example, we found that carbobenzoxy leucine chloromethyl ketone was
an irreversible inhibitor of poliovirus protease (143) and showed
also that it blocked viral protein cleavages and virus production
(154).

Inhibitors of viral protein cleavages are readily tested for
picornaviruses. The test can be made at a time after infection when
the cellular protein synthesis is completely inhibited and only
virus proteins are labeled with radioactive amino acids. The viral
proteins are size-separated in polyacrylamide and any inhibition of
cleavage can readily be detected. Variations on the assay involve
mixing purified viral protease with radioactive viral precursors
obtained from cell-free translation or with synthetic substrates
(157). The results may then be visualized by autoradiography of
dried gels (see figure 5). A spectrophotometer may also be used if
the cleaved substrate is chromogenic (157).

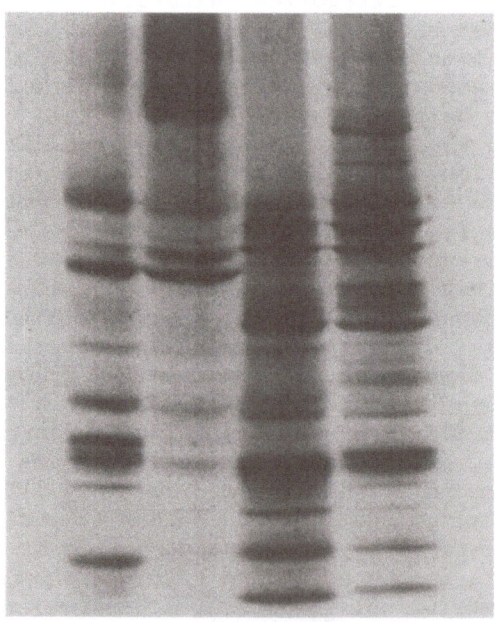

Figure 5. Picornavirus protein synthesis and inhibition of viral
protein cleavages. Lane 1: Rhinovirus type 1A proteins in
infected Hela cells. Lane 2: Effect of $2x10^{-4}$ molar zinc
chloride on HRV-1A processing. Lane 3: Poliovirus type 2. Lane
4: Addition of 10^{-4} molar idoacetamide to poliovirus-infected
cells.

We have recently found that certain natural protease inhibitors
prevent viral protein cleavages (158). Human alpha-2-macroglobulin

or rat alpha-1-macroglobulin blocked poliovirus protease action in
extracts, and covalently bound the enzyme. This finding has been
useful in characterizing the viral protease, and suggests that some
naturally-occurring inhibitors in serum or tissues may play a role
in limiting viral infection or pathology. The chemistry of the
alpha macrogobulins is interesting and involves the proteolytic
activation of a thiol-ester in the inhibitor. The activated
inhibitor then reacts covalently with a nucleophilic residue on the
protease. Since the activation of the inhibitor involves a major
change in protein conformation, it seems unlikely that low molecular
weight mimics could be synthesized as "suicide" inhibitors of the
viral proteases. A different approach could be to attach peptide
masking groups to toxins. If the peptide were removed by
viral-specific protease action, a toxic, short-lived species could
be generated inside the infected cell. Such an approach was
successful for in situ destruction of tumor cells expressing
collagenase (159).

     Some proteases have cofactors which are required for activity.
An interesting example is the recA protease of E. coli which
requires single-stranded polydeoxynucleotides and ATP in order to
make a specific cleavage in gene repressor proteins (160). The
protease associated with picornavirus maturation and generates the
internal structural peptides as viral RNA enters the capsid may have
similar characteristics. Potential inhibitors of the enzyme might
include polynucleotide analogues. A useful overview of possible
control points in proteolytic reactions of clinical significance is
available (162). From an additional standpoint, inhibitors of viral
proteases are appealing; the recent sequence studies of poliovirus
genomes show a high degree of conservation of the enzyme among
different poliovirus serotypes and their attenuated derivatives
(131). We confirmed this by determining the very low frequency
($<10^{-7}$) of spontaneous mutants resistant to bona fide inhibitors of
poliovirus protease.

Glycosylation of Togavirus Structural Proteins

     The envelope proteins of alphavirus virions are glycosylated.
E1 and E2 contain from 8 to 14% carbohydrate, and E3 is almost 50%
carbohydrate (rev. by Simons in 3). The polysaccharides are linked
by so-called N-glycosylation to the viral protein asparagine
residues. There are two classes of oligosaccharides present; the
high mannose type and the complex type (162), both of which
assemble during transport of the glycoprotein precursors through the
Golgi apparatus or other membrane systems of the host. Inhibitors
of glycosylation, such as 2-deoxyglucose or tunicamycin, can prevent
transport and assembly of togavirus proteins to a degree (3). It is
clear, however, that under-glycosylated virus particles which are
formed have approximately normal specific infectivities. The
stability of the virions lacking carbohydrate as well as
antigenicity may be somewhat altered.

The alphaviruses serve as a useful model for synthesis of
glycoproteins. However, the viruses appear to utilize host enzyme
systems entirely for carrying out these complex events. The
carbohydrate attachment sequences in the precursor proteins are
asparagine-X-ser (or thr) (104), a standard cellular sequence used
as a cellular signal for carbohydrate addition, and the type and
amount of sugar added is dependent mainly on the host cell rather
than the particular togavirus. Taken together, the data do not
provide much basis for selective inhibition of viral glycosylation.
It can be argued that little is known about the regulation of
carbohydrate addition (160), and that subtle, unrecognized sequence
differences between virus and host glycoprotein structures may be
used to advantage in the future. An interesting sidelight is that
the amount of the terminal sialic acid on the oligosaccharide chains
determines the ability of Sindbis virus to activate the alternate
pathway of complement. The lower the abundance of sialic acid on
the virion, the better the activation of the host immune system
(163). This result suggests that a treatment which reduced the
amount of sialic acid on the virus might be beneficial during
viremia. In conclusion to this section, it appears to us that
inhibitors of viral glycosylation are likely to have similar and
unacceptable effects on cellular glycoproteins.

Other Post-Translational Modifications

The envelope proteins of alphaviruses are modified in yet
another way during passage through the cell interior to the surface.
Envelope proteins 2 and 3 of Sindbis virus receive covalently
esterified fatty acids (164). This was shown by labeling infectd
cells with radioactive palmitic acid. The reaction probably occurs
as the envelope proteins are moving through the Golgi apparatus, ten
to twenty minutes after their synthesis is completed. Accessibility
studies indicate that the fatty acid moiety is located in a portion
of the glycoprotein located within the viral membrane. Thus the
function of the fatty acid may well be to position and then anchor
the protein-membrane complex by hydrophic interactions. Serine
hydroxyls are a common site for acylation by fatty acids in cellular
membrane proteins, and this can convert a moderately hydrophilic
site into a lipophilic one (rev. in 165). At this stage it would
seem that selective inhibition of acylation of viral glycoproteins
would be difficult to achieve.

Protein kinase activity is well-known with RNA and DNA tumor
viruses, and appears to be a virus-specific activitiy of unknown
function. An interesting recent study on FMDV found protein kinase
released or activated in disrupted FMDV virions (166). The
specificity is unclear, but apparently cellular proteins as well as
viral can be substrates. There has been documentation that two of

FMDV structural proteins are phosphorylated (168). It has not yet
been proven that the enzyme is virally-coded, although highly
purified virions provide a good starting point for further
purification and characterization (167).

MORPHOGENESIS

     The association of viral proteins with each other and with viral
RNA to yield infectious virus particles is a complex and fascinating
set of reactions, of fundamental interest in cellular as well as
molecular biology. The various stages of the process are driven by
virus-specific physical-chemical (and possibly enzymatic) events
which offer possibilities for intervention.

     Some virus systems offer excellent possibilities for studying
morphogenesis in vitro, with puirified viral components. The RNA
plant viruses offer excellent models in this regard. In general,
animal viruses do not offer faithful assembly systems in extracts,
and thus their morphogenesis has usually been studied in infected
cells, with accompanying difficulties in interpretation. Excellent
reviews of morphogenesis of picornaviruses and alphaviruses are
available (134 and Brown in 3). Since there are no confirmed
reports of in vitro reconstitution of members of either virus group,
final conclusions on the details of assembly and maturation must
await additional studies.

     In the alphaviruses, assembly begins by association of core
protein with 42S genome RNA in the cytoplasm. It has been suggested
that regions of the core protein amino-terminus, rich in prolines
and basic amino acids, combine with viral RNA (169) to form a
nucleus for condensation. In infected cells, each 42S genome-length
viral RNA combines with 300 molecules of core protein and moves to
the cell periphery to become enveloped. The 26S mRNA of the virus,
although it contains the 3' third of the genome, is never found in
cores or virions. By comparison, there is packaging of subgenomic
DI RNAs which lack most of the genome, but possess 5' and 3'
terminal sequences. Taken together, the tentative model for
alphaviruses assembly is condensation of positive-charged regions of
the core protein with specific 5' sequences of the genome. This
model suggests that mimics of arg/lys rich core peptides or 5'
genome nucleotide sequences will block assembly at an early stage.

     A problem with this naive view arises from in vitro studies on
assembly of Sindbis core proteins with nucleic acid (170). The data
support the view that little specificity is involved, but rather
that the reaction is limited by the concentration of reactants, so
that heterologous RNAs (and DNA) can enter Sindbis virus core-like
structures at concentrations which are not unreasonable for infected
cell cytoplasm. Probably the simple reason why 26S virus mRNA or

cellular messages are not packaged is that their concentration is
too low, or they are physically sequestered from the core protein.
Similar results were also obtained with Semliki Forest virus
polypeptides under modified conditions.

The picornaviruses use two different assembly pathways. Those
which do not form empty capsids, eg cardioviruses and FMDV, assemble
from structural protomers and RNA much as the alphaviruses. By
comparison, the enteroviruses and rhinoviruses assemble first into
14S subunits containing five protomers. Twelve of these then combine
into an empty capsid before complexing viral RNA. It is satisfying
that clusters of basic amino acids occur in structural proteins of
both picornaviruses and togaviruses.

If these interactions are relatively simple ones, driven by
protein-RNA ionic bonds or protein-protein hydrophobic or
-S-S-bonding, then little hope exists of specific intervention.
However, these are some clues that viral enzymes may also
participate. First, specific polypeptide bond cleavage occurs in
the final association of picornavirus structural proteins with viral
RNA and alphavirus cores with envelope proteins, and the protease
may be virus-coded. Second, a factor, probably a protein, has been
detected in the poliovirus assembly pathway (171). The factor
associates with viral structural protein and promotes assembly. It
is not present in uninfected cells, requires expression of the viral
genome, is sensitive to trypsin and is not 14S particles per se
(134).

Dichloropyrimidines have been found to selectively block
maturation cleavages of poliovirus NCVP1a and VP0 (172). These
"morphogenesis" inhibitors prevent accumulation of 14S and 80S
particles in infected cells, but do not prevent in vitro assembly
of14S protomers to the 80S capsids. The action of the
chloropyrimidines contrasts the actions of the thiopyrimidine S-7
and thiouracil which bind to native virions preventing host mediated
eclipse and uncoating, as already discussed. However, like
thiouracil (96), their actin is antagonized by L-cysteine.

Further studies in the morphogenesis area should focus on
characterizing purified factors required for assembly, to see if
they are "scaffolding" proteins or enzymes (proteases or kinases).
Accompanying the maturation of poliovirus are two additional
reactions of interest. One is the cleavage and attachment of the
genome-linked peptide to viral RNA. This has been dealt with in
previous sections, but the role of VPg in assembly should not be
discounted. It would be interesting to examine the intraction of
purified VPg with viral RNA and/or empty capsids. The other
reaction is the protein cleavage which generates the VP4 peptides,
of which sixty copies are present inside the virion. These peptides
are acidic in overall charge, and are formed proteolytically as

viral RNA associates with coat proteins. Curiously, the amino
terminal of the peptides is chemically modified or blocked during
maturation. One assessment is that pyro-glutamic acid is the
blocking group (Scraba in 2), but its origin and role in virus
assembly is obscure. The function of the VP4 peptides appears to be
to help retain the viral RNA in a condensed form, since other
interactions between RNA and poliovirus or rhinovirus structural
proteins are weak or non-existant. The overall negative charge of
VP4 peptides is thought to keep RNA compact by charge repulsion.

Togavirus assembly proceeds in a more complicated fashion than
the picornaviruses. The envelope proteins pass through the internal
membrane systems of the host, and are glycosylated and acylated.
Eventually they accumulate in the cell membrane. The pathway to the
cell surface of these proteins is blocked by monensin and other
Na^+/K^+ ionophores (173,174) as well as uncouplers of energy
metabolism (174). Disruption of cytoskeletal structures affected
transport of the viral glycoproteins, but to a smaller extent than
ionophores.

.   Once in place, the anchored envelope proteins combine with the
RNA-core protein complex at the inner side of the cell membrane. A
complex is formed between core protein, membrane and PE2, a
precursor to envelope protein 2. As this occurs, a final
proteolytic event cleaves PE2, and the released fragment is not
packaged within the mature virus which buds from the membrane.
There is some data suggesting the core protein may have proteolytic
activity (rev. by Brown in 3) and may cleave PE2 as it associates
with it. The cleaved E2 then undergoes a dramatic conformational
alteration perhaps directing the budding phenomenon. Inhibition of
the proteolytic processing of PE2 may be a useful approach for
preventing alphavirus maturation.

FUTURE DIRECTIONS

Our understanding of the structure and function of the small,
positive-strand RNA viruses is rapidly approaching that of RNA plant
viruses and bacteriophages (175 for a recent review). While this is
intellectually satisfying, it also holds the promise of designed
inhibitors of viral-specific processes. There are targets in
various stages of replication, including early events, viral RNA and
protein synthesis, protein processing and morphogenesis. It remains
to be seen whether the detailed knowledge will be exploited in terms
of useful antivirals in the clinic.

Rapid progress in sequencing viral genomes is providing exciting
new information with which rational inhibitors of viral polymerases
and processing enzymes may be designed. In our view, this will be
the route for circumvention of the antigenic diversity of their

small RNA viruses, and we have emphasized strategies for inhibition of viral enzymes rather than for synthesis of new vaccines. Despite these new opportunities, the design of antiviral drugs remains a very challenging task. The next several years will be exciting for virologists studying togaviruses and picornaviruses, and also for clinicians seeking to treat disesases caused by these pathogens.

ACKNOWLEDGEMENTS

The sequence of the 3' terminus of a rhinovirus RNA was provided by our colleagues, A. A. Cordova and R. Colonno of the Du Pont Virology group, and J. C. Kauer provided expert guidance on the crown ether compounds. Ms. V. Graham prepared the manuscript for publication.

REFERENCES

1.   F. Horsfall and I. Tamm, "Viral and Rickettsial Infections of Man", Lippincott, Philadelphia (1965).
2.   R. Perez-Bercoff, "The Molecular Biology of Picornaviruses", Plenum Publishing Co., New York (1979).
3.   R. Schlesinger, "The Togaviruses", Academic Press, New York, (1980).
4.   A. Ziemiecki and H. Garoff, J. Mol. Biol. 122:259 (1978).
5.   M. F. G. Schmidt, M. Bracha, and M. J. Schlesinger, Proc. Natl. Acad. Sci. USA, 76:1687 (1979).
6.   D. M. Corothers and H. Metzger, Immunochemistry 9:341 (1972).
7.   E. Fries and A. Helenius, Eur. J. Biochem. 97:213 (1979).
8.   J. White and A. Helenius, Proc. Natl. Acad. Sci. USA 77:3273 (1980).
9.   J. J. Mooney, J. M. Dalrymple, C. R. Alving, and P. K. Russell, J. Virol. 15:225 (1975).
10.   A. C. Chanas, E. A. Gould, J. C. S. Clegg, and M. G. R. Varma, J. gen. Virol. 58:37 (1982).
11.   M. N. Waxham and J. S. Wolinsky, Virology 126:194 (1983).
12.   A. C. Chanas, D. S. Ellis, S. Stamford, and E. A. Gould, Antiviral Research 2:191 (1982).
13.   R. R. Rueckert, in: "Comparative Virology", K. Maramorosch and E. Kurstak, eds., Academic Press, New York, (1971).
14.   R. L. Crowell, and B. J. Landau, in: "Comprehensive Viriology Vol. 18", H. Fraenkel-Conrat and R. R. Wagner, eds., Plenum Publishing Corp. (1983).
15.   P. Boulanger and K. Lonberg-Holm, in: "Virus Receptors Part 2", K. Lonberg-Holm and L. Philipson, eds., Chapman and Hall, London (1981).
16.   S. J. Barteling, F. Wagenaar, and A. L. J. Gielkens, J. gen. Virol. 62:357 (1982).
17.   K. Lonberg-Holm and L. Philipson, in: "Cell Membranes and Viral

Envelopes Vol. 2", H. A. Blough and J. M. Tiffancy, eds., Academic Press, London, (1980).

18.   G. F. Van de Woude, Virology 31:436 (1967).

19.   B. Baxt and H. L. Bachrach, Virology 116:391 (1982).

20.   C. E. Cords, C. G. James, and L. C. McLaren, J. Virol. 15:244 (1975).

21.   K. Lonberg-Holm, L. B. Gosser, and E. J. Shimshick, J. Virol. 19:746 (1976).

22.   M. Chow and D. Baltimore, Proc. Natl. Acad. Sci. USA 79:7518 (1982).

23.   G. A. Lund, B. R. Ziola, A. Salmi, and D. G. Scraba, Virology 78:35 (1977).

24.   K. Strohmaier, R. Franze, and K. H. Adam, J. gen. Virol. 59:295 (1982).

25.   J. L. Bittle, R. A. Houghten, H. Alexander, T. M. Shinnick, et al., Nature 298:30 (1982).

26.   O. R. Kaaden, K. H. Adam, and K. Strohmaier, J. gen. Virol. 34:397 (1977).

27.   R. Thorpe, P. D. Minor, A. Mackay, G. C. Schild, and M. Spitz, J. gen. Virol. 63:487 (1982).

28.   E. A. Emini, B. A. Jameson, A. J. Lewis, G. R. Larsen, and E. Wimmer, J. Virol. 43:997 (1982).

29.   A. D. Osterhaus, A. L. van Wezel, G. van Steenis, A. G. Hazendonk, and G. Drost, Dev. Biol. Standard 50:221 (1982).

30.   P. D. Minor, G. C. Schild, J. Bootman, D. M. A. Evans, et al., Nature 301:674 (1983).

31.   B. Mandel, Adv. Virus Res. 23:205 (1978).

32.   B. D. Korant, K. Lonberg-Holm, F. H. Yin, and J. Nobel-Harvey, Virology 63:384 (1975).

33.   B. Mandel, Virology 44:554 (1971).

34.   V. Chlumecka, P. D'Obrenan and J. S. Colter, J. gen. Virol. 35:425 (1977).

35.   B. Mandel, Virology 69:500 (1976).

36.   J. De Sena and D. L. Jarvis, Can. J. Microbiol. 27:1185 (1981).

37.   S. J. Klebanoff and M. E. Belding, J. Inf. Diseases, 129:345 (1974).

38.   H. Fraenkel-Conrat, J. Biol. Chem. 217:373 (1955).

39.   R. L. Ward, J. Virol. 26:299 (1978).

40.   R. L. Ward, C. S. Ashley, Appl. Environ, Microbiol. 36:191 (1978).

41.   J. W. Huggins, P. B. Jahrling, W. Rill, and C. D. Linden, J. gen. Virol. 64:149 (1983).

42.   J. S. Pierce, E. G. Strauss, and J. H. Strauss, J. Virol. 13:1030 (1974).

43.   S. C. Marker, D. Connelly, and P. B. Jahrling, J. Virol. 21:981 (1977).

44.   S. C. Marker, and P. B. Jahrling, Arch. Virol. 62:53 (1979).

45.   C. Gottlieb, S. Kornfeld, and S. Schlesinger, J. Virol. 29:344 (1979).

46.   J. A. Maassen and C. Terhorst, Eur. J. Biochem. 115:153 (1981).

46a. A. Helenius, B. Morein, E. Fries, K. Simons, P. Robinson, V. Schirrmacker, C. Terhorst, and J. L. Strominger, Proc. Natl. Acad. Sci. USA 75:3846 (1978).

47.   P. Boulanger and L. Philipson, in: "Virus Receptors Part 2," K. Lonberg-Holm and L. Philipson eds., Chapman and Hall, London (1981).

48.   P. R. McClintock, L.C. Billups, and A. L. Notkins, Virology 106:261 (1980).

49.   K. Lonberg-Holm, in: "Virus Receptors Part 2," K. Lonberg-Holm and L. Philipson, eds., Chapman and Hall, London (1981).

50.   T. Morishima, P. R. McClintock, G. S. Aulakh, L. C. Billups, and A. L. Notkins, Virology 122:461 (1982).

51.   T. Morishima, P. R. McClintock, L. C. Billups, and A. L. Notkins, Virology 116:605 (1982).

52.   M. Tardieu, R. L. Epstein, and H. L. Weiner, Int. Rev. Cytol. 80:27 (1982).

53.   A. T. H. Burness and I. U. Pardoe, J. gen. Virol. 55:275 (1981).

54.   R. L. Crowell and B. J. Landau, in: "Receptors and Human Diseases", A. G. Bearn and P. W. Choppin eds., Josiah Macy Jr. Foundation, New York, (1979).

55.   K. Lonberg-Holm, R. L. Crowell, and L. Philipson, Nature 259:679 (1976).

56.   V. Svensson, R. Persson, and E. Everitt, J. Virol. 38:70 (1981).

57.   K. Lonberg-Holm, J. gen. Virol. 28:313 (1975).

58.   J. S. Colter, J. B. Campbell, and L. R. Hatch, J. Cell. Comp. Physiol. 65:229 (1965).

59.   J. Lenard and D. K. Miller, Cell 28:5 (1982).

60.   M. Marsh, E. Bolzau and A. Helenius, Cell, 32:931 (1983).

61.   A. Helenius and M. Marsh, Ciba Foundation Symposium 92:59 (1982).

62.   K. S. Matlin, H. Reggio, A. Helenius, and K. Simons, J. Mol. Biol. 156:609 (1982).

63.   M. Marsh, A. Helenius, J. Mol. Biol. 142:439 (1980).

64.   M. Marsh, K. Matlin, K. Simons, H. Reggio, J. White, J. Kartenbeck, and A. Helenius, Cold Spring Harbor Symp. Quant. Biol. USA, 46:835 (1982).

65.   J. White, K. Matlin, and A. Helenius, J. Cell Biol. 89:674 (1981).

66.   K. Matlin, H. Reggio, A. Helenius, and L. Simons, J. Cell Biol. 91:601 (1981).

67.   P. J. Talbot and D. E. Vance, Can. J. Biochem. 58:1131 (1980).

68.   A. Helenius, J. Kartenbeck, K. Simons, and E. Fries, J. Cell Biol. 84:404 (1980).

69.   M. C. Willingham and I. Pastan, Cell 21:67 (1980).

70.   R. G. W. Anderson, M. S. Brown, and J. L. Goldstein, Cell 10:351 (1977).

71.   B. Tycko and F. R. Maxfield, Cell 28:643 (1982).

72.   S. Ohkuma and B. Poole, Proc. Natl. Acad. Sci. USA 75:3327 (1978).

73.   F. R. Maxfield, J. Cell Biol. 95:676 (1982).

74. M. Forgac, L. Chantley, B. Wiedenmann, L. Altstiel, and D. Branton, Proc. Natl. Acad. Sci. USA 80:1300 (1983).
75. N. J. Dimmock, J. Gen. Virol. 59:1 (1982).
76. J. J. Skehel, P. M. Bayley, E. B. Brown, S. R. Martin, M., et al., Proc. Natl. Acad. Sci. USA 79:968 (1982).
77. J. White, J. Kartenbeck, and A. Helenius, J. Cell Biol. 87:264 (1980).
78. P. Vaannanen and L. Kaariainen, J. gen. Virol. 46:467 (1980).
79. A. Helenius, M. Marsh, J. White, J. gen. Virol. 58:47 (1982).
80. M. Marsh, J. Wellsteed, H. Kern, E. Harms, and A. Helenius, Proc. Natl. Acad. Sci. USA 79:5297 (1982).
81. D. K. Miller and J. Lenard, Proc. Natl. Acad. Sci. USA 78:3605 (1981).
82. P. Fuchs and A. Levanon, Arch. Virol. 56:163 (1978).
83. C. D. Richardson, A. Scheid, and P. W. Choppin, Virology 105:205 (1980).
84. M. S. Brown, R. G. W. Anderson, and J. L. Goldstein, Cell 32:663 (1983).
85. K. Lonberg-Holm and N. M. Whiteley, J. Virol. 19:857 (1976).
86. A. C. King, L. Hernaez-Davis, and P. Cautrecasas, Proc. Natl. Acad. Sci. USA 77:3283 (1980).
87. C. Tietze, P. Schlesinger, and P. Stahl, Biochem. Biophys. Res. Commun. 93:1 (1980).
88. K. Lonberg-Holm and J. Nobel-Harvey, J. Virol. 12:819 (1973).
89. H. Ishitsuka, Y. T. Ninomiya, C. Ohsawa, M. Fujiu, and Y. Suhara, Antimicrob. Agents Chemother. 22:617 (1982).
90. J. J. McSharry, L. A. Caliguiri, and H. J. Eggers, Virology 97:307 (1979).
91. L. A. Caliguiri, J. J. McSharry, and G. W. Lawrence, Virology 105:86 (1980).
92. K. S. Kim, V. J. Sapienza, and R. I. Carp, Antimicrob. Agents Chemother. 18:276 (1980).
93. J. J. McSharry and F. Panic, in: "Chemotherapy of Viral Infections", P. E. Came and L. A. Caliguiri, eds., Springer-Verlag, Berlin (1982).
94. K. Lonberg-Holm, L. B. Gosser, and J. C. Kauer, J. gen. Virol. 27:329 (1975).
95. H. J. Eggers, Virology 78:241 (1977).
96. F. M. Steel and F. L. Black, J. Virol. 1:653 (1967).
97. P. Pohjanpelto, Virology 6:472 (1958).
98. M. Pons, Virology 22:253 (1964).
99. G. K. Hirst, Int. Poliomyelitis Conf. 5th, Copenhagen, pp 11-12 (1960).
100. M. L. Fenwick and P. D. Cooper, Virology 18:212 (1962).
101. C. C. Halsted, D. S. Y. Seto, J. Simkins, and D. H. Carver, Virology 40:751 (1970).
102. D. H. Carver and D. S. Y. Seto, J. Virol. 2:1482 (1968).
103. D. Robberson, G. Thornton, M. Marshall, and R. Arlinghaus, Virology 116:454 (1982).
104. H. Garoff, A. Frischauf, H. Lehrach, and H. Delius, Nature

288:236 (1980).

105. C. Rice and J. Strauss, Proc. Nat. Acad. Sci. USA 78:2062 (1981).

106. J. Ou, E. Strauss, and J. Strauss, Virology 109:281 (1981).

107. S. Monroe, J. Ou, C. Rice, S. Schlesinger, E. Strauss and J. Strauss, J. Virol. 41:153 (1982).

108. B. Robertson, D. Morgan, D. Moore, M. Grubman, J. Card, T. Fischer, G. Weddell, D. Dowbenko and D. Yansara, Virology 126:614 (1983).

109. H. Riedel, H. Lehrach and H. Garoff, J. Virol. 42:725 (1982).

110. J. Ou, C. Rice, L. Dalgarno, E. Strauss and J. Strauss, Proc. Nat. Acad. Sci. USA 79:5235 (1982).

111. G. Wengler, G. Wengler, and H. Gross, Virology 123:273 (1982).

112. N. Stebbing, Pharmac. Ther. 6:291 (1979).

113. E. DeClercq, Biochem. J. 205:1 (1982).

114. M. Ranki and L. Kaariainen, Virology 98:298 (1979).

115. D. Tuschall, E. Hiebert and J. Flanegan, J. Virol. 44:209 (1982).

116. M. Baron and D. Baltimore, J. Biol. Chem. 257:12,359 (1982).

117. A. Traub, B. Diskin, H. Rosenberg and E Kalmar, J. Virol. 18:375 (1976).

118. N. Kitamura, B. Semler, P. Rothberg, G. Larsen, C. Adler, A. Dorner, E. Emini, R. Hanecak, J. Lee, S. van der Werf, C. Anderson and E. Wimmer, Nature 291:547 (1981).

119. J. Polatnik, R. Arlinghaus, J. Graves, and K. Cowan, Virology 32:609 (1967).

120. B. Semler, R. Hanecak, L. Dorner, C. Anderson, and E. Wimmer, Virology 126:624 (1983).

121. M. Baron and D. Baltimore, J. Virol. 43:969 (1982).

122. B. Korant in "Proteases and Biological Control". E. Reich, D. Rifkin and E. Shaw, eds. Cold Spring Hbr. Lab., NY (1975).

123. M. Baron and D. Baltimore, Cell 30:74 (1982).

124. E. Wimmer, Cell 28:199 (1982).

125. Y. Becker, in "Monographs in Virology", Vol. 11 Karger, Basel (1976).

126. U.S. Patents 3997,565; 4,024,158; 4,104,275.

127. D. Tershak, F. Yin and B. Korant, in "Handbook of Pharmacol. and Exp. Therap." L. Calaguiri and P. Came, eds. Springer, Berlin (1982).

128. P. Cooper, in "Comprehensive Virology" 9:133, Plenum, NY (1977).

129. B. Korant, Virology 81:25 (1977).

130. K. Saunders and A. King, J. Virol. 42:389 (1982).

131. A. Nomoto, T. Omata, H. Toyoda, S. Kuge, H. Horie, Y. Yataoka, Y. Genba, Y. Nakano, and N. Imma, Proc. Nat. Acad. Sci. USA 79:5793 (1982).

132. G. Koch and D. Richter (editors) "Biosynthesis, Modification and Processing of Cellular and Viral Polyproteins", Academic Press, NY (1980).

133. R. Monckton and E. Westaway, J. Gen. Virol 63:227 (1982).

134. J. Putnak and B. Phillips, Microb. Rev. 45:287 (1981).

135. J. Bell, M. Hunkapiller, L. Hood and J. Strauss, Proc. Nat. Acad. Sci. USA 75:2722 (1978).
136. J. Bell, C. Rice, M. Hunkapiller, and J. Strauss, Virology 119:255 (1982).
137. A. Makoff, C. Paynter, D. Rowlands, and J. Bootroyd, Nucl. Acids. Res. 10:8285 (1982).
138. J. Boothroyd, T. Harris, D. Rowlands, and P. Lowe, Gene 17:153 (1982).
139. P. Minor, G. Schild, J. Bootman, D. Evans, M. Ferguson, P. Reeve, M. Spitz, G. Stanway, A. Cann, R. Hauptman, L. Clarke, R. Mountford and J. Almond, Nature 301:675 (1983).
140. E. Herbert and M. Muhler, Cell 30:1 (1982).
141. J. Langner, P. Bohley, H. Kirschke, B. Wiederanders and B. Korant, Europ. J. Biochem. 125:21 (1982).
142. M. Jacobson, J. Asso and D. Baltimore, J. Mol. Biol. 49:657 (1970).
143. B. Korant, N. Chow, M. Lively and J. Powers, Proc. Nat. Acad. Sci. USA 76:2992 (1979).
144. H. Pelham, Europ. J. Biochem. 85:457 (1978).
145. B. Korant, Acta Biol. Med. Germ. 36:1565 (1977).
146. C. Shih, N. Naseer and D. Shih, J. Virol. 42:1127 (1982).
147. R. Cancedda and M. Schlesinger, Proc. Nat. Acad. Sci. USA 71:1843 (1974).
148. G. Aliperti and M. Schlesinger, Virology 90:366 (1978).
149. C. Lawrence and R. Thach, J. Virol. 15:918 (1975).
150. A. Palmenberg, M. Pallansch and R. Rueckert, J. Virol 32:770 (1979).
151. A. Gorbalenya, Y. Svitkin, and V. Agol., Biochem. Bphys. Res. Commun. 98:952 (1981).
152. R. Hanecak, B. Semler, C. Anderson and E. Wimmer, Proc. Nat. Acad. Sci. 79:3973 (1982).
153. A. Palmenberg and R. Rueckert, J. Virol 41:244 (1982).
154. B. Korant in "Design of Inhibitors of Viral Functions" K. Gauri, editor, Academic Press, NY (1981).
155. B. Korant, B. Butterworth, and J. Kauer, Nature 248:588 (1974).
156. B. Korant and B. Butterworth, J. Virol. 18:298 (1976).
157. B. Korant, N. Chow, M. Lively, and J. Powers, Ann NY Acad. Sci. (334:304.
158. B. Korant and K. Lonberg-Holm, Acta. Biol. Med. Germ. 40:1481 (1981).
159. M. Marquisee and J. Kauer, J. Med. Chem. 21:1188 (1978).
160. N. Craig and J. Roberts, J. Biol. Chem. 256:8039 (1981).
161. A. Barrett in "Enzyme Inhibitors as Drugs" M. Sandler, ed. Univ. Park Press, Baltimore (1980).
162. P. Hsieh, M. Rosner and P. Robbins, J. Biol. Chem. 258:2555 (1983).
163. R. Hirsch, D. Griffin and J. Winkelstein, Proc. Nat. Acad. Sci. USA 80:548 (1983).
164. M. Schmidt and M. Schlesinger, J. Biol. Chem. 255:3334 (1980).
165. M. Schmidt, Virology 116:327 (1982).

166. M. Grubman, B. Bat, J. LaTorre, and H. Bachrach, J. Virol. 39:455 (1981).
167. M. Grubman, J. Virol. 44:1102 (1982).
168. J. LaTorre, M. Grubman, B. Baxt, and H. Bachrach, Proc. Nat. Acad. Sci. USA 77:7444 (1980).
169. H. Garoff, A. Frischauf, K. Simons, H. Lehrach, and H. Delius, Proc. Nat. Acad. Sci. USA 77:6376 (1980).
170. G. Wengler, U. Boege, G. Wengler, H. Rischoff, and K. Wahn, Virology 118:401 (1982).
171. C. Drescher-Lincoln, J. Putnak, and B. Phillips, Virology 126:301 (1983).
172. P. LaColla, M. Corrias, M. Marongiu and A. Pani in "Design of Inhibitors of Viral Functions K. Gauri, ed. Academic Press, NY (1981).
173. D. Johnson and M. Schlesinger, Virology 103:407 (1980).
174. L. Kaariainen, K. Hashimoto, I. Saraste, I. Virtanen and K. Penttinen, J. Cell Biol. 87:783 (1980.
175. V. Agol Prog. Med. Virol. 26:119 (1980).
176. J. Lacal and L. Carrasco, Eur. J. Biochem. 127:359 (1982).

TARGETS IN NEGATIVE STRAND RNA VIRUSES (EXCLUDING ORTHO- AND PARAMYXOVIRUSES)

David H.L. Bishop

Department of Microbiology
Unviersity of Alabama in Birmingham
Birmingham, AL 35294

INTRODUCTION

The negative strand RNA virus families include, in addition to the Orthomyxoviridae and Paramyxoviridae, the Rhabdoviridae, Arenaviridae and Bunyaviridae. The viruses are termed negative stranded because the virion RNA does not itself function as an mRNA, rather, in virus infected cells, complementary sequence mRNA species are transcribed from the virion RNA by the RNA polymerase (Baltimore et al., 1970).

All negative strand RNA viruses have an envelope derived from the lipid membranes of the host cell in which the viruses originated. In addition, the viruses have one or more species of glycoprotein protruding from the envelope and forming an exterior fringe of "spikes" on the surface of the virus particle. Encompassed by the envelope, the viral RNA is in the form of nucleocapsid(s) in association with structural protein(s) and viral coded RNA polymerase molecules.

The different genetic properties and strategies of infection of three families of negative sense RNA viruses (Rhabdoviridae, Arenaviridae, Bunyaviridae) will be described below after a brief review of the member viruses of each family that are of concern to man and his economy.

THE RHABDOVIRIDAE

This family of viruses includes more than 50 recognized virus serotypes which either multiply only in vertebrates, or only in arthropods, or both in arthropods and veterbrates. In addition, several other rhabdoviruses have been isolated from plants, or arthropods that feed on plants. Biological, or mechanical, vectors for particular rhabdoviruses include aphids, leafhoppers, mosquitoes, sandflies, tabanids, midges and mites (Matthews, 1982).

Some rhabdoviruses are transmitted either directly, or indirectly, from one infected animal to another. An example is rabies virus which is usually transmitted by bite, but can also be transmitted by aerosol infection (e.g., in a cave harboring numerous infected bats). Although no rhabdoviruses are known to be transmitted congenitally (vertically) in plants, or vertebrates, Sigma virus is transmitted that way in Drosophila. It is probable that other arthropod-borne rhabdoviruses are transmitted transovarially allowing them to overwinter and persist in nature.

Rhabdoviruses that infect vertebrates and are of concern to man and his economy include rabies virus, vesicular stomatitis virus, VSV, bovine ephemeral fever virus (BEFV), and the fish rhabdoviruses, Egtved, infectious hematopoietic necrosis virus (IHNV), pike fry rhabdovirus and spring viremia of carp virus. The fish rhabdoviruses are frequently a problem in hatcheries and can cause recurrent epidemics commensurate with the migration patterns of the fish. VSV in the Americas and BEFV in Australasia are important, albeit non-lethal, pathogens of horses and, or cattle.

The most serious rhabdovirus pathogen of man and domestic animals is rabies virus. Discussions of the host range and infection course of rabies virus in an infected animal have been presented elsewhere (Brown and Crick, 1979). For the purpose of this review the salient feature that makes rabies an important and difficult target for antiviral therapy is the fact that shortly after the virus is introduced to a wound site, it replicates in cells (often myocytes) and eventually gains access to neurons serving those cells. The progression of rabies virus through the extended processes of a nerve cell for the most part appears to involve replicating nucleocapsids and their gradual diffusion (?) within the cytoplasm of infected neurons, although virus particles are produced at cellular membranes. Crossing synapses, or the delivery of virus from one cell to another within the central nervous system, or the brain, may well involve budding virus particles. The presence of cytoplasmic inclusion bodies (composed of viral nucleocapsids) in brain cells of an infected animal (Negri bodies) is a characteristic sign of infection in approximately 75% of usual reservoir host species. In the U.S.A. these include skunks, foxes and bats, in Europe predominantly foxes.

It is probable that the sequestering of rabies virus to the nervous system and its particular replicative progress, renders the virus a difficult target for the host humoral, or cell mediated, immune system. In man the infection course may take weeks to several months. Infections involving facial sites of entry (wounds or from transplanted cornea) have a shorter incubation period (a few weeks). Preventive rabies vaccination of dogs and cats that is required in certain countries provides an effective measure of protection both for domestic animals and for man. Treatment of recently infected man with immune sera provides some measure of protection. In model animal systems interferon is also efficacious.

The Structure of Rhabdoviruses

Other than the plant rhabdoviruses which are frequently bacilliform in shape and up to 380 nm long (Matthews, 1982), most rhabdoviruses are bullet-shaped (60-70 nm wide, 160-180 nm in length).

Rabies virus has 5 structural proteins, a glycoprotein (G, $65\text{-}80\times10^3$ daltons), a nucleocapsid protein (N, $58\text{-}62\times10^3$ daltons), a large transcriptase component (L, approx. 190×10^3 daltons) and two proteins (MI or NS, apparent size $35\text{-}40\times10^3$ daltons and M2, $22\text{-}25\times10^3$ daltons). As mentioned above, the glycoprotein is embedded in the viral envelope and forms the outer fringe of projections seen on the surface of virus particles. These five proteins represent the total number of proteins coded by the virus.

The number of protein molecules per virion are of the order of 20-150 (L), 1,500-2,000 (G), 1,500-2,000 (N), 900-1,000 (MI, NS) and 1,600-1,700 (M2). VSV is similar in composition although the number and sizes of the 5 VSV protein species (named L, G, N, NS and M) are slightly different (Gallione et al., 1981; Rose and Gallione, 1981). The NS protein of VSV is phosphorylated and migrates with an apparent size of $39\text{-}50\times10^3$ daltons (depending on the gel system) although from NS mRNA analyses it has an actual size of around 25×10^3 daltons (Gallione et al., 1981).

The nucleotide sequence has almost been completed for VSV. The overall organization of the genetic information along the viral RNA involves the following arrangment (Abraham and Banerjee, 1976; Ball and White, 1976):

(3') Short non-coding region - N - NS - M - G - L - short non-coding region (5')
      ("leader")

The 3' end of the VSV viral RNA has the sequence $_{HO}$UpGpCp... The 5' end has a short complementary sequence terminating with the residues ...pGpCpAppp (Hefti and Bishop, 1975). Thus, unlike several positive strand viruses, the viral RNA does not have a capped 5' terminus (or covalently attached 5' protein), or a 3' poly A sequence.

The intergenic regions of VSV have a common sequence, as do the 3' and 5' ends of the 5 complementary mRNA species (Rose, 1980):

(3')	...AUACUUUUUUUNAUUGUCNNUAG...	(5')	vRNA strand
(5')	...UAUG-poly(A)$_n$ mRNA #1	AACAGNNAUC... mRNA #2	(3') mRNA species

Each mRNA is essentially proportional in size to the protein for which it codes (Knipe et al., 1975). The 5' ends of the mRNA species are capped and variously methylated (Banerjee et al., 1977). The first U of the UUGUC sequence serves as template for the initial nucleotide of the mRNA species. The septanucleotide stretches of U residues serve as templates for poly A addition (presumably by slippage) at the 3' ends of the complementary mRNA species. The poly A sequence can vary from less than a hundred to several hundred residues in length. Thus, there is a dinucleotide (NA) that is apparently not found in the mature form of the mRNA species (although whether it is found in an mRNA precursor will be discussed below). A similar dinucleotide between the leader and beginning of the N gene is also not transcribed.

The virion ribonucleocapsid consists of a single copy of viral RNA (Huang and Wagner, 1966), multiple copies of N protein and a few molecules of the transcriptase components (NS+L). The N protein is arranged over the entire RNA molecule in a manner that protects the nucleic acid against ribonuclease digestion but allows the bases to be read by the transcriptase. N protein is not permanently displaced from the template during mRNA transcription. Whether it is temporarily displaced is not known.

In the virus particle the ribonucleocapsid is organized in a helix of approximately 35 turns suggesting that there are specific interactions between the individual turns of the helix, possibly involving N-N protein reactions, although M protein may also play a role. How M, a basic protein, is involved in the virus organization is not clear. One possibility is that it interacts in a non-covalent association with the nucleocapsid on the one hand and the carboxy terminal portion of G protein molecules on the other (and, or the internal surface of the lipid envelope).

The majority of each viral glycoprotein molecule, including its amino terminus, is located on the outside of the virus particle. The carboxy terminus of G protein is, as mentioned, located on the inside. There is a hydrophobic, transmembranal, sequence of carboxy proximal amino acids that holds the G protein in the envelope (Rose et al., 1980). The G protein is important for virus infectivity since after its removal by proteolytic enzymes, the resulting "spikeless" particles exhibit low infectivity (Wagner, 1975). Proteolytic enzymes do not remove the transmembranal, or carboxy terminal, portions of G protein. G protein elicits and interacts with neutralizing antibodies. Antibodies raised to other viral proteins do not neutralize virus infectivity.

In summary, most rhabdovirus particles are bullet-shaped, 60-70 nm in diameter, 160-180 nm long and composed of an internal, helical, single ribonuclecapsid associated with transcriptase enzyme molecules and matrix protein. The internal components of rhabdoviruses are bounded by a lipid envelope in which are embedded glycoprotein molecules. The envelope is derived from the membrane site of maturation of the virus.

The Infection Process of Rhabdoviruses

The overall infection process of rhabdoviruses, as understood from tissue culture and in vitro analyses, involves the following sequence of events: adsorption, penetration, primary mRNA transcription, mRNA translation, RNA replication, and secondary transcription, translation and replication. Viral morphogenesis involves the budding of virus particles (usually from the cell surface plasma membrane) and the acquisition of a lipid envelope. Depending on the virus, host cell and temperature of incubation, progeny virus particles can be detected between 1 and 2 hours post-infection with yields of up to 10^5 particles per cell by the end of the infection cycle.

Concomitant with a productive infection is inhibition of host macromolecular syntheses. Although progeny virus can be obtained from enucleated cells (Follett et al., 1974), indicating that the nucleus is not required, recent studies indicate that transcription products (specifically "leader" RNA) may enter the nucleus and function to disturb normal host macromolecular syntheses (Weck and Wagner, 1979; Kurilla et al., 1982).

The processes involved in adsorption and penetration of a cell are poorly understood. The location of the receptor binding site on the viral G protein, or nature of the receptors on cells targeted by viruses, is not known. In some manner the viral envelope fuses with the cell plasma membrane (or membrane of a cytoplasmic vacuole if the virus is phagocytized). This results in the deposition of the ribonucleocapsid and associated transcriptase enzymes into the cytoplasm.

Primary Transcription

With the availability of nucleoside triphosphates in the infected cell the viral nucleocapsid directs the synthesis of mRNA species. The process is termed primary transcription to differentiate it from later transcriptional events. Primary transcription is repetitive yielding several hundred copies of each mRNA species (Flamand and Bishop, 1973, 1974).

Since virus particles contain transcriptase enzyme molecules, the transcription process can be studied in vitro. In vivo analyses and virus mutant analyses have substantiated the in vitro results. The salient features of primary transcription are the sequential and repetitive syntheses of leader and the 5 capped and methylated mRNA species (Rhodes and Banerjee, 1976). Sequential RNA synthesis has been demonstrated in time-course reaction analyses (Iverson and Rose, 1982) and from the results of ultraviolet light inhibition studies (Abraham and Banerjee, 1976; Ball and White, 1976). By the latter analyses it has been shown that with progressively higher doses of ultraviolet light there is inhibition of the transcription of L, then L+G then L+G+M, then L+G+M+NS, then L+G+M+NS+N (etc.), i.e. the target sizes of the various mRNA species are not proportional to their lengths.

Various models for the mechanism of mRNA synthesis have been proposed. The currently favored model postulates that there is a single entry site for transcriptase molecules at the 3' end of the template (Emerson, 1982) and stop/restart transcription of the individual mRNA species. An alternate model postulates multiple entry sites for the polymerase and attenuated transcription until the preceeding genes are transcribed (Testa et al., 1980).

The leader transcript consists of some 47 nucleotides and starts with the sequence pppApCpGp... The mature form of each mRNA species has a conserved 5' sequence with an inverted "capping" guanosine that is methylated at position 7, and additional methyl residues attached to the 2' 0 of the first and (variably) second nucleotide as well as (again variably) to the first and second adenosine residues (Rose, 1975), viz:

$$m^7 GpppmA(m)p(m)A(m)pCpApGpNpNpApUpCp...$$

The 3' end of each mRNA species terminates in a variable length poly A sequence. As mentioned, the poly A sequence is templated by 7 U residues of the intergenic region and presumably is produced by slippage of the transcriptase on the U residues. The 5' end of each mRNA is templated by the UpUpGpUpCp... sequence.

Studies with incomplete transcription reaction mixtures and S-adenosyl-L-methionine (SAM) indicate that capping and methylation are early events. It has been shown (Abraham et al., 1975) that during the capping process one phosphate of the bridge derives from the adenosine nucleotide, the other 2 from the capping GTP precursor (i.e., Gpp-pApApC...). The methyl groups come from SAM. Their addition can be inhibited by S-adenosyl homocysteine.

Although the end results of the transcription process are the completion of an uncapped and unpolyadenylated leader RNA and 5 capped, methylated, polyadenylated mRNA species, a variety of incomplete products are also obtained in vitro (Testa et al., 1980). These include capped, methylated, or unmethylated, or partially methylated products and uncapped, unmethylated, short oligonucleotides with 5' triphosphates that have sequences that are identical to the ends of certain mRNA species (i.e., pppApCpApGpNpNpApUpCp...). Whether these products are artifacts of the in vitro system is not known. The synthesis of these uncapped, unmethylated, short oligonucleotides is not sensitive to ultraviolet light inhibition to the same extent as the individual mRNA species, or even the capped oligonucleotides (Piwnica-Worms and Keene, 1983). This evidence can be interpreted to indicate entry of the polymerase at different mRNA initiation sites and attenuated transcription.

Another postulated mechanism of transcription involves the synthesis of readthrough, polycistronic, precursors of the mRNA species (Banerjee et al., 1977). Evidence that supports this model includes the identification of transcripts containing an internal poly A sequence and N plus NS mRNA sequences (Herman et al., 1978). If such transcripts function as precursors to individual, mature, mRNA species, then processing must occur involving at least endonuclease cleavage and (possibly) removal of the intergenic dinucleotide. Although readthrough transcripts are found in low amounts, recent studies on in vivo paramyxovirus transcription (T. Morrison personal communication) have identified such products in larger numbers. However, it is not clear for rhabdoviruses, or paramyxoviruses, to what extent readthrough transcripts and, or independent initiation of the various mRNA species are artifacts of the transcription process.

Genetic studies with VSV have established that there are 5 complementation groups (i.e., minimally 5 gene products). Temperature sensitive (ts), conditional lethal, mutants representing each of the 5 groups have been used to analyse the different steps of the infection process. Members of Group I, for which there are the most number of ts mutants, exhibit 2 phenotypes. At the non-permissive temperature, some support primary transcription, others do not (Flamand and Bishop, 1973). In vitro reconstruction experiments with purified L and NS proteins plus nucleocapsids that contain solely RNA and N protein, have established that both L, NS and nucleocapsids are required for mRNA transcription (Emerson and Yu, 1975). Using for such reconstruction experiments the L protein of a Group I ts mutant plus NS and nucleocapsids (RNA-N) from wild-type virus, it has been shown that transcription is temperature sensitive, i.e., Group I mutants have L protein defects (Hunt et al., 1976). Considering the size of L, the frequency of isolation of Group I mutants supports this conclusion. Based on these observations it has been concluded that some Group I mutants are able to make mRNA from the parental nucleocapsid templates because they have a thermostable transcriptase. Mutants that do not make mRNA from the parental template have a thermolabile transcriptase. Since mutant viruses are grown at permissive temperatures and therefore have an active transcriptase at that temperature, thermostability can vary depending on the particular ts mutant. No Group I ts mutant at a non-permissive temperature makes mRNA at amplified rates during the infection time-course (secondary transcription, see later). Low, but reproducible, levels of intracistronic complementation between certain Group I mutants have been observed at non-permissive temperatures. This has been interpreted to indicate that multiple copies of the L protein form the functional transcriptase.

Mutants representing the other 4 Groups all make mRNA by primary transcription. Although both NS and N are integral components of the active transcription reaction, all the available mutants corresponding to these gene products (Groups II and IV) support primary transcription at non-permissive temperatures, some, but not all, Group II mutants even support secondary transcription. Others, and all the Group I and IV mutants, do not. Although one might expect to obtain thermolabile mutants of Groups II and IV, it may be that not enough mutants have been obtained to detect some that make thermolabile proteins.

Translation, RNA Replication and Secondary Transcription

In a normal infection process primary transcription is followed 30-45 min later by an amplified rate of mRNA synthesis (Flamand and Bishop, 1974). This increased rate is termed secondary transcription. Inhibitors of protein synthesis (puromycin or cycloheximide) block secondary transcription. Inhibitors of DNA synthesis, or DNA-directed RNA synthesis (actinomycin D or a-amanitin) neither effect transcription, replication or the overall infection process. The amplified transcription rates account for the eventual production of some hundred thousand copies of each mRNA species in the infected cell by 4-8 hours post-infection (Flamand and Bishop, 1974).

As discussed above, all mutants representing complementation Groups I and IV (and most of the Group II mutants) do not support secondary transcription at non-permissive temperatures (Flamand and Bishop, 1974). These results agree with the conclusion that secondary transcription is the product of RNA replication and that RNA replication occurs in infected cells only with the availability of newly synthesized viral proteins. Mutants of complementation Group III (M protein) and Group V (G protein) make amplified levels of mRNA at non-permissive temperatures (Flamand and Bishop, 1974), indicating that their gene products are not involved in the steps of mRNA transcription or RNA replication; rather, they are important in later stages of viral morphogenesis.

The switch from transcription to replication must involve at least the availability of N protein since in infected cells newly formed plus and minus sense nucleocapsids are in the form of RNA-N protein complexes (Simonsen et al., 1979). Whether alternate phosphorylated forms of the NS protein are involved in the switch is not clear. NS protein is found abundantly in infected cells and was initially considered as a non-structural (hence NS) protein (Wagner, 1975). In the infected cell it can be found in at least 2 phosphorylated states, NS1 and NS2 (Kingsford and Emerson, 1979). However their exact functions are unknown either in transcription, or RNA replication.

The difference between transcription and replication is that an exact copy of the viral RNA is transcribed in its entirety into plus sense RNA. It has been postulated that newly synthesized N protein associates with a nascent plus sense RNA (presumably in the leader region) and catalyses the further addition of N protein. This stabilizes the transcription complex so that the usual attenuation of leader synthesis does not occur and the enzyme reads through into the N gene and similarly through and beyond each of the intergenic regions. Evidently during RNA replication no slippage occurs on the intergenic uridylate sequences. What recognition sequences are involved in RNA-N protein interactions, and how they are specific for VSV RNA species in infected cells is not known.

From the point of view of the overall strategy of replication, one of the interesting aspects of negative strand virus RNA replication is that transcription and RNA replication may take place concomitantly since they both transcribe plus sense RNA in the same direction. For positive sense viruses translation of the parental viral RNA (one of the initial events in the infection process) must be halted during RNA replication since replication of the parental RNA strand is countercurrent to translation of that RNA species.

Virus Morphogenesis

Rhabdovirus particles are produced by a process of budding through a cellular membrane. Frequently this involves the outer plasma membrane of an infected cell allowing the virus to directly escape and either infect a neighbouring cell, or circulate systemically, or be excreted (as in the case of rabies virus released from cells of the salivary gland). On occasion, depending on the virus, cell, and tissue, viruses bud into intracellular vacuoles through an internal membrane. The process of budding involves 2 important stages. First, the transportation of viral glycoprotein to the site of morphogenesis (e.g., the outer surface of the plasma membrane). Second, the recognition of glycoproteins at this site by other viral components.

Studies using in vivo and in vitro mRNA translation, as well as ts mutants of VSV viral glycoprotein have elucidated some of the steps in the transportation of G protein to the plasma membrane (Knipe et al., 1977). Also informative have been studies which document that in epithelial cells G protein is eventually inserted into particular domains of the plasma membrane (leading to polarity of virus maturation, Roth et al., 1979). Translation of the G mRNA appears to occur predominantly on polysomes associated with membranes of the endoplasmic reticulum (Grubman et al., 1975). As in the case of other membrane and secreted proteins, the amino terminal sequence of the VSV G protein has a "signal" or "leader" sequence of hydrophobic amino acids (Rose 1977; Lingappa et al., 1978; Irving et al., 1979; Chatis and Morrison, 1979; Rose and Gallione, 1981) that allows the protein to be inserted through the hydrophobic membrane. Such "leaders" are removed before chain termination (Blobel and Dobberstein, 1975). Commensurate with entry of the glycoprotein into the internal lumen of the endoplasmic reticulum, the protein is glycosylated (Rothman and Lodish, 1977) and eventually becomes anchored in the membrane by virtue of another hydrophobic sequence of amino acids that is proximal to the carboxy terminus of the protein (Rose et al., 1980; Rose and Gallione, 1981). Transportation of the G protein through the Golgi apparatus of the infected cell leads to the eventual positioning of the glycoprotein on the plasma membrane surface.

Little is known about the envelopement process. It is supposed that the carboxy terminus of the viral glycoprotein (which, as mentioned, is located on the cytoplasmic side of the membrane) is recognized by the viral M protein and, or the viral nucleoprotein that is associated with the nucleocapsid. Presumably the helical configuration of the nucleocapsid determines the ultimate size and shape of the nascent virus particle. How viruses are released from the membrane is not known.

THE ARENAVIRIDAE

Although there are only relatively few members of this family (Matthews, 1982), several are human pathogens. These include Lassa virus, the etiologic agent of Lassa fever in West Africa, Junin and Machupo viruses, the etiologic agents of (respectively) Argentine and Bolivian hemorrhagic fevers and lymphocytic choriomeningitis virus (LCM) which is found in both Europe, Africa, Asia and the Americas. Other than LCM virus, the continental distribution of arenaviruses is limited, probably because the viruses have restricted natural host ranges (Johnson et al., 1973). Lassa, Junin and Machupo viruses cause severe hemorrhagic disease with various tissues and organs involved. Nosocomial Lassa virus infections, involving patients and hospital staff, have been documented and are a major concern in patient management. Childrens' rodent pets are frequently a source of LCM virus infection if the animals are shedding virus.

One of the particular attributes that makes arenaviruses a problem is their ability to develop persistent infections in animals. Persistently infected animals shed virus in urine and saliva over a long period and are usually viremic with little or no neutralizing antibodies. The development of persistence appears to depend on a variety of factors, including the age and genetic background of an animal as well as the dose and route of infection (Rawls et al., 1981). Frequently, persistence is the outcome of a congenitally acquired infection or one that is initiated shortly after birth. Infections acquired later in life are usually self-limiting (with transient viremias) and induce neutralizing antibodies. The development of viral persistence appears to be related to the maturity of the host's immune system.

Other than Tacaribe virus, which came from a fruit eating bat, the normal hosts of arenaviruses are rodents (Matthews, 1982). Transmission from animal to animal, or to man, is either by aerosol, or via contamination.

The Structure of Arenaviruses

Arenaviruses are enveloped and pleomorphic in shape. Most of the molecular and genetic studies of arenaviruses have been undertaken with LCM virus and a South American arenavirus, Pichinde. Frequently virus particles seen in negatively stained electron micrographs are roughly spherical in shape ranging in diameter from 50 to 300 nm. The viruses have surface projections that consist of 1 or 2 species of glycoprotein (G1, 44-72×10^3 daltons and G2, 34-40×10^3 daltons) that protrude 5-10 nm from the virus envelope. These projections are club-shaped and, compared to other enveloped viruses, relatively few in number (Murphy and Whitfield, 1975). The arrangement of the 2 glycoproteins in the envelope is not known. They are required for virus infectivity (Vezza et al., 1977). Within the virus particle a variable number of electron-dense granules are often apparent (Pfau, 1974). These structures are host ribosomes (Farber and Rawls, 1975; Vezza et al., 1978). Their function is unknown. Their inclusion in virus particles, although characteristic, may be fortuitous. It has been shown using viruses containing ts ribosomes that ribosomes are not needed to establish a productive infection (Leung and Rawls, 1977).

The viral genetic information is resident in 2 species of single stranded, negative sense, RNA, designated on the basis of molecular weight differences as L (large) and S (small). They have sizes of $2.1-3.2 \times 10^6$ daltons (L) and $1.1-1.6 \times 10^6$ daltons (S). The RNAs are in the form of 2 convoluted, apparently circular, nucleocapsids (Palmer et al., 1977; Vezza et al., 1977); each consists of an individual strand of RNA, nucleoprotein (N, $60-70 \times 10^3$ daltons) and small amounts of a large protein, believed to be a transcriptase component (L, $180-200 \times 10^3$ daltons) (Harnish et al., 1981). Other proteins have also been observed in virus preparations (Vezza et al., 1977), their function or location is unknown. RNA polymerase activities have been detected in virus extracts (Carter et al., 1974; Leung et al., 1979).

The 3' ends of the L and S RNA species are homologous and conserved for LCM, Tacaribe and Pichinde viruses (Auperin et al., 1982). Whether the 5' ends are complementary to the 3' ends, or conserved between arenaviruses, is not known.

Since the viral genetic information resides in 2 species of RNA, new virus genotype combinations can be obtained by RNA segment reassortment. High frequency reassortment of RNA segments has been documented for Pichinde (Vezza and Bishop, 1977; Vezza et al., 1980) and LCM viruses (Kirk et al., 1980; Romanowski and Bishop, 1983) by dual homologous virus infections using either ts mutants, or distinguishable wild-type virus isolates. No interspecies reassortment has yet been detected between Pichinde and LCM viruses (Romanowski and Bishop, 1983). Diploid viruses containing more than 1 viral S RNA have been identified by genetic experiments (Romanowski and Bishop, 1983). Whether diploidy, or reassortment between distinguishable virus serotypes, occurs in nature is not known. The ability to produce intertypic reassortants between different arenavirus topotypes has allowed assignment of the viral gene products to the 2 genome RNA species (Vezza et al., 1980; Harnish et al., 1983). From such analyses it has been shown that the L RNA codes for the L protein; the S RNA for N and the 2 glycoproteins. In infected cells the glycoproteins are initially translated as a precursor polypeptide (GPC, Buchmeier et al., 1978).

It has been shown from analyses of intertypic reassortant arenaviruses obtained from parent viruses with different disease phenotypes, that the S RNA codes for the major determinants of virulence. Whether this attribute resides in the viral glycoproteins, or is a function of the N protein, is not known.

Recent studies involving DNA cloning and sequencing (Auperin and Bishop, 1983) indicate that the gene order on the Pichinde S RNA species is:

3'   non-coding end - N - intergenic region - GPC - non-coding end   5'

The intergenic region of Pichinde S RNA is remarkable in that following the 2 stop codons that terminate the N gene there is an 18 base pair hairpin that is followed immediately afterwards by an AUG codon that initiates the GPC gene (Auperin and Bishop, 1983). The function of the hairpin in modulating transcription or translation is not known. The strategy of the arenavirus L RNA species has not been determined.

The mRNA species of arenaviruses have not been characterized. Whether there are discrete mRNA species transcribed for each S RNA gene product (e.g., N mRNA and GPC mRNA species), or spliced mRNA species, or a polycistronic mRNA, is not known.

The Infection Process of Arenaviruses

The infection process of arenaviruses is understood only in outline. By analogy with other negative sense RNA viruses the sequence of events is believed to involve adsorption, penetration, primary mRNA transcription, translation, RNA replication, secondary transcription, translation and replication. Viral morphogenesis occurs by budding from cellular plasma membranes.

The individual stages of an acute infection, or the steps involved in the development of a persistent infection (Pedersen, 1979), are not well characterized. In culture, arenaviruses induce much less cytopathology than that observed in a rhabdovirus infection. Frequently persistently infected cells outgrow dying cells in a culture; sometimes even during a plaque assay leading to "bullseye" plaques. Whether the establishment of a persistently infected cell is due to the presence of defective interfering virus is not known.

The replication of arenaviruses is not inhibited by antagonists of DNA synthesis such as 5-bromodeoxyuridine, or 5-iododeoxyuridine (Pfau, 1974). Actinomycin D and α-amanitin reportedly reduce virus yields (Buck and Pfau, 1969), whether they also reduce mRNA yields is not clear. Enucleation inhibits the growth of arenaviruses (Banerjee et al., 1976).

Arenavirus morphogenesis takes place at the outer plasma membrane of an infected cell. How the viral glycoproteins get there, what processing is involved and how they interact with other viral components during envelopement remain subjects for future investigation.

In summary, arenaviruses are pleomorphic, enveloped viruses that have an outer layer of surface glycoproteins and internal components that include ribosomes and 2 nucleocapsids containing the viral genetic information. Arenaviruses have a virion transcriptase and are capable of genetic reassortment. The viruses frequently induce persistent infections in animals and in tissue culture. Little is known about their molecular biology.

THE BUNYAVIRIDAE

The Bunyaviridae is comprised of 4 defined genera of viruses in addition to several unassigned virus serogroups and serotypes (Bishop et al., 1980). In general, members of the Bunyavirus genus are mosquito transmitted, members of the Nairovirus and Uukuvirus genera are transmitted by ticks, while members of the Phlebovirus genus are transmitted by sandflies (Bishop and Shope, 1979; Bishop et al., 1980). Most of the viruses productively infect only a narrow range of arthropods. The reason is not known but may be a genetic attribute of the virus. For some viruses it has been demonstrated that they can be transmitted transovarially allowing them to overwinter and persist in nature.

Although many of the Bunyaviridae members are of no impact or consequence to man's ecomomy, some cause fatal human disease, others are pathogens of domestic animals. Examples among the bunyaviruses include the mosquito transmitted La Crosse (LAC) virus that is endemic to parts of the U.S.A. and on occasion causes fatal human meningoencephalitis; as well as Akabane virus that is often responsible for abortion and congenital abnormalities among sheep and cattle in Australasia. Phleboviruses that are important human pathogens include the sandfly fever viruses and Rift Valley fever virus. In Africa the latter virus is a pathogen of both man and domestic animals. Crimean-Congo hemorrhagic fever virus, a member of the Nairovirus genus, is an important human pathogen in Eastern Europe and parts of Asia and Africa. Nairobi sheep disease virus, another nairovirus, is an animal pathogen in certain regions of Africa. One of the unassigned bunyaviruses that is a major human pathogen and does not have an arthropod vector is Hantaan virus, the etiologic agent of Korean hemorrhagic fever.

The Bunyavirus genus is comprised of some 15 serogroups of viruses as well as a few unassigned viruses (Bishop et al., 1980). The viruses are enveloped, 90-120 nm in diameter, and possess oyter glycoprotein spikes composed of 2 species of glycoprotein (G1, 115×10^3 daltons and G2, 35×10^3 daltons) that are required for virus infectivity. The arrangement of the viral glycoproteins is not known. The viral envelope encompasses 3 nucleocapsids. Each nucleocapsid is circular, consisting of a single unique species of negative sense, end-hydrogen bonded, RNA, designated on the basis of size differences, as large (L), medium (M) and small (S). The average sizes of the 3 RNA species are 2.9×10^6 daltons (L, range 2.7-3.1×10^6), 2.0×10^6 daltons (M, range $1.8 - 2.3 \times 10^6$), and 0.45×10^6 daltons (S, range $0.28 - 0.50 \times 10^6$). In the nucleocapsid each RNA is associated with many copies of a nucleoprotein (N, 22×10^3 daltons) and a few molecules of a large protein ($160 - 200 \times 10^3$ daltons) that is believed to be a transcriptase component. The 3' ends of bunyavirus L, M, and S RNA species have conserved sequences that, at least for LAC and the related snowshoe hare virus, are complementary to their respective 5' ends.

The conserved bunyavirus sequences differ from the conserved ends of phleboviruses, uukuviruses, or nairoviruses.

In genetic studies, it has been shown that members of the California serogroup of bunyaviruses are able to form recombinant viruses by RNA segment reassortment (Gentsch and Bishop, 1976). Intertypic recombinant viruses have been isolated from nature (Klimas et al., 1981). Analyses of reassortant bunyaviruses obtained using different parents have established that the bunyavirus S RNA codes for the N protein and a non-structural protein, NS_C (Gentsch and Bishop, 1978; Fuller and Bishop, 1982). These 2 proteins are read from overlapping reading frames in the viral complementary mRNA (Bishop et al., 1982). The M RNA codes for both viral glycoproteins, G1 and G2, and another non-structural protein, NS_M (Gentsch and Bishop, 1979; Fuller and Bishop, 1982). It is believed that the L RNA codes for the large protein identified in virus preparations.

It has been shown that the viral glycoproteins induce, and interact with, neutralizing antibodies (Gentsch et al., 1980). The M RNA gene products (the glycoproteins) are not only major determinants of virus virulence (Shope et al., 1981) but also determine the ability of the viruses to produce productive infections in mosquitoes, and be disseminated to vertebrates (Beaty et al., 1981a, 1982). Mutations in either the M or L RNA species can result in attenuation of virus virulence (Rozhon et al., 1981).

No reassortants have been obtained from mixed virus infections involving ts mutants of the California serogroup bunyaviruses and members of the Bunyamwera or Group C serogroups (or with members of other genera). If this observation holds true for all the bunyavirus serogroups, it would indicate that reassortment may be limited in potential, perhaps only to related viruses of a serogroup. Although dual simultaneous infection of a permissive arthropod by genetically compatible bunyaviruses can lead to reassortant virus formation (Beaty et al., 1981b), sequential infection does not; rather the superinfecting virus is inhibited (Beaty et al., 1983). If the superinfecting virus belongs to another bunyavirus serogroup (or family) it is not inhibited. The reason for the particular viral interference is not known.

Members of the Phlebovirus genus have 3 unique viral RNA species with apparent sizes of 2.7×10^6 daltons (L, range $2.6-2.8 \times 10^6$), 2.0×10^6 daltons (M, range $1.8-2.2 \times 10^6$), and 0.75×10^6 daltons (S, range $0.7-0.8 \times 10^6$). The average sizes of the 3 major phlebovirus virion polypeptides are 25×10^3 (N, range $20-30 \times 10^3$), 55×10^3 daltons (G2, range $50-60 \times 10^3$) and 63×10^3 daltons (G1, range $55-70 \times 10^3$) (Robeson et al., 1979). High frequency genetic recombination has been obtained with ts mutants of Punta Toro phlebovirus (G. Robeson and D.H.L. Bishop, unpublished data).

Studies with Uukuniemi virus (a member of the Uukuvirus genus) have shown that it has 3 unique RNA species and 3 major viral polypeptides that are similar in size and number to those of phleboviruses (Pettersson et al., 1977, von Bonsdorff & Pettersson, 1975). Circular viral nucleocapsids, end hydrogen bonded viral RNA species and the presence of a virion transcriptase have been described for Uukuniemi virus (Ranki and Pettersson, 1975; Pettersson and Kaariainen, 1973; Pettersson and von Bonsdorff, 1975).

Members of the Nairovirus genus have 3 unique viral RNA species, the largest of which is apparently some 70% larger than the L RNA of bunyaviruses, phleboviruses, or uukuviruses (Clerx and Bishop, 1981). The values for the 3 viral RNA species of Qalyub (QYB) nairovirus are 4.7×10^6 daltons (L), 1.75×10^6 daltons (M) and 0.75×10^6 daltons (S). The sizes of the 3 major QYB virion polypeptides are 40×10^3 daltons (G2), 53×10^3 daltons (N) and 75×10^3 daltons (G1). The N polypeptide is twice the size of the N polypeptides of bunyaviruses, uukuviruses, or phleboviruses. Intracellular 115×10^3 and 85×10^3 dalton glycopolypeptide precursors to the QYB viral G1 and G2 species have been identified by pulse-chase experiments (Clerx and Bishop, 1981). Intracellular N protein and a 200×10^3 dalton large polypeptide have also been identified in QYB infected cells. Similar data has been obtained for other members of the genus. Further characterization of the strategy of infection developed by nairoviruses is required to determine if their inclusion in the family is justified.

The Infection Process of Bunyaviruses

As in the case of arenaviruses the infection process of bunyaviruses is understood only in broad terms. Adsorption and penetration are mediated by the viral glycoproteins. Their removal, or interaction with neutralizing antibodies, blocks attachment to cell receptors and penetration. Following entry into the cell the viral genome is transcribed by the virion polymerase (primary transcription). This process is not inhibited by cycloheximide or puromycin (Vezza et al., 1979) although either drug prevents the onset of secondary transcription. Presumably this occurs by inhibiting translation which in turn is required for RNA replication. Recent analyses indicate that bunyaviruses may require primer oligonucleotides for mRNA synthesis (unpublished data). Neither actinomycin D, nor a-amanitin, inhibit transcription although the former drug reduces virus yields (Vezza et al., 1979). The reason is not known. There are conflicting reports concerning the effect of enucleation on virus replication (Goldman et al., 1977; Pringle, 1980). Viral morphogenesis is unique by comparison with that of rhabdoviruses, arenaviruses, orthomyxoviruses and paramyxoviruses. Like other enveloped viruses bunyaviruses mature by budding, however envelopment takes place in the Golgi complexes leading to virus particle accumulation in intracellular vacuoles. How viruses are released from infected cells is not known.

In summary, bunyaviruses are spherical, enveloped, negative sense, RNA viruses that have a genome consisting of 3 species of unique, single-stranded, RNA. Related viruses are capable of genetic reassorment. Most members of the family are transmitted by select ranges of arthropods. Some use mosquito vectors, others ticks, others sandflies, others gnats. Several of the member viruses are human pathogens, others are animal pathogens. Viral morphogenesis involves budding at intracellular membranes of the Golgi complex. The molecular biology of the viruses is poorly understood.

UNIQUE ASPECTS OF RHABDOVIRUS, ARENAVIRUS AND BUNYAVIRUS REPLICATION AS TARGETS FOR ANTIVIRAL THERAPY

The negative stranded rhabdoviruses, arenaviruses and bunyaviruses replicate in the cytoplasm of infected cells. Their transcription into mRNA species is unaffected by inhibitors of DNA synthesis, or DNA-directed RNA synthesis.   Each has a viral coded, virus specific, RNA polymerase. Nucleotide analog antagonists specific for these polymerases have not yet been identified. Viral specific structures and events in an infected cell that might be considered targets for selective antiviral therapy include the the transcriptase, the specific recognition of viral nucleoprotein for its own species of RNA and the events involved in interaction of viral proteins in morphogenesis. As yet each of these processes are poorly understood.

ACKNOWLEDGEMENTS

This work was support by contract DAMD-17-78-C-8017 from the U.S. Army Medical Research and Development Command and NIH grants AI-15400 and AI-14183.

REFERENCES

Abraham, G., and Banerjee, A.K., 1976, Sequential transcription of the genes of vesicular stomatitis virus, Proc. Natl. Acad. Sci., U.S.A., 73:1504.

Abraham, G., Rhodes, D.P., and Banerjee, A.K., 1975, The 5' terminal structure of the methylated mRNA synthesized in vitro by vesicular stomatitis virus, Cell, 5:51.

Auperin, D.D., and Bishop, D.H.L., 1983, Molecular cloning, nucleotide sequence analyses, and predicted gene products, including the N protein of the S RNA of Pichinde arenavirus, J. Virol., in press.

Auperin, D., Compans, R.W., and Bishop, D.H.L., 1982, Nucleotide sequence conservation at the 3' termini of the virion RNA species of New World and Old World arenaviruses, Virology, 121:200.

Ball, L.A., and White, C.N., 1976, Order of transcription of genes of vesicular stomatitis virus, Proc. Natl. Acad. Sci., U.S.A., 73:442.

Baltimore, D., Huang, A.S., and Stampfer, M., 1970, Ribonucleic acid synthesis of vesicular stomatitis virus. II. An RNA polymerase in the virion, Proc. Natl. Acad. Sci., U.S.A., 66:572.

Banerjee, A.K., Abraham, G., and Colonno, R.J., 1977, Vesicular stomatitis virus: mode of transcription, J. Gen. Virol., 34:1.

Banerjee, S.N., Buchmeier, M., and Rawls, W.E., 1976, Requirement of cell nucleus for the replication of an arenavirus, Intervirol., 6:190.

Beaty, B.J., Bishop, D.H.L., Gay, M., and Fuller, F., 1983, Interference between bunyaviruses in Aedes Triseriatus mosquitoes, Virology, 127:83.

Beaty, B.J., Holterman, M., Tabachnick, W., Shope, R.E., Rozhon, E.J., and Bishop, D.H.L., 1981a, Molecular basis of bunyavirus transmission by mosquitoes: The role of the M RNA segment, Science, 211:1433.

Beaty, B.J., Rozhon, E.J., Gensemer, P., and Bishop, D.H.L, 1981b, Formation of reassortant bunyaviruses in dually-infected mosquitoes. Virology, 111:662.

Beaty, B.J., Miller, B.R., Shope, R.E., Rozhon, E.J., and Bishop, D.H.L., 1982, Molecular basis of bunyavirus per os infection of mosquitoes: The role of the M RNA segment. Proc. Natl. Acad. Sci., U.S.A., 79:1295.

Bishop, D.H.L., and Shope, R. E., 1979, Bunyaviridae, Comp. Virol., vol. 14, p.1, H. Fraenkel-Conrat, and R.R. Wagner, eds., Plenum Press, New York.

Bishop, D.H.L., Calisher, C., Casals, J., Chumakov, M.P., Gaidamovich, S.Ya., Hannoun, C., Lvov, D.K., Marshall, I.D., Öker-Blom, N., Pettersson, R.F., Porterfield, J.S., Russell, P.K., Shope, R.E., and Westaway, E.G., 1980, Bunyaviridae, Intervirol., 14:125.

Bishop, D.H.L., Gould, K.G., Akashi, H., and Clerx-van Haaster, C.M, 1982, The complete sequence and coding content of snowshoe hare bunyavirus small (S) viral RNA species, Nucl. Acids Res., 10:3703.

Blobel, G., and Dobberstein, B., 1975, Transfer of proteins across membrane I. Presence of proteolytically processed and unprocessed nascent immunoglobulin light chains in membrane-bound ribosomes of murine myeloma, J. Cell Biol., 67:835.

Brown, F., and Crick, J., 1979, Natural history of the rhabdoviruses of vertebrates and invertebrates. in: "Rhabdoviruses" vol. I, p.2, D.H.L. Bishop, ed., CRC Press, Boca Raton.

Buchmeier, M.J., Elder, J.H., and Oldstone, M.B.A., 1978, Protein structure of lymphocytic choriomeningitis virus: Identification of the virus structural and cell associated polypeptides, Virology, 89:133.

Buck, L.L., and Pfau, C.J., 1969, Inhibition of lymphocytic choriomeningitis virus replication by actinomycin D and 6-azauridine, Virology, 37:698.

Carter, M.F., Biswal, N., and Rawls, W.E., 1974, Polymerase activity of Pichinde virus, J. Virol., 13:577.

Chatis, P.A., and Morrison, T.G., 1979, Vesicular stomatitis virus glycoprotein is anchored to intracellular membranes near its carboxyl end and is proteolytically cleaved at its amino terminus, J. Virol., 29:957.

Clerx, J.P.M., and Bishop, D.H.L., 1981, Qalyub Virus, a member of the newly proposed Nairovirus genus (Bunayviridae), Virology, 108:361.

Emerson, S.U., 1982, Reconstitution studies detect a single polymerase entry site on the vesicular stomatitis virus genome, Cell, 31:635.

Emerson, S.U., and Yu, Y.-H., 1975, Both NS and L proteins are required for in vitro RNA synthesis by vesicular stomatitis virus, J. Virol., 15:1348.

Farber, F.E., and Rawls, W.E., 1975, Isolation of ribosome-like structures from Pichinde virus, J. Gen. Virol., 26:21.

Flamand, A., and Bishop, D.H.L., 1973, Primary in vivo transcription of vesicular stomatitis virus and temperature sensitive mutants of five VSV complementation groups, J. Virol., 12:1238.

Flamand, A., and Bishop, D.H.L., 1974, In vivo synthesis of RNA by vesicular stomatitis virus and its mutants, J. Mol. Biol., 87:31.

Follett, E.A.C., Pringle, C.R., Wunner, W.H., and Skehel, J.J., 1974, Virus replication in enucleated cells: Vesicular stomatitis virus and influenza virus, J. Virol., 13:394.

Fuller, F., and Bishop, D.H.L., 1982, Identification of viral coded non-structural polypeptides in bunyavirus infected cells, J. Virol., 41:643.

Gallione, C.J., Greene, J.R., Iverson, L.E., and Rose, J.K., 1981, Nucleotide sequences of the mRNA's encoding the vesicular stomatitis virus N and NS proteins, J. Virol., 39:529.

Gentsch, J., and Bishop, D.H.L., 1976, Recombination and complementation between temperature-sensitive mutants of the bunyavirus, snowshoe hare virus, J. Virol., 20:351

Gentsch, J., and Bishop, D.H.L., 1978, Small viral RNA segment of bunyaviruses codes for viral nucleocapsid protein, J. Virol., 28:417.

Gentsch, J.R., and Bishop, D.H.L., 1979, M viral RNA segment of bunyaviruses codes for two unique glycoproteins: G1 and G2, J. Virol., 30:767.

Gentsch, J.R., Rozhon, E.J., Klimas, R.A., El Said, L.H., Shope, R.E., and Bishop, D.H.L., 1980, Evidence from recombinant bunyavirus studies that the M RNA gene products elicit neutralizing antibodies, Virology, 102:190.

Goldman, N., Presser, I., and Sreevalsan, T., 1977, California encephalitis virus: Some biological and biochemical properties, Virology, 76:352.

Grubman, M.J., Moyer, S.A., Bannerjee, A.K., and Ehrenfeld, E., 1975, Subcellular localization of vesicular stomatitis messenger RNA's, Biochem. Biophys. Res. Commun., 62:531.

Harnish, D., Dimock, K., Leung, W-C., and Rawls, W.E., 1981, Immunoprecipitable polypeptides in Pichinde virus infected BHK-21 cells. in: "The replication of negative strand viruses", p. 23, D.H.L. Bishop and R.W. Compans, eds., Elsevier, North-Holland, New York.

Harnish, D.G., Dimock, K., Bishop, D.H.L., and Rawls, W.E., 1983, Gene mapping in Pichinde virus: assignment of viral polypeptides to genomic L and S RNAs, J. Virol., 46:638.

Hefti, E., and Bishop, D.H.L., 1975, The 5' nucleotide sequence of vesicular stomatitis viral RNA, J. Virol., 15:90.

Herman, R.C., Adler, S., Lazzarini, R.A., Colonno, R.J., Banerjee, A.K., and Westphal, H., 1978, Intervening polyadenylate sequences in RNA transcripts of vesicular stomatitis virus, Cell, 15:587.

Huang, A.S., and Wagner, R.R., 1966, Comparative sedimentation coefficients of RNA extracted from plaque-forming and defective particles of vesicular stomatitis virus, J. Mol. Biol., 22:381.

Hunt, D.M., Emerson, S.U., and Wagner, R.R., 1976, RNA⁻ temperature-sensitive mutants of vesicular stomatitis virus: L-protein thermosensitivity accounts for transcriptase restriction of group I mutants, J. Virol., 18:596.

Irving R.A., Toneguzzo, F., Rhee, S.H., Hofmann, T., and Ghosh, H.P., 1979, Synthesis and assembly of membrane glycoproteins: Presence of leader peptide in nonglycosylated precursor of membrane glycoprotein of vesicular stamatitis virus, Proc. Natl. Acad. Sci., U.S.A., 76:570.

Iverson, L.E., and Rose, J.K., 1982, Localized attenuation and discontinuous synthesis during vesicular stomatitis virus transcription, Cell, 23:477.

Johnson, K.M., Webb, P.A., and Justines, G., 1973, Biology of Tacaribe-complex viruses. in: "Lymphocytic choriomeningitis virus and other arenaviruses" p. 241, F. Lehmann-Grube, ed., Springer, Berlin.

Kingsford, L., and Emerson, S.U., 1979, Transcriptional activities of different phosphorylated species of NS protein purified from vesicular stomatitis virus and cytoplasm of infected cells, J. Virol., 33:1097.

Kirk, W.E., Cash, P., Peters, C.J., and Bishop, D.H.L., 1980, Formation and characterization of an intertypic lymphocytic choriomeningitis recombinant virus, J. Gen. Virol., 51:213.

Klimas, R.A., Thompson, W.H., Calisher, C.H., Clark, G.G., Grimstad, P.R., and Bishop, D.H.L., 1981, Genotypic varieties of La Crosse virus isolated from different geographic regions of the continental United States and evidence for a naturally occurring intertypic recombinant La Crosse virus, Am. J. Epidem., 114:112.

Knipe, D., Rose, J.K., and Lodish, H.F., 1975, Translation of individual species of vesicular stomatitis viral mRNA, J. Virol., 15:1004.

Knipe, D.M., Baltimore, D., and Lodish, H.F., 1977, Maturation of viral proteins in cells infected with temperature-sensitive mutants of vesicular stomatitis virus, J. Virol., 21:1149.

Kurilla, M.G., Piwnica-Worms, H., and Keene, J.D., 1982, Rapid and transient localization of the leader RNA of vesicular stomatitis virus in the nuclei of infected cells, Proc. Natl. Acad. Sci., U.S.A., 79:5240.

Leung, W-C., and Rawls, W.E., 1977, Virion-associated ribosomes are not required for the replication of Pichinde virus, Virology, 81:174.

Leung, W-C., Leung, M.F.K.L., and Rawls, W.E., 1979, Distinctive RNA transcriptase, polyadenylic acid polymerase, and polyuridylic acid polymerase activities associated with Pichinde virus, J. Virol., 30:98.

Lingappa, V.R., Katz, F.N., Lodish, H.F., and Blobel, G., 1978, A signal sequence for the insertion of a transmembrane glycoprotein, J. Biol. Chem., 253:8667.

Matthews, R.E.F., 1982, "Classification and nomenclature of viruses," Karger, Basel.

Murphy, F.A., and Whitfield, S.G., 1975, Morphology and morphogenesis of arenaviruses, Bull. WHO, 52:409.

Palmer, E.L., Obijeski, J.F., Webb, P.A., and Johnson, K.M., 1977, The circular segmented nucleocapsid of an arenavirus-Tacaribe virus. J. Gen. Virol., 36:541.

Pedersen, I.R., 1979, Structural components and replication of arenaviruses, Adv. Virus. Res., 24:277.

Pettersson, R., and Kaariainen, L., 1973, The ribonucleic acids of Uukuniemi virus, a noncubical tickborne arbovirus, Virology, 56:608.

Pettersson, R., and von Bonsdorff, C.-H., 1975, Ribonucleoproteins of Uukuniemi virus are circular, J. Virol., 15:386.

Pettersson, R., Hewlett, M.J., Baltimore, D., and Coffin, J.M., 1977, The genome of Uukuniemi virus consists of three unique RNA segments, Cell, 11:51.

Pfau, C.J., 1974, Biochemical and biophysical properties of the arenaviruses, Prog. Med. Virol., 18:64.

Piwnica-Worms, H., and Keene, J.D., 1983, Sequential synthesis of small capped RNA transcripts in vitro by vesicular stomatitis virus, Virology, 125:206.

Pringle, C.R., 1980, Enucleation as a technique in the study of virus-host interactions, Curr. Topics Micro. Immunol., 76:50.

Ranki, M., and Pettersson, R., 1975, Uukuniemi virus contains an RNA polymerase, J. Virol., 16:1420.

Rawls, W.E., Chan, M., and Gee, S., 1981, Mechanisms of persistence in arenavirus infections: a brief review, Can. J. Microbiol., 27:568.

Rhodes, D.P., and Banerjee, A.K., 1976, 5'-Terminal sequence of vesicular stomatitis virus mRNA's synthesized in vitro, J. Virol., 17:33.

Robeson, G., El Said, L.H., Brandt, W., Dalrymple, J., and Bishop, D.H.L., 1979, Biochemical studies on the Phlebotomus fever group viruses (Bunyaviridae family). J. Virol., 30:339.

Romanowski, V., and Bishop, D.H.L., 1983, The formation of arenaviruses that are genetically diploid, Virology, 126:87.

Rose, J.K., 1975, Heterogenous 5'-terminal structures occur on vesicular stomatitis virus mRNAs, J. Biol. Chem., 250:8098.

Rose, J.K., 1977, Nucleotide sequences of ribosome recognition sites on messenger RNAs of vesicular stomatitis virus, Proc. Natl. Acad. Sci., U.S.A., 74:3672.

Rose, J.K., 1980, Complete intergenic and flanking gene sequences from the genome of vesicular stomatitis virus, Cell, 19:415.

Rose, J.K., and Gallione, C.J., 1981, Nucleotide sequences of the mRNA's encoding the vesicular stomatitis virus G and M proteins determined from cDNA clones containing the complete coding regions, J. Virol., 39:519.

Rose, J.K., Welch, W.J., Sefton, B.M., Esch, F.S., and Ling, N.C., 1980, Vesicular stomatitis glycoprotein is anchored in the viral membrane by a hydrophobic domain near the COOH-terminus, Proc. Natl. Acad. Sci., U.S.A., 77:3884.

Roth, M.G., Fitzpatrick, J.P., and Compans, R.W., 1979, Polarity of influenza and vesicular stomatitis virus maturation in MDCK cells: lack of a requirement for glycosylation of viral glycoproteins, Proc. Natl. Acad. Sci., U.S.A., 76:6430.

Rothman, J.E., and Lodish, H.F., 1977, Synchronized transmembrane insertion and glycosylation of a nascent membrane protein, Nature (Lond.), 269:775.

Rozhon, E.J., Gensemer, P., Shope, R.E., and Bishop, D.H.L., 1981, Attenuation of virulence of a bunyavirus involving an L RNA defect and isolation of LAC/SSH/LAC and LAC/SSH/SSH reassortants, Virology, 111:125.

Shope, R.E., Rozhon, E.J., and Bishop, D.H.L., 1981, Role of the middle-sized bunyavirus RNA segment in mouse virulence, Virology, 114:273.

Simonsen, C.C., Batt-Humphries, S., and Summers, D.F., 1979, RNA synthesis of vesicular stomatitis virus-infected cells: in vivo regulation of replication, J. Virol., 31:124.

Testa, D., Chanda, P.K., and Banerjee, A.K., 1980, Unique mode of transcription in vitro by vesicular stomatitis virus, Cell, 21:267.

Vezza, A.C., and Bishop, D.H.L., 1977, Recombination between temperature sensitive mutants of the arenavirus, Pichinde, J. Virol., 24:712.

Vezza, A.C., Cash, P., Jahrling, P., Eddy, G., and Bishop, D.H.L., 1980, Arenavirus recombination: the formation of recombinants between prototype Pichinde and Pichinde Munchique viruses and evidence that arenavirus S RNA codes for N polypeptide, Virology, 106:250.

Vezza, A.C., Clewley, J.P., Gard, G.P., Abraham, N.Z., Compans, R.W., and Bishop, D.H.L., 1978, The virion RNA species of the arenaviruses Pichinde, Tacaribe and Tamiami, J. Virol., 26:485.

Vezza, A.C., Gard, G.P., Compans, R.W., and Bishop, D.H.L., 1977, Structural components of the arenavirus Pichinde, J. Virol. 23:776.

Vezza, A.C., Repik, P.M., Cash, P., and Bishop, D.H.L., 1979, In vivo transcription and protein synthesis capabilities of bunyaviruses: wild-type snowshoe hare virus and its temperature-sensitive Group I, Group II and Group I/II mutants, J. Virol., 31:426.

von Bonsdorff, C.-H., and Pettersson, R., 1975, Surface structure of Uukuniemi virus, J. Virol., 16:1296.

Wagner, R.R., 1975, Reproduction of rhabdoviruses, Comp. Virol., vol. 4, p.1, H. Fraenkel-Conrat, and R.R. Wagner, eds., Plenum Press, New York.

Weck, P.K., and Wagner, R.R., 1979, Transcription of vesicular stomatitis virus is required to shut off cellular RNA synthesis, J. Virol., 30:410.

TARGETS FOR THE DESIGN OF ANTIVIRAL AGENTS: TARGETS IN

ORTHOMYXOVIRUSES

J. J. Skehel and D. C. Wiley

National Institute for Medical Research, Mill Hill
London, NW7, UK; and Department of Biochemistry
Harvard University, Cambridge, Mass. 0238, USA

To initiate replication influenza viruses bind to sialic acid
residues of membrane glycoconjugates and are taken into intracellular vesicles by endocytosis. Fusion of virus membranes and
endosomal membranes occurs and virus transcription complexes are
transferred to cell nuclei where virus messenger RNAs are synthesized. The translation of these messengers leads to the production
of the proteins involved in RNA replication, in virus assembly, and
in the release of progeny virus particles from the cell by budding
at the plasma membrane. Each of these processes in the replication
cycle will be considered in turn to indicate possible targets for
inhibitors of virus production.

Receptor binding

It has been known for some time that influenza viruses bind
to cells by interacting with membrane receptor molecules containing
sialic acid (reviewed in 1) and that the virus component involved is
the haemagglutinin glycoprotein of the virus membrane. This interaction varies in detailed specificity for different influenza
viruses and comparitive analyses of the haemagglutinins of viruses
which demonstrate preferences for binding to carbohydrate side
chains containing neuraminic acid in either $\alpha2,6$ or $\alpha2,3$ linkage
with penultimate residues, have been used to obtain information
concerning the location of the receptor binding site on the
haemagglutinin. The haemagglutinin of the 1968 Hong Kong influenza
virus (H3N2) recognizes sialyloligosaccharides with the terminal
sequence SA $\alpha2,6$ Gal- and variants selected for their ability to
grow in the presence of non-immune horse serum, a rich source of
SA $\alpha2,6$ Gal- containing glycoprotein inhibitors of haemagglutinin-
cell membrane interaction, specifically recognize sialyloligo-

saccharides with the sequence SA $\alpha 2,3$ Gal-. Comparisons of the
amino acid sequences of wild type and variant haemagglutinins
deduced from the nucleotide sequences of their RNA genes indicates[2]
that they differ only at residue 226 of the HA_1 polypeptide chain.
At this position the wild type haemagglutinin contains leucine
(codon CUG) variants which recognize SA $\alpha 2,3$ Gal- linkages contain
glutamine (codon CAG), and a variant which bound to sialyloligo-
saccharide with either SA $\alpha 2,6$ Gal- or SA $\alpha 2,3$ Gal- linkages contained methionine (codon AUG) at residue 226. These observations
of changes in receptor binding specificity as a consequence of amino
acid substitutions at residue 226 support the proposition that the
sialic acid binding site is a surface pocket at the distal end of
the haemagglutinin molecule.[3] This proposal was originally based
on the presence of the conserved residues tyrosine 98, tryptophan
153, histidine 183, glutamic acid 190, and leucine 194, in this
pocket and the location of amino acid 226 in this site (Figure 1)
is consistent with a role for this residue in receptor binding.
Further analyses of the structure of haemagglutinins with different
binding specificities and of haemagglutinin-sialyllactose complexes
should provide a detailed description of the receptor binding site
and the basis of receptor binding specificities. They may also
indicate the feasibility of specific inhibition of virus adsorption.

Membrane Fusion

        Binding of virus particles to their receptors is followed by
endocytosis and the next stage in virus infection involves fusion
of virus and endosomal membranes. In vitro, haemolysis and membrane
fusion by influenza viruses have been found to be optimum at about
pH 5.0, the pH of endosomes and also to involve the haemagglutinin.
[4,5,6,7] The mechanism of haemagglutinin-mediated fusion is not
known but a number of observations indicate that at pH 5.0
haemagglutinin structure is modified and the molecule displays
properties which suggest that in the process of fusion it may interact directly with the endosomal membranes. Specifically, on incubation at pH 5.0, soluble haemagglutinin released from virus
particles by bromelain digestion,[8] acquires the ability to bind to
lipid vesicles, to bind detergent, or to aggregate in lipid and
detergent-free solutions. The molecule also becomes suceptible to
proteolysis but CD analyses indicate that the structural modifications which these observations imply, do not involve gross
conformational changes; they are more consistent with the relative
movement of molecular domains which retain their individual secondary structure.[9] The region of the molecule responsible for the
pH 5.0-specific aggregation and lipid interactions appears to be
the NH_2-terminus of HA_2 (Figure 1). This is the site at which the
proteolytic processing of the haemagglutinin precursor, necessary
for fusion activation, occurs.[7,10,11] It is highly conserved,
hydrophobic and is released from pH 5.0-haemagglutinin aggregates
by thermolysin digestion which reverses the aggregation.[12] For

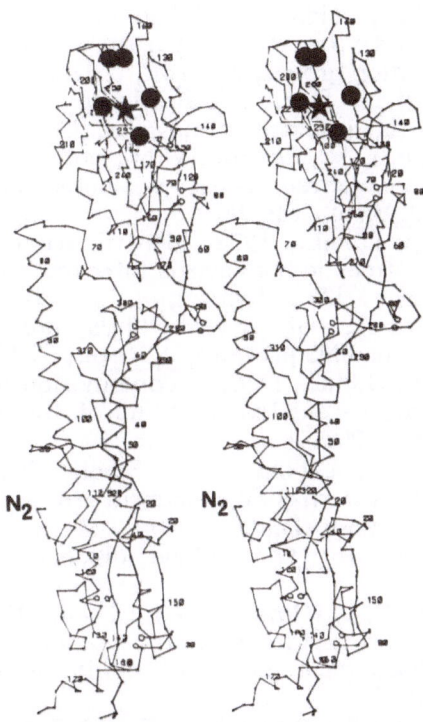

Fig. 1.   An α-carbon tracing of a haemagglutinin subunit which
          shows conserved amino acids defining the sialic acid
          binding site, ●, and amino acid residue 226 which
          changes in binding mutants ★. N₂ denotes the amino
          terminus of HA₂ involved in virus entry.

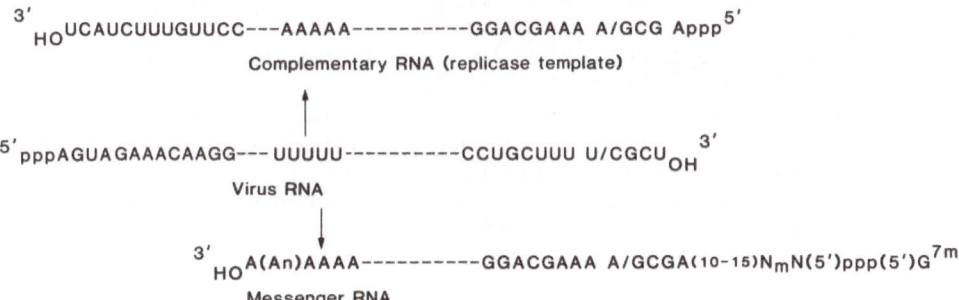

Fig. 2.   Terminal structures of influenza virus RNAs and their
          transcripts.

Flamand, A., and Bishop, D.H.L., 1973, Primary in vivo transcription of vesicular stomatitis virus and temperature sensitive mutants of five VSV complementation groups, J. Virol., 12:1238.

Flamand, A., and Bishop, D.H.L., 1974, In vivo synthesis of RNA by vesicular stomatitis virus and its mutants, J. Mol. Biol., 87:31.

Follett, E.A.C., Pringle, C.R., Wunner, W.H., and Skehel, J.J., 1974, Virus replication in enucleated cells: Vesicular stomatitis virus and influenza virus, J. Virol., 13:394.

Fuller, F., and Bishop, D.H.L., 1982, Identification of viral coded non-structural polypeptides in bunyavirus infected cells, J. Virol., 41:643.

Gallione, C.J., Greene, J.R., Iverson, L.E., and Rose, J.K., 1981, Nucleotide sequences of the mRNA's encoding the vesicular stomatitis virus N and NS proteins, J. Virol., 39:529.

Gentsch, J., and Bishop, D.H.L., 1976, Recombination and complementation between temperature-sensitive mutants of the bunyavirus, snowshoe hare virus, J. Virol., 20:351

Gentsch, J., and Bishop, D.H.L., 1978, Small viral RNA segment of bunyaviruses codes for viral nucleocapsid protein, J. Virol., 28:417.

Gentsch, J.R., and Bishop, D.H.L., 1979, M viral RNA segment of bunyaviruses codes for two unique glycoproteins: G1 and G2, J. Virol., 30:767.

Gentsch, J.R., Rozhon, E.J., Klimas, R.A., El Said, L.H., Shope, R.E., and Bishop, D.H.L., 1980, Evidence from recombinant bunyavirus studies that the M RNA gene products elicit neutralizing antibodies, Virology, 102:190.

Goldman, N., Presser, I., and Sreevalsan, T., 1977, California encephalitis virus: Some biological and biochemical properties, Virology, 76:352.

Grubman, M.J., Moyer, S.A., Bannerjee, A.K., and Ehrenfeld, E., 1975, Subcellular localization of vesicular stomatitis messenger RNA's, Biochem. Biophys. Res. Commun., 62:531.

Harnish, D., Dimock, K., Leung, W-C., and Rawls, W.E., 1981, Immunoprecipitable polypeptides in Pichinde virus infected BHK-21 cells. in: "The replication of negative strand viruses", p. 23, D.H.L. Bishop and R.W. Compans, eds., Elsevier, North-Holland, New York.

Harnish, D.G., Dimock, K., Bishop, D.H.L., and Rawls, W.E., 1983, Gene mapping in Pichinde virus: assignment of viral polypeptides to genomic L and S RNAs, J. Virol., 46:638.

Hefti, E., and Bishop, D.H.L., 1975, The 5' nucleotide sequence of vesicular stomatitis viral RNA, J. Virol., 15:90.

Herman, R.C., Adler, S., Lazzarini, R.A., Colonno, R.J., Banerjee, A.K., and Westphal, H., 1978, Intervening polyadenylate sequences in RNA transcripts of vesicular stomatitis virus, Cell, 15:587.

Huang, A.S., and Wagner, R.R., 1966, Comparative sedimentation coefficients of RNA extracted from plaque-forming and defective particles of vesicular stomatitis virus, J. Mol. Biol., 22:381.

virus RNAs. The AAUAAA sequence commonly found near the 3'-termini
of cellular messenger RNAs which may signal polyadenylation[23] is
not present in virus transcripts and this distinction may provide
a target for specific inhibition of virus messenger RNA termination.

In contrast to virus messenger RNAs the transcripts which
function as templates in genome RNA synthesis are complete trans-
scripts without terminal modifications. They contain pppAp-residues
at their 5'-termini and -pUOH at their 3'-termini (Figure 2) and
are considered for these reasons to be synthesized as separate
molecules and not to be derived by processing of a precursor.[24]

The proteins involved in RNA synthesis have been deduced from
genetic experiments to include the three largest virus gene products
PB_1, PB_2 and PA of molecular weights about 90,000; the nucleoprotein,
molecular weight about 50,000; and the NS_1 protein product of the
smallest virus gene with a molecular weight of about 25,000.[25]
However, only PB_1 and PB_2 have been assigned to specific roles in
transcription. In experiments involving cross-linking of virus
proteins to primers and to transcripts produced in vitro PB_2 was
found to bind specifically to RNAs containing methylated cap 1
structures ($^m7G(5')ppp(5')Nm-$) and, therefore, appears to be involved in the priming of messenger RNA synthesis.[26,27] In similar
experiments PB_1 has been shown to bind to the first nucleotide
added to primers and, therefore, has been proposed to participate
in the initiation of transcription which may include covalent
linkage of the initiating nucleotide to the protein.[28] The results
of studies of temperature sensitive mutants are consistent with
these roles for PB_1 and PB_2 in transcription and also suggest that
PB_2 may be the nuclease which generates the capped primers.[29]
The properties of this nuclease are unusual and it could be a target
for inhibition by antiviral compounds. In vitro and presumably
in vivo nuclease action results in the formation of oligonucleotides
between 10 and 14 nucleotides long which contain 3'-terminal
hydroxyl groups. Substrate recognition appears to be defined by the
requirements for a cap 1 structure at the 5'-terminus and analyses
of the 3'-terminal sequences of primers used in vivo indicate that
-pCpAOH may be the most common 3'-terminal dinucleotide.[30]
Inhibition of the binding of PB_2 to capped oligonucleotides and
presumably of nuclease action can be achieved in vitro by cap
analogues such as $^m7GpppG_m$, ^m7GpppG, and ^m7GTP[27] and the nuclease
is also inhibited by the uncapped ribopolymers poly (S^4U) and
poly (AG).[19] Inhibition by these oligonucleotides which also
prevent the process of transcript elongation in vitro suggests that
the nuclease may be blocked by interactions at sites additional to
those involved directly in cap binding and such information may
eventually be important for drug design.

Less is known of the enzymology of template complementary RNA
synthesis. Before the transcriptase complex of the infecting virus

can produce these complete transcripts, virus protein synthesis is
required and the continuous requirement for this synthesis concomitant with both template RNA and virus RNA syntheses[31] has
prevented the development of systems for the study of these
processes in vitro. From studies with temperature sensitive mutants
the large polypeptide PA is involved in virus RNA synthesis and
since the 5'-nucleotides of virus RNAs are triphosphates and the
3'-terminal nucleotides have 3'OH groups, like template complementary RNAs the virus RNAs appear to be produced independently. This
conclusion is supported by estimates of the sensitivity of the
synthesis of each RNA to UV irradiation which indicate that the
target size is proportional to the size of the individual RNAs.[32]

Virus Assembly

The assembly of the eight separate nucleocapsids and the
production of virus particles containing a representative of each
one are poorly understood processes. There is evidence from
analyses of the yields of reassorted viruses which are produced from
mixed infections that assembly is a random process[33,34] but there
is a report[35] of non-random segregation in such experiments
particularly of the three largest RNAs and this suggests that
specific interactions between individual nucleocapsids may occur
during genome assembly.

Virus assembly also involves the production of a virus membrane
and there is a considerable amount of information on the synthesis
and structure of the two glycoproteins of this membrane, the
haemagglutinin and the neuraminidase.[3,36] The properties of the
third virus specified membrane component, MP, are not as well
defined. The molecule appears to have an affinity for lipid[37,38]
and organic solvents such as chloroform[39] and in virus assembly
may interact directly with the lipids of the cytoplasmic face of
the plasma membrane. It also appears likely that MP can recognize
the areas of the plasma membrane which contain the virus glycoproteins[40] and this may involve interaction with the regions of
these molecules which project into the cytoplasm (for review see 41).
However, from studies on the protein composition of viruses produced
as a result of mixed infections with a number of membrane-containing
RNA viruses it seems that such MP-glycoprotein interactions are not
highly specific since pseudotypes containing the MP of one virus and
the glycoproteins of another are often isolated. The interaction
between MP and nucleocapsids which may trigger the process of membrane budding appeared in similar experiments to be virus specific.
[43] The sensitivity of a number of viruses to amantadine hydrochloride which was found to co-vary with the gene for MP[43,44] may
reflect an influence of this compound on MP function at this stage
of virus replication. It is, however, also possible that the dissociation of MP from the infecting transcriptase complex in the

initial stages of infection is the process affected by amantadine
and the importance of this step in virus uncoating remains to be
determined.

Biosynthesis and insertion of the haemagglutinin into the
plasma membrane occurs as for many secreted and membrane proteins,
following association of an amino-terminal signal peptide with the
endoplasmic reticulum. During transfer through the membrane compartments of the cell involved in glycoprotein export the molecule
is glycosylated at about seven sites each of which involves the
formation of an N-glycosidic linkage with asparagine. None of
these processes including acylation near the carboxy-terminus of
HA_2[45] appear to be specific for virus membrane glycoproteins and
probably as a consequence do not represent specific antiviral
targets. Neuraminidase biosynthesis is somewhat different. In
this case the amino-terminal region with mediates the initial
association between nascent polypeptide and membranes is not
removed as are signal peptides but remains as the site through
which the completed glycoprotein is attached to the plasma
membrane.[46] It is not at present obvious whether or not this
unusual though not unique feature could be exploited to inhibit
virus replication.

At the plasma membrane both glycoproteins are involved in the
formation of infectious virus particles. In the case of the
haemagglutinin cleavage of the precursor glycoprotein occurs with
the consequent formation of the HA_1 and HA_2 components of the
haemagglutinin subunit required in infectious virus particles[47,48]
For influenza viruses isolated from humans this processing reaction
involves removal of an arginine residue from the carboxyl terminus
of HA_1 in addition to polypeptide chain cleavage but the enzymes
involved have not been identified. Their inhibition is a possible
antiviral target. The neuraminidase has a specific role in the
formation and release of virus particles from infected cells which
involves removal of sialic acid residues from both viral and
cellular glycoproteins and glycolipids. As a consequence the membranes of the infected cells no longer contain receptors for the
virus haemagglutinin and the release of virus particles can proceed.
Failure of this receptor destroying mechanism in cells infected at
non-permissive temperatures with temperature-sensitive neuraminidase
mutants or in infected cells incubated in inhibitors of neuraminidase action[49,50] leads to the binding and accumulation of virus
particles at the cell surface. Inhibition of neuraminidase action
is, therefore, a target for antiviral drugs and since the three
dimensional structure of the enzyme has been determined and the
active site identified[36,51], a logical approach to their design
may eventually be possible.

In conclusion, all of the stages in the replication cycle
considered here have features which appear to make them susceptible

to inhibitors of virus production. It is clear that for the
processes for which most detailed information is available,
particularly with regard to the structure of the components involved in specific reactions, possible targets are more readily
appreciated. Whether this knowledge can be translated into the
preparation of appropriately specific inhibitors remains to be
seen.

Acknowledgements

        We acknowledge support from research grants NATO No. 0215/83,
NIH A113654 and NSF PC-771398.

References

1.   A. Gottschalk, in: "The Viruses" Vol 3, F. M. Burnet and
      W. M. Stanley, eds., Academic Press (1959) pp. 51-61.
2.   G. N. Rogers, J. C. Paulson, R. S. Daniels, J. J. Skehel,
      I. A. Wilson and D. C. Wiley, Nature (1983) in press.
3.   I. A. Wilson, J. J. Skehel and D. C. Wiley. Nature 289:368
      (1981).
4.   J. M. Whie and A. Helenius, Proc. Natl. Acad. Sci. USA
      77:3273 (1980).
5.   P. Vaananen and L. Kaariainen, J. gen. Virol. 46:467 (1980).
6.   T. Maeda and S. Ohnishi, Febs Lett. 122:283 (1980).
7.   R. T. C. Huang, R. Rott and H.-D. Klenk, Virology 110:243
      (1981).
8.   C. M. Brand and J. J. Skehel, Nature New Biol. 238:145 (1972).
9.   J. J. Skehel, P. M. Bayley, E. B. Brown, S. R. Martin,
      M. D. Waterfield, J. M. White, I. A. Wilson and D. C. Wiley,
      Proc. Natl. Acad. Sci. 79:968 (1982).
10.   J. M. White, K. Martin and A. Helenius, J. Cell. Biol. 89:674
      (1981).
11.   J. J. Skehel and M.D. Waterfield, Proc. Natl. Acad. Sci.,
      72:93 (1975).
12.   R. S. Daniels, A. R. Douglas, J. J. Skehel, M. D. Waterfield,
      I. A. Wilson and D. C. Wiley, in: "The Origin of Pandemic
      Influenza Viruses", C. M. Chu and W. G. Laver, eds., (1982)
      in press.
13.   C. J. Galloway, G. E. Dean, M. Marsh, G. Rudnick and I. Mellman,
      Proc. Natl. Acad. Sci. 80:3334 (1983).
14.   J. W. McCauley and B. W. J. Mahy, Biochem. J. 211:281 (1983).

15. J. J. Skehel and A. J. Hay, Nucleic Acids Res. 5:1207 (1978).
16. J. S. Robertson, Nucleic Acids Res. 6:3745 (1979).
17. U. Desselberger, V. R. Racionello, J. J. Zazra and P. Palese, Gene 8:315 (1980).
18. I. Ulmanen, B. A. Broni and R. M. Krug. Proc. Natl. Acad. Sci. 78:7355 (1981).
19. S. J. Plotch, M. Bouloy, I. Ulmanen and R. M. Krug, Cell 23:847 (1981).
20. R. Dhar, R. M. Chanock and C. J. Lai, Cell 21:495 (1980).
21. A. J. Caton and J. S. Robertson, Nucleic Acids Res. 8:2591 (1980).
22. J. S. Robertson, M. Schubert and R. A. Lazzarini, J. Virol. 38:157 (1981).
23. N. Proudfoot, Nature 298:516 (1982).
24. A. J. Hay, J. J. Skehel and J. McCauley, Virology 116:517 (1982).
25. C. Scholtissek, Adv. Genet. 20:1 (1979).
26. Ulmanen, I., Broni, B. and Krug, R.M. Proc. Natl. Acad. Sci., 78:7355 (1981).
27. D. Blaas, E. Patzelt and E. Kuechler, Nucleic Acids Res. 10:4803 (1982).
28. M. A. Horisberger, Virology 120:279 (1982).
29. I. Ulmanen, B. Broni and R. M. Krug, J. Virol. 45:27 (1983).
30. A. R. Beaton and R. M. Krug, Nucleic Acids Res. 9:4423 (1981).
31. A. J. Hay and J. J. Skehel, Br. Med. Bull. 35:47 (1979).
32. G. L. Smith and A. J. Hay, Virology, 118:96 (1982).
33. G. K. Hirst, Cold Spring Harb. Symp. Quant. Biol. 27:303 (1962).
34. W. G. Laver and J. C. Downie, Virology 70:105 (1976).
35. M. D. Lubeck, P. Palese and J. L. Schulman, Virology 95:269 (1979).
36. J. N. Varghese, W. G. Laver and P. M. Colman, Nature 303:35 (1983).
37. A. Gregoriades and B. Frangione, J. Virol. 40:323 (1981).
38. D. J. Bucher, I. G. Kharitonenkov, J. A. Zakomiridiu, V. B. Grigoriev, S. M. Klimenko and J. F. Davis, J. Virol. 36:586 (1980).
39. A. Gregoriades, Virology 54:369 (1973).
40. A. J. Hay and J. J. Skehel, in: "Negative Strand Viruses", B. W. J. Mahy and R. D. Barry, eds., Academic Press, N.Y., 2:635 (1975).
41. C. M. Ward, Current Topics in Microbiol. and Immunol. (1982).
42. J. Zavada, J. gen. Virol. 63:15 (1982).
43. A. J. Hay, N. C. T. Kennedy, J. J. Skehel and G. Appleyard, J. gen. Virol. 42:189 (1979).
44. M. D. Lubeck, J. L. Schulman and P. Palese, J. Virol. 28:710 (1978).
45. M. F. Schmidt, Virology 116:327 (1982).
46. J. Blok, G. M. Air, W. G. Laver, C. W. Ward, G. G. Lilley, E. F. Woods, C. M. Roxburgh and A. S. Inglis, Virology 119:109 (1982).

47.   H.-D. Klenk, R. Rott, M. Orlich and J. Blodorn, Virology
      68:426 (1975).
48.   S. G. Lazarowitz and P. W. Choppin, Virology 68:440 (1975).
49.   P. Palese, T. Tobita and M. Ueda, Virology 61:397 (1974).
50.   D. Bucher and P. Palese, in: "Influenza Viruses and Influenza",
      E. D. Kilbourne, ed., Academic Press, N.Y., pp 84-123 (1975).
51.   P. M. Colman, J. N. Varghese and W. G. Laver, Nature 303:41
      (1983).

BIOLOGIC AND CHEMOTHERAPEUTIC FORAYS

INTO THE FIELD OF UNCONVENTIONAL VIRUSES

Paul Brown

Laboratory of Central Nervous System Studies
National Institute of Neurologic and Communicative
  Disorders and Stroke, Building 36, Room 5B25
National Institutes of Health, Bethesda, Maryland, 20205

INTRODUCTION: HISTORICAL REMARKS

The subject of this presentation is a group of agents which, because of their unusual biological behavior and common pathology, have been termed "unconventional viruses". They are responsible for two diseases of man and two diseases of animals that are characterized by incubation periods of up to 10 or more years, followed by subacute to chronic, invariably fatal neurologic illnesses, with a distinctive brain histopathology of diffuse spongiosis, astrocytosis, and neuronal loss, in the absence of any inflammatory changes.

The first of these diseases to be recognized was scrapie, a widespread and commercially important disease of sheep and goats, which was described in England at least as early as the beginning of the eighteenth century (1). Speculation about the likelihood of an infectious etiology began to appear in the nineteenth century veterinary literature, and by 1924 several unsuccessful transmission attempts had been reported (2). The French investigators Cuille and Chelle in 1936 finally succeeded in transmitting the disease to sheep by intraocular inoculation of infected spinal cord tissue from a sheep with natural scrapie (3). Subsequent adaptation of the virus to mice in 1961 (4), and most recently to hamsters, with a comparatively rapid disease evolution and high brain infectivity titers (5), has made scrapie the most easily studied of the unconventional virus diseases, and much of the biological data presented here is based on studies of this prototype virus.

A second veterinary disease, transmissible mink encephalopathy, which is clinically and pathologically very similar to
scrapie, was first described in 1965 in two outbreaks traced to a
possibly scrapie-contaminated food source, and is thought likely
to represent scrapie virus infection in a new host (6).

Meanwhile, in the early 1920's two German neuropsychiatrists,
Creutzfeldt and Jakob, reported a total of six probably heterogeneous neurological diseases, of which two were certainly
examples of the disease (CJD) that came to be named after them
(7-9); and in 1957, Gajdusek and Zigas reported the epidemic
occurrence of a fatal cerebellar ataxia called kuru, that was
limited to the Fore linguistic group and their neighbors in the
Eastern Highlands of Papua New Guinea (10). Two years later, in
1959, astute observations by the neuropathologist Klatzo and the
veterinarian Hadlow changed the entire course of thinking about
these diseases. Klatzo pointed out the pathological similarity
between kuru and CJD (11), and Hadlow not only drew attention to
the similarities between kuru and scrapie, but further suggested
that kuru be inoculated into primates for long-term observation
(12). Gajdusek and his colleagues consequently resumed their
previously unrewarding attempts to show an infectious cause for
kuru, which was finally transmitted to the chimpanzee in 1966 (13)
and followed by similar success with CJD in 1968 (14). Some human
strains of CJD have now also been transmitted, with much reduced
incubation times, to guinea pigs, rats, and mice, greatly
facilitating experimental study of the agent (15).

CJD occurs with equal frequency in men and women, most often
between 55 and 75 years of age. Although by no means a stereo-
typed illness, it typically evolves as a progressive dementia with
associated cerebellar or visual abnormalities and, later, extrapyramidal signs with myoclonus or other abnormal movements, and a
characteristic electroencephalographic pattern of pseudoperiodic
slow wave activity. Most patients are dead within three to six
months of the onset of symptoms (16).

The significance of these "unconventional virus" diseases to
human medicine certainly does not lie in their public health
importance: from epidemiologic evidence, the two veterinary diseases probably do not affect man; kuru is currently disappearing
even from its exotic localization in Papua New Guinea; and CJD,
while worldwide in distribution, annually affects no more than one
per million people, and does not appear to be increasing. Rather,
their significance is implicit in the causative agents themselves,
which may represent a new class of microorganisms with novel
mechanisms of replication, and which may conceivably cause other,
more common illnesses of man which are at present of unknown
etiology.

PATHOGENESIS: ENTRY, TRANSIT, EXIT

Since the clinical and pathological abnormalities of these
diseases are restricted to the central nervous system, the primary
target for any practical application of chemotherapy must be virus
that is already present and replicating within neuronal cells.
Knowledge of how virus attains these cells may also have eventual
chemotherapeutic importance if a biologic marker (such as antibody)
is discovered which can identify infected individuals during the
pre-clinical stage of disease.

Experimentally, infection can be regularly initiated by means
of inoculation, either directly into the brain, or by a variety of
peripheral routes including intravenous, intraperitoneal, intramuscular, subcutaneous, subgingival, and intraocular (3,17-20).
Virus can also be transmitted orally (21), but not, apparently,
via respiratory or venereal routes.

Studies of natural scrapie in sheep have shown an initial
appearance of virus in tonsils, suprapharyngeal lymph nodes, and
intestine, implying an oral route of infection (22). Natural kuru
and mink encephalopathy may also be transmitted orally, but it is
impossible to exclude the alternative possibility of virus entry
by subcutaneous or mucosal routes through epidermal breaks in the
gums, conjunctivae, or skin. CJD in man has been iatrogenically
transmitted from a corneal transplant (23) and from virus-
contaminated stereotactic EEG electrodes (24), and may also have
been transmitted from neurosurgical (25,26) and dental procedures
(27), but the natural mode of presumed inter-human transmission
is entirely unknown (28).

Inside the host, studies of both natural and experimental
scrapie, and of experimental CJD, have established the cardinal
role of the spleen, and particularly the blastoid B-cell
component, as the most important early site of virus replication
(22,29,30). Virus is demonstrable in the spleen as early as
several weeks after infection, well before it can be detected in
nervous tissue, and long before the appearance of clinical
disease. Interestingly, this is true even after direct intra-
cerebral inoculation, which suggests the importance of blood-borne
virus in the early phase of infection. Later during the incubation period, virus can also be isolated from a variety of other
visceral tissues, as well as from brain, spinal cord, and
peripheral nerves (22,29,30).

The question of how virus reaches the central nervous system
is not yet resolved. Viremia throughout most of the course of
infection has been documented in experimental CJD (30,31). However, sequential infectivity titrations in experimental scrapie
are most consistent with initial penetration of splenic virus via

Table 1. Virus Isolations in Tissues from CJD Patients

Tissue	Number Positive	Number Tested
Brain	218	331[a]
CSF	3	23
Cord	3	8
Eye	3	7
Kidney	3	28
Liver	3	28
Lung	2	7
Lymph node	2	13
Spleen	1	30
Skeletal muscle, heart, adipose tissue, marrow, adrenal, thyroid, prostate, testes, and peripheral nerve	0	25
Blood/lymphocytes	0	7
Urine	0	16
Feces	0	5
Saliva	0	8

[a]Includes some histologically unverified cases.

sympathetic nerves to the thoracic spinal cord, with subsequent spread upward to the brain and downward to the lumbar cord, dorsal root ganglia, and peripheral nerves (32,33). It has also been shown that following unilateral intraocular inoculation, scrapie virus attains the central nervous system by retrograde spread along the optic nerve to the contralateral superior colliculi and thence to the rest of the brain (19). Thus, there is little question that neural spread can occur under experimental conditions, but it is not known whether it also occurs in natural infections. The dilemma is reminiscent of the early studies on the pathogenesis of poliomyelitis.

The sequential distribution of virus during the incubation period of human CJD is not known, but by the time symptoms have appeared, virus can be isolated from many of the same visceral tissues that are affected in scrapie. The distribution and proportion of successful virus isolations from patients in the clinical phase of CJD are presented in Table 1. It should be noted that although virus has been shown to invade and replicate or at least persist in several visceral tissues, it has been isolated from these tissues far less regularly in man than in experimental animal models. As with animals, infectivity titers in visceral tissues are much lower than in brain. Isolation of virus from human blood or lymphocytes has not yet been accomplished, even by transfusing primates with entire units of whole blood from terminally ill CJD patients; nor has virus been isolated from saliva, urine, or feces, which while comforting to medical personnel caring for patients, fails to illuminate the possible means by which CJD is ordinarily acquired.

Summarizing what is known of the pathogenesis of these diseases, it can be said that with the exception of human CJD, for which only iatrogenic, tissue-penetrating transmission has been documented, natural infection by the unconventional viruses appears to occur by the oral route. The spleen is an early and principal extraneural site of infection, and invasion of other visceral and central nervous system tissues may occur either by blood-borne or neural spread of virus. Bodily secretions and excretions have not as yet been shown to contain virus.

CELLULAR BIOLOGY: WHERE IS THE VIRUS? (WHERE IS IT NOT)?

Studies of the intracellular location of virus have been hindered not so much by a failure of detection as by an embarrassment of riches - some degree of infectivity is demonstrable almost everywhere within the cell. In the 1960's, shortly after scrapie was adapted to mice, there appeared from Hunter's laboratory in England the first of a succession of papers aimed at defining the intracellular distribution of virus in brain tissue from terminally ill mice (34).

By the simple technique of differential centrifugation of a sucrose-brain homogenate, several relatively crude subcellular fractions were obtained for infectivity titrations. Although virus was found in all fractions, approximately 99% of the total infectivity was associated with particulate material. Later experiments using discontinuous sucrose gradients, with electron microscopic examination of the resulting interface material, showed that nuclei, myelin, and nerve ending particles were only minimally infective, and that the still comparatively impure mitochondrial and microsomal fractions contained the bulk of recoverable virus (35). However, further studies indicated a lack of association between infectivity and mitochondria or lysosomes, and a stong association with all membrane-containing preparations, irrespective of the concentration of other subcellular organelles (36).

More precise localization was subsequently made possible by
the use of enzyme markers limited to, or concentrated in, various
subcellular organelles and membranes, with comparison of enzyme
activities and virus infectivity in subfractions of the initial
density gradient preparations (37). The results showed a clear
association of scrapie infectivity with enzyme markers found in
cell membranes and endoplasmic reticulum, but could not differentiate between the two entities.

The same approach was then used on a continuous cell culture
originating from scrapie-infected mouse brain, which supports low-
grade replication of the virus in vitro. Fractions were obtained
which by analysis of enzyme marker activities (5'-nucleotidase and
cytochrome c reductase) consisted mostly of either plasma membrane
or endoplasmic reticulum, and infectivity was found to correlate
best with the nucleotidase plasma membrane marker (38).

Strong independent support for these physico-chemical
conclusions about the plasma membrane location of virus has come
from histologic study of brain tissue. It has been recognized for
some time that spongiosis constitutes the morphologic hallmark of
all the unconventional virus diseases, and that the glial proliferation and neuronal degeneration also common to these diseases
are secondary processes. This spongiosis consists of irregular
accumulations of intracellular vacuoles that by electron microscopy can be seen to be bounded by membrane, and which frequently
also contain amorphous fragments of membranous material.

Recently, a remarkable microscopic study by Beck et al.(39)
of the sequential changes observed during the pre-clinical stage
of experimental kuru in the spider monkey, has provided irrefut-
able evidence that these vacuoles originate from detachment of the
inner leaf of abnormal, multilaminar patches of plasma membranes
of adjacent neuronal dendrites or of dendrites and presynaptic
terminals (Fig. 1). Preliminary observations show the same
abnormalities in experimental CJD and scrapie.

During the course of the incubation period, the number of
multilaminar membrane patches was found to decrease as vacuolar
development progressed, and none were seen in the single animal
sacrificed at the symptomatic stage of disease. Moreover, the
initial appearance of multilamination followed an even earlier
profusion of cytoplasmic organelles called coated vesicles. The
function of these vesicles is uncertain, but in normal embryonic
development they are thought to be involved in synaptogenesis,
budding off from the Golgi cisternae and migrating toward the
plasma membrane, with which they merge to contribute material to
synaptic attachment sites. Morphologic evidence of hyperactive
Golgi complexes, endoplasmic reticula, and protein synthesis was
also observed early in the incubation period.

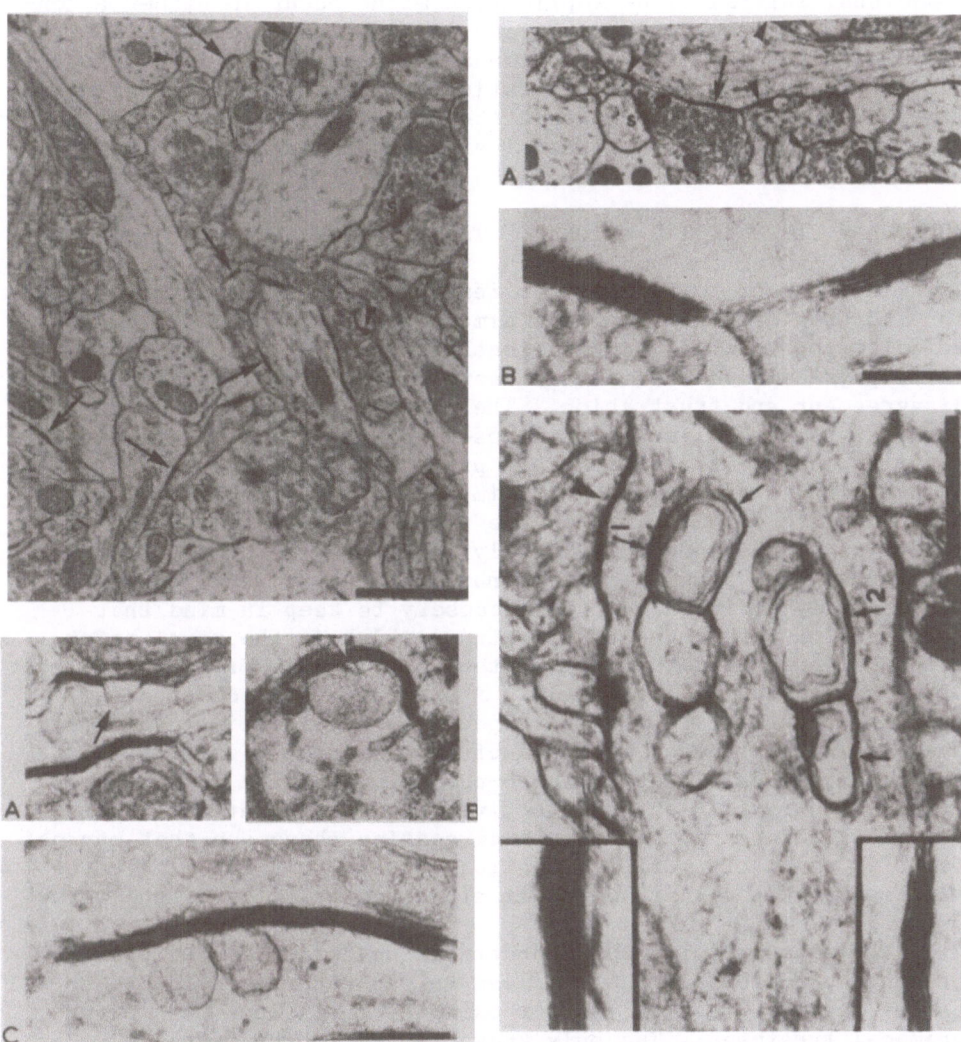

Fig. 1. Electron microscopy of experimental kuru in the monkey.
         Top left (13 weeks after inoculation): arrows indicate
         abnormal configurations of plasma membranes (ACPMs).
         Top right (same animal): higher magnifications of ACPMs.
         Bottom left (A, 4 weeks; and B and C, 13 weeks after
         inoculation): formation of vacuoles from innermost
         leaflet of ACPMs. Bottom right (13 weeks after inocu-
         lation): confluent vacuoles containing ACPMs.
         Original transparencies of these photographs were very
         kindly furnished by Mrs. Elizabeth Beck and Professor
         P. M. Daniel, whose article should be consulted for
         additional details (39).

Thus, initial virus replication might occur in either a non-membranous, intracellular location, or in plasma membrane loci, from which it could set in motion a train of metabolic events that provoke reduplication and separation of the internal plasma membrane layer, to culminate in the picture of intracellular vacuolation seen in the clinical stage of disease.

MOLECULAR BIOLOGY: THE NATURE OF THE BEAST

In contrast to the more or less harmonious and straight-forward progression of studies aimed at defining the tissue and cellular location of unconventional viruses, efforts to determine their physical and chemical characteristics have been marked by disagreement and frustration. The crucial problem in all of this work has been the limitation imposed, first, by the comparative impurity of even the most highly processed virus-containing tissue preparation; and second, by the imprecision of the infectivity assay used to detect virus in the preparation under analysis. The purity of a substance is evidently defined by reference to other substances, and thus when it is noted that a preparation is "extensively purified", it is necessary to keep in mind that purification with respect to protein, which is the usual reference, has never exceeded 10% of the total sample protein, and most often has fallen well below 1%. A similar degree of success applies to purification with respect to nucleic acids and lipids, and nothing whatsoever is known of other possibly important chemical compounds that could co-purify with virus infectivity. Imperfect purification of infective tissue preparations is also responsible for aggregation-disaggregation phenomena that may not occur in similarly purified control tissues, and can play havoc with apparent physical and chemical characteristics of the virus.

The second limitation, concerning detection of the virus, presents a curious logical impasse to the interpretation of so-called negative results, as for example, resistance to a given chemical treatment. The only known assay for the presence of an unconventional virus is the transmission of disease to animals, and the reproducibility of virus titrations has about a one logarithm variance. Thus, if there are 10^8 infective units before treatment and 10^7 units after treatment, the difference cannot be considered significant, despite a 90% reduction in virus infectivity, with the result that conclusions about the nature of the virus are being drawn from observations on the behavior of a minority subpopulation of resistant particles (40). The difficulty is compounded when the amount of virus is estimated from the length of the incubation period of the transmitted disease rather than actually measured by dilution end-point infectivity titrations. Although equations describing a precise relationship between inoculation dose and length of incubation period have been

published (41), some workers have noted a variance of up to two
logarithms between such estimates and dilution end-point titers
(42). Moreover, it is not always known whether chemical treatments of inocula destroy agent titer or affect the expression of
disease, and thus the operational estimate of titer in the
inoculated animal (38).

These caveats notwithstanding, a respectable body of knowledge has accumulated about the physical and chemical nature of
the unconventional viruses, based largely on work with the scrapie
agent. Early studies showed that it passed through filters of
50 nm but not 27 nm pore size, a finding that has been repeatedly
verified, and establishes an upper limit of between 30 and 50 nm
for an infective particle (38). Aggregation and membrane linkage
phenomena allow for the possibility that its minimal (monomeric)
size may be even smaller.

Density gradient centrifugation experiments have yielded
peak infectivity fractions with buoyant density values ranging
between 1.14 and 1.34 g/cm^3, but most often showing a relatively
broad band of infectivity centering around a density 1.20 g/cm^3
(43,44). Such variability probably results from technical differences in sample purification and centrifugation protocols, and in
any case primarily reflects the density of virus-associated
membrane fragments (44). A recent rate-zonal centrifugation
experiment in sucrose gradients showed infectivity to be spread
over a range of particle sizes with sedimentation constants of
about 40S to more than 500S (45). Similarly treated homogenates
subjected to nuclease and protease digestion, and then used in an
agarose gel electrophoresis revealed a forward wave of infectivity
occurring in the same region as control DNA fragments of about
10^6 daltons (46). However, it should be remembered that size
estimates of the scrapie agent based on comparison to migration of
DNA fragments in electrophoretic gels may not be valid (40).

An altogether different approach to determining the size of
infective particles has been taken by investigators using ionizing
radiation to inactivate the scrapie virus. In this method, the
virus preparation is subjected to precisely controlled high
voltage electron or gamma radiation, and energy-infectivity dose
response curves are produced which can be used to estimate the
molecular weight corresponding to a calculated target size.
Estimates of the molecular weight of the scrapie infective
particle in such calculations have ranged from 64,000 to 150,000
daltons (47,48), and the infective particles of kuru and CJD
viruses have been found to be at least as small (49).

While this methodology cicumvents some problems of impurity
or heterogeneity of the virus preparations, it, too, has serious
limitations. The calculation of target size is based on a complex

physical model of the inactivation process, and estimation of molecular weight requires a further assumption for the density of the irradiated molecule. The estimated molecular weights given above were based on the density of nucleic acid. If instead of nucleic acid the estimate were based on the density of glyco-protein, the value of 64,000 daltons would increase to about 100,000 daltons (48), and would be still higher using the density of a lipid-containing macromolecule.

An alternative to target size calculation is a direct compar-ison of the radiation inactivation constant of scrapie virus to values obtained for other viruses or macromolecules of known size. When comparison is made with the inactivation constants of single and double-stranded nucleic acids from different viruses, the molecular size of a hypothetical scrapie nucleic acid is found to range from 75,000 to 1,600,000 daltons, or approximately 10 times greater than estimates made with the same data by target size calculations (40).

Further limitations on conclusions drawn from irradiation sensitivity data are the unknown but probably high degree of molecular aggregation present in the tissue samples, and the unknown degree of molecular repair occurring in response to radiation damage. Such repairs need not be restricted to nucleic acid mechanisms, but might involve ionic re-associations following irradiation of a complex crystalline-gel membrane structure. Both aggregation and repair phenomena would have the effect of causing an underestimation of the true size of the inactivated particle.

Considering the technical limitations of these filtration, density gradient, and irradiation data, it would be unwise to hazard a more radical summary statement than that the size of the non-aggregated infectious component of unconventional viruses probably lies within or below the size range of the smallest known conventional viruses ($\leq$30 nm, 40S, 10^6 daltons).

If their size does lie within the range of small conventional viruses, it may be asked why they have not been seen by electron microscopy. There have in fact been scattered reports in the literature of the presence of ordinary virions, and most recently, of spiroplasma organisms in brains from patients with CJD, but these must be considered fortuitous passengers of the disease in view of the failure to observe similar structures in hundreds of other well-studied cases of both human and experimental CJD, and the fact that all such microorganisms would be destroyed by physical and chemical treatments that do not affect the transmissible agent of CJD.

A variety of virus-like or particulate structures have also been observed in brain tissues from scrapie-infected mice and

Table 2. The Effect of Various Physical and Chemical
Treatments on Scrapie Virus Infectivity

Treatment	Loss of Infectivity
Nucleases	None
Proteases	None to nearly complete
Carbohydrases	None
Lipases	None
Organic solvents	Minimal to complete
Detergents	Minimal to complete
Oxidizing agents	Partial (variable results)
Alkylating agents	Minimal
Chaotropic agents	Partial to complete (some are reversible)
Carbethoxylation	Complete (reversible by hydroxylamine)
Psoralen photoreaction	None
Mercaptoethanol + SDS	Partial
pH >12	Partial to complete
Heat:	
100°C	Partial
121°C	Complete or nearly complete
Irradiation:	
Ultraviolet (254 nm)	Minimal at usual dose $(D_{37}25 \text{ kJ/m}^2$
Ionizing	Minimal at usual dose $(D_{37}45 \text{ kGy})$

sheep which do not occur in normal animals (50,51). One type of
particle of about 30 nm size with a circular or rectilinear
profile has been repeatedly noted to lie within post-synaptic cell
processes, but there is no evidence that it represents the scrapie
agent, rather than a pathologic change of cell organelles in
affected tissues. The same may be said of the abnormal, 4-6 nm
helical filaments seen in detergent-treated extracts of scrapie
and CJD brain tissue (52).

If the physical structure of the unconventional viruses
remains a mystery, their chemical composition has been the single
most disputed subject in the whole of unconventional virology.
All of the data are based on comparative susceptibilities to
physical or chemical attack (Table 2), and conclusions inferred
about virus composition are particularly embarrassed by the
imperfect purification of virus-containing tissue preparations.

Critical reviews of these studies have recently been published by
Millson et al. (38), Latarjet (48), and Prusiner (41).

The crucial question concerns the presence or absence of
viral nucleic acid. Neither DNA nor RNA extracted from scrapie-
affected brain tissue, either alone or in combination with normal
membrane components, has produced disease in inoculated animals
(53-55). Digestion of crude and partially purified tissue samples
with a variety of nucleases, has, with the exception of a single
unverified experiment (56), repeatedly failed to affect scrapie
infectivity (38,41,57). Chemical or photochemical attack by
agents to which nucleic acids are particularly sensitive,
including divalent zinc cations, psoralens, and hydroxylamine, do
not reduce scrapie infectivity (41). Resistance to ultraviolet
irradiation at 254 nm wave length is over three times greater than
for any known biologically active particle which depends for
replication on the integrity of nucleic acid, and the action
spectrum of ultraviolet irradiation does not resemble that of a
nucleic acid (48,58). Finally, the presence of oxygen dramatically increases the sensitivity of scrapie virus to ionizing
radiation, whereas it almost always protects against damage to
nucleic acid (59).

Such uniformly negative evidence might be supposed to have
convinced all but the most skeptical critics that nucleic acids do
not figure among the essential components of the unconventional
viruses. A similar dossier of molecular size and chemical
characteristics has been inferred for the replicating cytopatho-
genic material from lysates of Naegleria amebae (60). There is,
however, an understandably massive reluctance to relinquish the
doctrine of nucleic acid-mediated replication mechanisms, and
although alternative mechanisms have been proposed that involve
replicating membrane fragments, polysaccharides, or proteins, they
remain hypotheses unburdened by experimental evidence. Moreover,
biologic precedents exist for the objection that chemical and
radiation attack may simply be nullified by the presence of protective tissue constituents, as for example occurs with viroids
(61). This proposal will be impossible to disprove until such
time as complete purification of the virus has been accomplished.

The situation with regard to a protein component necessary
for infectivity, while not unequivocal, is better than that for
nucleic acid. Concentrations of SDS, urea, and phenol that
denature most proteins have been repeatedly shown to be effective
inactivating agents (38). Early studies of the effects of
proteolytic enzymes on scrapie infectivity yielded variable
results, but one successful pronase digestion of acetone-extracted
brain suggested the importance of a lipid-protected protein
component of infected membranes (38). Subsequent experiments
using brain extracts incubated with SDS before treatment with

either pronase or 2-mercaptoethanol also showed partial to nearly
complete inactivation of scrapie infectivity (57,62), and a series
of publications from Prusiner's laboratory has accumulated further
evidence for the role of protein in the expression of scrapie
infectivity. Using more highly purified preparations than in
earlier experiments where Proteinase K was ineffective, significant titer reductions were observed to follow treatment with
chaotropic ions and diethyl pyrocarbonate, as well as with
Proteinase K (63-65). It is curious and somewhat confusing to
note that independent confirmation of Proteinase K susceptibility
has recently been shown not to depend on pre-treatment of membrane
fractions with detergents, and not to be correlated with the
degree of proteolysis (42).

    None of this evidence can be taken to prove that scrapie
virus actually contains a protein, since it has never been shown
whether the protein necessary for infectivity is a component of
the virus or of the host; it is even possible, in the role of
devil's advocate, to argue that none of the evidence is fully
conclusive for the involvement of a protein. The claim that
hydroxylamine-reversible inactivation by diethyl pyrocarbonate
favors an action on protein (64) is not warranted in the absence
of any information about similar reversibility of inactivated
nucleic acid. More generally, the effect of the various chemicals
and enzymes may be to modify or prevent particle aggregation that
could be necessary for infectivity. Inactivation by molar
concentrations of chaotropic ions, for example, does not occur
without an accompanying freeze-thaw cycle (63). Enzymes might
lyse tissue proteins that prevent aggregation, or might simply
adsorb to a virus subunit to prevent aggregation, without
proteolytic action, which could explain the lack of correlation
between the degree of proteolysis and reduction of infectivity
observed with Proteinase K digestion (42).

    Furthermore, unconventional viruses have long been known to
show exceptional resistence to heat, formaldehyde, and other
alkylating agents that denature most proteins (38), and the
atypical oxygen sensitization of ionizing radiation damage,
mentioned in regard to nucleic acid, also applies to protein (59).
It therefore seems unlikely, if we are dealing with a protein,
that it has a simple structure; indeed, the degree of oxygen
sensitivity to ionizing radiation damage shown by scrapie virus
preparations has been noted to be higher than for any biological
or biochemical system except the release of bound enzymes by
intact lysosomes; and the action spectrum to ultraviolet radiation
most closely resembles that of endotoxin, a lipopolysaccharide-
protein complex, and of peroxidase, a glycoprotein (48,59).

    In this regard, it is interesting to recall the biochemical
evidence of abnormal glycoprotein, polysaccharide, and lipid

(especially ganglioside) metabolism of brain tissue (66-71), as
well as the histochemical demonstration of accumulated cholesterol
esters and glycoproteins in neuronal membranes in scrapie and CJD
(72-74).   Thus, after years of intensive search for the essential
scrapie nucleic acid or protein, we may have come round full
circle to the notion, originally based on the inactivating effects
of organic solvents and detergents, that scrapie infectivity
depends not on any individual macromolecule, but rather on a
complex structure composed mostly of lipid and polysaccharide
buried within the host cell membrane (38,75), and that resolution
of the problem may depend upon "rigorous and patient application
of the techniques of membrane biochemistry" (38).

CHEMOTHERAPEUTIC TARGETS: SHOTS IN THE DARK

     What can be said about therapeutic targets in a group of
agents for which the natural mode of disease transmission and
pathogenesis are either unknown or uncertain, which cannot be
visualized by electron microscopy, induce no detectable immune
response, and are so embedded in host cell membranes as to defy
purification; for which the structure is unknown, the chemical
composition debated, the mechanism of replication a mystery, and
which show such an unlikely resistence to physical and chemical
inactivation that their very definition as viruses is in question?

     The answer is that the chemotherapy of unconventional viruses
has been completely empirical.   Several drugs which have shown
some effect on conventional viruses have been tried in patients
with CJD, and Table 3 summarizes the few therapeutic attempts that
have been reported in the literature.

     The first report appeared in 1971, using the drug amantadine,
chosen because of its demonstrated anti-viral action and ability
to cross the blood-brain barrier (76).   The patient had moderately
advanced symptoms and a steadily progressive course, but within
two days of the administration of amantadine showed a dramatic
change for the better that persisted for the next ten days, when
treatment was withdrawn.   Within three days the patient relapsed,
and treatment was restarted with improvement and clinical
stabilization that lasted nine months, when the patient died of
sepsis (76,77).

     Two further cases treated with amantadine were reported in
1973 (77).   The first patient had a rapidly progressive illness
with advanced symptoms when treatment was begun.   Within four days
there appeared a remarkable improvement that was sustained in a
fluctuating manner for the next two months, when a steady
deterioration occurred that ended in death three weeks later.
Autopsy examination confirmed the clinical diagnosis of CJD.

The second case had a 20-month history of emotional and mental impairment, ataxia, and tremors. On admission to the hospital she was severely demented, unable to stand, walk, or care for herself, and had extrapyramidal rigidity and ankle clonus. Symptomatic improvement began within six days of the initiation of treatment, progressing to complete normality over the next three months. Treatment was stopped after seven months, and the patient was still well when last seen five years later (78). It must be noted that the long duration of this patient's illness before treatment was begun, without either myoclonus or a characteristic electroencephalogram, and no histopathologic verification, raise a question about the clinical diagnosis of CJD.

Nevertheless, as might be expected, these three case reports caused a stir of excitement among neurologists, and led to numerous further trials of amantadine in CJD patients. The results have been disappointing. Apart from one subsequent case in whom a partial improvement was sustained for five years (78), published reports have not validated these earlier successes (79-81), nor has treatment been successful in any of several additional unpublished cases known to us.

Other anti-viral drugs have also been used in a small number of CJD patients. Idoxurydine was reported to have no effect on two patients with advanced disease (79,82), and human interferon, given either parenterally or intrathecally, failed to influence the course of disease in three other patients (83,84). During the past two years, the antimoniotungstate HPA-23 has been used on a number of patients in France, usually without effect, but in two still living patients with as yet histologically unverified CJD, administration of the drug was associated with either a marked slowing or arrest of the previously rapid symptomatic progression of disease (85). A recent report from Italy also describes a patient with rapidly progressive disease in whom the administration of methisoprinal coincided with clinical stabilization and prolongation of an agonal stage of illness over an 18-month period (86). Finally, an attempt to treat CJD with vidarabine has recently been published in which one patient had no response, one patient showed a modest improvement, and one patient, described in detail, had a remarkable history of temporary symptomatic remissions during each of five therapeutic courses with the drug, but went on to die nine months later (87).

The results of empirical chemotherapeutic studies of CJD and scrapie in experimental animals have also been uneven, despite the advantageous use of drug treatment schedules beginning at the time of virus inoculation, rather than at the onset of symptomatic disease, as must be done in humans. Interferon and interferon inducers have been repeatedly shown to be without effect on experimental scrapie or CJD (88-92). Treatment with the antibiotics

Table 3.   Reported attempts to treat human cases of CJD

Author	Date	Diagnostic Basis	Clinical Condition before Treatment
Braham (76)	1971	Clinical	Moderately advanced
Goldhammer et al (82)	1972	Biopsy	Terminal
Herishanu (79)	1973	Clinical	Moderately advanced
Sanders and Dunn (77)	1973	Autopsy	Advanced
Sanders and Dunn (77)	1973	Clinical	Moderately advanced
Ratcliffe et al (80)	1975	Autopsy	Advanced
Sanders (78)	1979	Autopsy	Moderately advanced
Ververken et al (83)	1979	Autopsy	Moderately advanced[a]
Scully et al (81)	1980	Biopsy	Early
Kovanen et al (84)	1980	Autopsy	Moderate
Kovanen et al (84)	1980	Biopsy	Early
Villa et al (86)	1982	Autopsy	Terminal
Furlow et al (87)	1982	Autopsy	Moderate

[a]Personal communication, Dr. H. Carton, Department of Neurology, University of Leuven, Leuven, Belgium.

Treatment Drug, Daily Dose, and Duration	Effect and Outcome
Amantadine, 600 mg po for 12 days, then 9 months	Improvement (X2), stable, then death after 9 months
Idoxuridine, 3 g iv for 5 days	No effect, deterioration, death after 2 weeks
Amantadine, 200 mg po for 1 week Amantadine, 600 mg po, and Idoxuridine, 6 g iv for 5 days	No effect, deterioration with septicemia, and death after 3 weeks
Amantadine, 200-600 mg po for 10 weeks	Fluctuating improvement, then death after 4 months
Amantadine, 100-200 mg po for 7 months	Continuous improvement, alive and well after 5 years
Amantadine, 1000 mg po for 2 weeks	No effect, deterioration, death after 2 weeks
Amantadine, 100-400 mg po for 5 years	Improvement, stable, then death after 5 years
Human fibroblast IF, 10^6 units intrathecally q2d for 1 week[a]	No effect, deterioration, death after 3 months[a]
Amantadine, 200-400 mg po for 8 months[b]	No effect, deterioration, death after 8 months[b]
Human leukocyte IF, 3 X 10^6 units sc for 4 months	No effect, deterioration, death after 7 months
Human leukocyte IF, 3 X 10^6 units sc for 13 months	No effect, progressive deterioration
Methisoprinal, 7 g po for 18 months	Stabilization, then death after 18 months.
Vidarabine, 15-20 mg/kg iv for 5-15 days (X5)	Improvement (X5), but death after 9 months

[b]Personal communication, Dr. Dennis M.D. Landis, Department of Neurology, Massachusetts General Hospital, Boston, USA.

thiamphenicol and rifampicin, and with the anti-viral agents
amantadine, cytosine arabinoside, and isoprinosine, has been
reported not to modify the outcome of experimental CJD in mice
(93). Amantadine, cytosine arabinoside, methisazone, and virazole
have also failed to modify experimental scrapie in mice (94).
Unpublished studies from our own laboratory have employed a wide
variety of drugs, including steroids and other hormones, vitamins,
and antimetabolites, as well as a number of antimicrobials active
against parasites, bacteria, and viruses (95). None had any
effect on either the length of the incubation period or the
mortality rate of CJD in mice or scrapie in hamsters.

    The only drugs shown to have any statistically significant
effect upon experimental unconventional virus diseases are HPA-23
and amphotericin B. HPA-23, so named by virtue of its position as
the 23rd in a series of Hetero Poly Anion compounds investigated
for their antiviral activity, is a three-lobed hexagonal crystal
with a molecular weight of 6800 daltons, in which a cloud of
negatively charged oxy-tungsten and antimonium atoms surrounds a
central sodium ion through six oxygen bridges. It is known to be
active against numerous conventional and oncogenic viruses by
means of nucleic acid polymerase inhibition, augmentation of
"killer cell" activity, and synergism with interferon. Electroencephalographic abnormalities occurring during and after drug
administration suggest that it also has a direct physico-chemical
effect on neuronal membranes (96). When administered in repeated
doses during the days following virus inoculation, the drug
markedly prolongs the incubation period, reduces the amount of
virus in brain, and lowers the mortality rate of experimental
scrapie in mice (94,96). As mentioned above, however, treatment
of human CJD, begun during the clinical phase of the disease, does
not appear to have achieved a comparable success.

    Amphotericin B belongs to the group of polyene antibiotics
with a ring structure that includes a hydrophobic region of
conjugated double bonds and a hydrophilic region of hydroxyl
groups with a mycosamine sugar moiety. Its action on fungi is
thought to be mainly due to a preferential binding to the ergo-
sterol present in fungal cell membranes, with alteration of
permeability functions and death of the cells. It is also known
to have an affinity for cholesterol, the principal sterol in
animal cell membranes. Preliminary experiments in our laboratory
have shown a significant, dose-dependant, prolongation of the
incubation period of scrapie in the hamster, when the drug is
administered weekly during the first four months following virus
inoculation (95). Also, there is an unverified but tantalizing
observation of an unusually long 12-month duration of illness in a
CJD monkey that received a course of treatment beginning after the
appearance of clinical signs (95). Further studies of the drug in
experimental CJD in mice and monkeys are in progress.

Moving from the empirical to the theoretical realm in a concluding attempt to formulate some ideas for the immediate future of chemotherapy, we may observe, first, that in terms of disease pathogenesis, the possible chemotherapeutic targets for an attack on the unconventional viruses might be situated at several points in the body. Attachment of virus to cells, or replication of virus within cells, might be blocked at the natural point of entry into the body, which is perhaps the gastrointestinal tract; or in the principal extraneural site of replication, which is the spleen. It could be important from the standpoint of chemotherapy to understand what critical metabolic differences distinguish the blastoid B-cell from the neurone, since virus replicates freely in both cell types, but in contrast to its destruction of the central nervous system, causes neither morphologic nor functional abnormalities in the spleen.

At a somewhat later stage of the incubation period (but still well before the onset of clinical illness), when virus gains access to the central nervous system, chemotherapy might be aimed at some aspect of the early metabolic alterations that have been demonstrated. In addition to the morphologic observations of binucleate neurones, endoplasmic reticulum activity, chromatolytic protein synthesis, and the appearance of multilaminar membranes and abnormal, paired helical filaments within the first few weeks of the incubation period (39,52,97), parallel derangements of membrane function have been revealed by cell fusion, spin label, and conventional electroencephalographic techniques (98-100), and glycoprotein and glucose energetic metabolism undergo a progressive alteration throughout the remainder of the incubation period (66,101).

For practical purposes, however, human disease can only be treated after the onset of symptoms, at a time when virus has already attained high titers in the central nervous system, and neurones are riddled with intracytoplasmic vacuoles. The many neurochemical aberrations known to be present at this stage of disease should be re-evaluated from the viewpoint of possible chemotherapeutic exploitation. In particular, detailed histochemical examination of plasma and vacuolar membranes might show specific differences between multilaminated foci and intervening areas of normal membrane. In this regard, the observation (39) that the plasma membrane abnormalities so prevalent during the incubation period decrease and may even disappear in the clinical stage of disease is especially intriguing, since it suggests at least the possibility that symptomatic disease might result from vacuole-induced cell derangements, and not from intact plasma membrane abnormalities. If the virus accompanied this morphologic transition, and if the physico-chemical structure of such vacuoles differed from that of the plasma membrane, it might be feasible to attack the virus without destroying the integrity of the cell.

Another feature of unconventional virus disease that could bear on eventual therapeutic strategies is the phenomenon of genetic regulation of disease expression. We obviously cannot breed humans with a view towards extending the incubation period of disease to beyond the normal expected life span, as may occur in some host-virus strain combinations in experimental scrapie infection of the mouse (102); but if we could elucidate the biochemical consequences of this genetic regulation, it might be possible by chemotherapy to create a resistant metabolic state in the normally susceptible human being.

Finally, is there anything to be learned from the limited success of the drugs HPA-23 and amphotericin B? Their common ability to effect cell membranes suggests that they may modify the expression of disease by interacting with virus on or within the membrane. The extreme electronegativity of HPA-23, and the amphiphilic structure of amphotericin B suggest the further possibility of an ionic or detergent-like activity on virus-membrane binding sites. It has been proposed that a disruptive interaction between polyene drugs and membrane requires that the membrane be in a disordered state (103): would it be altogether fanciful to imagine that in CJD the activity of amphotericin B might be preferentially directed at foci of abnormal membrane multilamination? Other drugs that show membrane activity could also be tried: for example, vancomycin, which inhibits glyco-peptide polymerization, or the polymyxins, which are thought to disorient the normal lipoprotein membrane bilayer, or even chelating agents such as L-alanosine and diethyldithiocarbamate, which have a pronounced, and reversible, effect on both the chemistry and structure of plasma membranes (104).

It is impossible to develop these ideas without more precise knowledge of the nature of the viruses under attack, but in the meantime, attention to genetic and organ-specific differences in disease expression, to metabolic and morphologic abnormalities of the central nervous system, and to the testing of membrane-active drugs, may provide an interim basis for the future direction of chemotherapeutic research in the field of unconventional viruses.

ACKNOWLEDGEMENT

The author is most grateful to Drs. Ralph Garruto, Marie-Claude Moreau-Dubois, and D. Carleton Gajdusek for their careful reading and criticism of this manuscript.

REFERENCES

1. McGowan, J.P., Scrapie in sheep, Scot. J. Agricult. 5:365 (1922).
2. Gaiger, S.H., Scrapie, J. Comp. Path. 37:259 (1924).
3. Cuille, J. and P.L. Chelle, La maladie dite tremblante du mouton est-elle inoculable?, C. R. Acad. Sci. (Paris) 203:1552 (1936).
4. Chandler, R.L., Encephalopathy in mice produced by inoculation with scrapie brain material, Lancet i:1378 (1961).
5. Kimberlin, R.H. and C.A. Walker, Characteristics of a short incubation model of scrapie in the golden hamster, J. Gen. Virol. 34:295 (1977).
6. Hartsough, G.R. and D. Burger, Encephalopathy of mink. I. Epizootic and clinical observations, J. Infect. Dis. 115:387 (1965).
7. Creutzfeldt, H.G., Über eine eigenartige herdförmige erkrankung des zentralnervensystems, Zeitschr. Ges. Neurol. Psychiat. 57:1 (1920).
8. Jakob, A., Über eine der multiplen sklerose klinisch nahestehende erkrankung des centralnervensystems (spastische pseudosklerose) mit bemerkenswertem anatomischem befunde, Med. Klin. 13:372 (1921).
9. Masters, C.L. and D.C. Gajdusek, The spectrum of Creutzfeldt-Jakob disease and the virus-induced subacute spongiform encephalopathies, in: "Recent Advances in Neuropathology", W.T. Smith and J.B. Cavanagh, eds., Churchill-Livingston, New York (1982).
10. Gajdusek, D.C. and V. Zigas, Degenerative disease of the central nervous system in New Guinea. The endemic occurrence of of "kuru" in the native population, N. Engl. J. Med. 257:974 (1957).
11. Klatzo, I., D.C. Gajdusek, and V. Zigas, Pathology of kuru, Lab. Invest. 8:799 (1959).
12. Hadlow, W.J., Scrapie and kuru, Lancet ii:289 (1959).
13. Gajdusek, D.C., C.J. Gibbs, Jr., and M. Alpers, Experimental transmission of a kuru-like syndrome to chimpanzees, Nature 209:794 (1966).
14. Gibbs, C.J., Jr., D.C. Gajdusek, D.M. Asher, M.P. Alpers, E. Beck, P.M. Daniel, and W.B. Matthews, Creutzfeldt-Jakob disease (spongiform encephalopathy): transmission to the chimpanzee, Science 161:388 (1968).
15. Gibbs, C.J., Jr., D.C. Gajdusek, and H. Amyx, Strain variation in the viruses of Creutzfeldt-Jakob disease and kuru, in: "Slow Transmissible Diseases of the Nervous System", Vol.2, S.B. Prusiner and W.J. Hadlow, eds., Academic Press, New York (1979).
16. Brown, P., F. Cathala, D. Sadowsky, and D.C. Gajdusek, Creutzfeldt-Jakob disease in France. II. Clinical characteristics of 170 cases dying during the decade 1968-1977, Ann. Neurol. 6:430 (1979).

17. Gajdusek, D.C. and C.J. Gibbs, Jr., Subacute and chronic
    diseases caused by atypical infections with unconventional
    viruses in aberrant hosts, in: "Perspectives in Virology",
    Vol.8, M. Pollard, ed., Academic Press, New York (1973).
18. Kimberlin, R.H. and C.A. Walker, Pathogenesis of mouse
    scrapie: effect of route of inoculation on infectivity
    titres and dose response curves, J. Comp. Pathol. 88:39
    (1978).
19. Fraser, H., Neuronal spread of scrapie agent and targeting
    lesions within the retino-tectal pathway, Nature 295:149
    (1982).
20. Carp, R.I., Transmission of scrapie by oral route: effect of
    gingival scarification, Lancet i:170 (1982).
21. Gibbs, C.J., Jr., H.L. Amyx, A. Bacote, C.L. Masters, and
    D.C. Gajdusek, Oral transmission of kuru, Creutzfeldt-Jakob
    disease and scrapie disease to nonhuman primates, J. Infect.
    Dis. 142:205 (1980).
22. Hadlow, W.J., R.E. Race, R.C. Kennedy, and C.M. Ecklund,
    Natural infection of sheep with scrapie virus, in: "Slow
    Transmissible Diseases of the Nervous System", Vol. 2,
    S.B. Prusiner and W.J. Hadlow, eds., Academic Press,
    New York (1979).
23. Duffy, P., J. Wolf, G. Collins, A. Devoe, B. Streeten, and
    D. Cowen, Possible person-to-person transmission of
    Creutzfeldt-Jakob disease, N. Engl. J. Med. 290:692 (1974).
24. Bernoulli, C., Danger of accidental person-to-person trans-
    mission of Creutzfeldt-Jakob disease by surgery, Lancet
    i:659 (1977).
25. Nevin, S., W.H. Mcmenemey, S. Behrman, and D.P. Jones,
    Subacute spongiform encephalopathy - a subacute form of
    encephalopathy attributable to vascular dysfunction
    (spongiform cerebral atrophy), Brain 83:519 (1960).
26. Foncin, J.F., J. Gaches, F. Cathala, and J. LeBeau, Trans-
    mission iatrogène possible de maladie de Creutzfeldt-
    Jakob avec atteinte des grains du cervelet, Rev. Neurol.
    (Paris) 136:280 (1980).
27. Matthews, W.B., Evidence for case-to-case transmission of
    Creutzfeldt-Jakob disease, J. Neurol. Neurosurg. Psychiat.
    45:235 (1982).
28. Brown, P., An epidemiologic critique of Creutzfeldt-Jakob
    disease, Epidemiol. Rev. 2:113 (1980).
29. Outram, G.W., The pathogenesis of scrapie in mice, in: "Slow
    Virus Diseases of Animals and Man", R.H. Kimberlin, ed.,
    North Holland, Amsterdam (1976).
30. Kuroda, Y., C.J. Gibbs, Jr., H.L. Amyx, and D.C. Gajdusek,
    Creutzfeldt-Jakob disease in the mouse, Infect. Immun.
    in press (1983).
31. Manuelidis, E.E., E.J. Gorgacz, and L. Manuelidis, Viremia in
    experimental Creutzfeldt-Jakob disease, Science 200:1069
    (1978).

32. Kimberlin, R.H. and C.A. Walker, Pathogenesis of mouse
    scrapie: evidence for neural spread of infection to the
    CNS, J. Gen. Virol. 51:183 (1980).
33. Kimberlin, R.H., H.J. Field, and C.A. Walker, Pathogenesis
    of mouse scrapie: evidence for spread of infection from
    central to peripheral nervous system, J. Gen. Virol.
    64:713 (1983).
34. Hunter, G.D., G.C. Millson, and R.L. Chandler, Observations
    on the comparative infectivity of cellular fractions
    derived from homogenates of mouse-scrapie brain, Res. Vet.
    Sci. 4:543 (1963).
35. Hunter, G.D. and G.C. Millson, The intracellular location of
    the agent of mouse scrapie, J. Gen. Microbiol. 34:319
    (1964).
36. Hunter, G.D. and G.C. Millson, Distribution and activation of
    lysosomal enzyme activities in subcellular components of
    normal and scrapie-affected mouse brain, J. Neurochem.
    13:375 (1966).
37. Millson, G.C., G.D. Hunter, and R.H. Kimberlin, An experi-
    mental examination of the scrapie agent in cell membrane
    mixtures. II. The association of scrapie activity with
    membrane fractions, J. Comp. Path. 81:255 (1971).
38. Millson, G.C., G.D. Hunter, and R.H. Kimberlin, The physico-
    chemical nature of the scrapie agent, in: "Slow Virus
    Diseases of Animals and Man", R.H. Kimberlin, ed., North
    Holland, Amsterdam (1976).
39. Beck, E., P.M. Daniel, A.J. Davey, D.C. Gajdusek, and
    C.J. Gibbs, Jr., The pathogenesis of transmissible spong-
    iform encephalopathy. An ultrastructural study, Brain
    105:755 (1982).
40. Rohwer, R.G., Scrapie: virus-like size and virus-like suscept-
    ibility to inactivation of the infectious agent, Nature
    in press (1983).
41. Prusiner, S.B., Novel proteinaceous infectious particles cause
    scrapie, Science 216:136 (1982).
42. Lax, A.J., G.C. Millson, and E.J. Manning, Involvement of
    protein in scrapie agent infectivity, Res. Vet. Sci. 34:155
    (1983).
43. Rohwer, R.G., P.W. Brown, and D.C. Gajdusek, The use of
    sedimentation to equilibrium as a step in the purification
    of the scrapie agent, in: "Slow Transmissible Diseases of
    the Nervous System", Vol.2, S.B. Prusiner and W.J. Hadlow,
    eds., Academic Press, New York (1978).
44. Brown, P., E.M. Green, and D.C. Gajdusek, Effect of different
    gradient solutions on the buoyant density of scrapie
    infectivity, Proc. Soc. Exper. Biol. Med. 158:513 (1978).
45. Prusiner, S.B., W.J. Hadlow, D.E. Garfin, S.P. Cochran, J.R.
    Baringer, R.E. Race, and C.M. Ecklund, Partial purification
    and evidence for multiple molecular forms of the scrapie
    agent, Biochem. 17:4993 (1978).

46. Prusiner, S.B., D.F. Groth, C. Bildstein, F.R. Masiarz, M.P. McKinley, and S.P. Cochran, Electrophoretic properties of the scrapie agent in agarose gels, Proc. Natl. Acad. Sci. (USA) 77:2984 (1980).

47. Alper, T., D.A. Haig, and M.C. Clarke, The exceptionally small size of the scrapie agent, Biochem. Biophys. Res. Commun. 22:278 (1966).

48. Latarjet, R., Inactivation of the agents of scrapie, Creutz-feldt-Jakob disease, and kuru by radiations, in: "Slow Trans-missible Diseases of the Nervous System", Vol.2, S.B. Prusiner and W. J. Hadlow, eds., Academic Press, New York (1978).

49. Gibbs, C.J., Jr., D.C. Gajdusek, and R. Latarjet, Unusual resistance to ionizing radiation of the viruses of kuru, Creutzfeldt-Jakob disease, and scrapie, Proc. Natl. Acad. Sci. (USA) 75:6268 (1978).

50. Fraser, H., The pathology of natural and experimental scrapie, in: "Slow Virus Diseases of Animals and Man", R.H. Kimberlin, ed., North Holland, Amsterdam (1976).

51. Narang, H.K., R.L. Chandler, and H.S. Anger, Further observa-tions on particulate structures in scrapie affected brain, Neuropathol. Appl. Neurobiol. 6:23 (1980).

52. Merz, P.A., R. Rohwer, R. Somerville, H.M. Wisniewski, C.J. Gibbs, Jr., and D.C. Gajdusek, Scrapie associated fibrils in human Creutzfeldt-Jakob disease, J. Neuropathol. Exper. Neurol. 42:327 (1983).

53. Marsh, R.F., J.S. Semancik, K.C. Medappa, R.P. Hanson, and R.R. Rueckert, Scrapie and transmissible mink encephalo-pathy: search for infectious nucleic acid, J. Virol. 13:993 (1974).

54. Ward, R.L., D.D. Porter, and J.G. Stevens, Nature of the scrapie agent: evidence against a viroid, J. Virol. 14:1099 (1974).

55. Hunter, G.D., S.C. Collis, G.C. Millson, and R.H. Kimberlin, Search for scrapie-specific RNA and attempts to detect an infectious DNA or RNA, J. Gen. Virol. 32:157 (1976).

56. Marsh, R.F., T.G. Malone, J.S. Semancik, W.D. Lancaster, and R.P. Hanson, Evidence for an essential DNA component in the scrapie agent, Nature 275:146 (1978).

57. Cho, H.J., Requirement of a protein component for scrapie infectivity, Intervirol. 14:213 (1980).

58. Alper, T., W.A. Cramp, D.A. Haig, and M.C. Clarke, Does the agent of scrapie replicate without nucleic acid?, Nature 214:764 (1967).

59. Alper, T., D.A. Haig, and M.C. CLarke, The scrapie agent: evidence against its dependence for replication on intrinsic nucleic acid, J. Gen. Virol. 41:503 (1978).

60. Dunnebacke, T.H. and F.L. Schuster, The nature of a cytopatho-genic material present in amebae of the genus Naegleria, Amer. J. Trop. Hyg. Med. 26:412 (1977).

61. Diener, T.O., Similarities between the scrapie agent and the agent of potato spindle tuber disease, Ann. Clin. Res. 5:268 (1973).

62. Somerville, R.A., G.C. Millson, and R.H. Kimberlin, Sensitivity of scrapie infectivity to detergents and 2-mercaptoethanol, Intervirol. 13:126 (1980).

63. Prusiner, S.B., D.F. Groth, M.P. McKinley, S.P. Cochran, K.A. Bowman, and K.C. Kasper, Thiocyanate and hydroxyl ions inactivate the scrapie agent, Proc. Natl. Acad. Sci. (USA) 78:4606 (1981).

64. McKinley, M.P., F.R. Masiarz, and S.B. Prusiner, Reversible chemical modification of the scrapie agent, Science 214:1259 (1981).

65. Prusiner, S.B., M.P. Mckinley, D.F. Groth, K.A. Bowman, N.I. Mock, S.P. Cochran, and F.R. Masiarz, Scrapie agent contains a hydrophobic protein, Proc. Natl. Acad. Sci. (USA) 78:6675 (1981).

66. Hunter, G.D., and G.C. Millson, Glycoprotein biosynthesis in normal and scrapie-affected mouse brain, J. Comp. Pathol. 83:217 (1973). .

67. Yu, R.K., R.W. Ledeen, D.C. Gajdusek, and C.J. Gibbs, Jr., Ganglioside changes in slow virus disease: analyses of chimpanzee brains infected with kuru and Creutzfeldt-Jakob agents, Brain Res. 70:103 (1974).

68. Ikuta, F., T. Kumanishi, T. Ohashi, and M. Koga, Studies on Creutzfeldt-Jakob disease - Is this disease a metabolic disorder? (In Japanese), Advanc. Neurol. Sci. 18:46 (1974).

69. Yu, R.K. and E.E. Manuelidis, Ganglioside alterations in guinea pig brains at end stages of experimental Creutzfeldt-Jakob disease, J. Neurol. Sci. 35:15 (1978).

70. Tamai, Y., H. Kojima, F. Ikuta, and T. Kumanishi, Alterations in the composition of brain lipids in patients with Creutzfeldt-Jakob disease, J. Neurol. Sci. 35:59 (1978).

71. Federico, A., P. Annunziato, and G. Malentacchi, Neurochemical changes in Creutzfeldt-Jakob disease, J. Neurol. 223:135 (1980).

72. Mackenzie, A., A.M. Wilson, and P.F. Dennis, Further observations on histochemical changes in scrapie mouse brain, J. Comp. Pathol. 68:489 (1968).

73. Bass, N.H., H.H. Hess, and A. Pope, Altered cell membranes in Creutzfeldt-Jakob disease, Arch. Neurol. (Chicago) 31:174 (1974).

74. Beck, E., P.M. Daniel, W.B. Matthews, D.L. Stevens, M.P. Alpers, D.M. Asher, D.C. Gajdusek, and C.J. Gibbs, Jr., Creutzfeldt-Jakob disease: the neuropathology of a transmission experiment, Brain 92:699 (1969).

75. Hunter, G.D., R.H. Kimberlin, G.C. Millson, and R.A. Gibbons, An experimental examination of the scrapie agent in cell membrane mixtures. I. Stability and physicochemical properties of the scrapie agent, J. Comp. Pathol. 81:23 (1971).

76. Braham, J., Jakob-Creutzfeldt disease: treatment by amantidine, Brit. Med. J. 4:212 (1971).

77. Sanders, W.L. and T.L. Dunn, Creutzfeldt-Jakob disease treated with amantidine, J. Neurol. Neurosurg. Psychiat. 36:581 (1973).

78. Sanders, W.L., Creutzfeldt-Jakob disease treated with amantidine, J. Neurol. Neurosurg. Psychiat. 42:960 (1979).

79. Herishanu, Y., Antiviral drugs in Jakob-Creutzfeldt disease, J. Amer. Geriat. Soc. 21:229 (1973).

80. Ratcliffe, J., A. Rittman, S. Wolf, and M.A. Verity, Creutzfeldt-Jakob disease with focal onset unsuccessfully treated with amantadine, Bull. Los Angeles Neurol. Soc. 40:18 (1975).

81. Case records of the Massachusetts General Hospital, R.E. Scully, J.J. Galdabini, and B.U. McNeely, eds., N. Engl. J. Med. 303:1162 (1980).

82. Goldhammer, Y., J.J. Bubis, I. Sarova-Pinhas, and J. Braham, J. Neurol. Neurosurg. Psychiat. 35:1 (1972).

83. Ververken, D., H. Carton, and A. Billiau, Intrathecal administration of interferon in MS patients, in: "Humoral Immunity in Neurological Diseases", D. Karcher, A. Lowenthal, and A.D. Strosberg, eds., Plenum Press, New York (1979).

84. Kovanen, J., M. Haltia, and K. Cantrell, Failure of interferon to modify Creutzfeldt-Jakob disease, Brit. Med. J. 280:902 (1980).

85. Dormont, D., F. Cathala, J.C. Chermann, and L. Court, personal communication (1983).

86. Villa, G., C. Caltagirone, and G. Macchi, Unusual clinical course in a case of Creutzfeldt-Jakob disease, Ital. J. Neurol. Sci. 2:155 (1982).

87. Furlow, T.W., R.J. Whitley, and F.J. Wilmes, Repeated suppression of Creutzfeldt-Jakob disease with vidarabine, Lancet ii:564 (1982).

88. Gresser, I. and I.H. Pattison, An attempt to modify scrapie in mice by the administration of interferon, J. Gen. Virol. 3:295 (1968).

89. Field, E.J., G. Joyce, and A. Keith, Failure of interferon to modify scrapie in the mouse, J. Gen. Virol. 5:149 (1969).

90. Worthington, M., Interferon system in mice infected with the scrapie agent, Infect. Immun. 6:643 (1972).

91. Allen, L.B. and K.W. Cochran, Acceleration of scrapie in mice by target-organ treatment with interferon inducers, Ann. N.Y. Acad. Sci. 284:676 (1977).

92. Salazar, A.M., C.J. Gibbs, Jr., D.C. Gajdusek, and R.A. Smith, Clinical usage of interferons in central nervous system disorders, in: "Handbook of Experimental Pharmacology: Interferons and Their Applications", Vol.71, P. Came and W. Carter, eds., Springer, Heidelberg (1983).

93. Tateishi, J., Antibiotics and antivirals do not modify experimentally-induced Creutzfeldt-Jakob disease, J. Neurol. Neurosurg. Psychiat. 44:723 (1981).

94. Kimberlin, R.H. and C.A. Walker, Antiviral compound effective against experimental scrapie, Lancet ii:591 (1979).

95. Salazar, A.M., H.L. Amyx, C.J. Gibbs, Jr., and D.C. Gajdusek, manuscript in preparation (1983).

96. Chermann. J.C., D. Dormont, R. Liderean, and F. Herodin, Action d'un tungstoantimoniate sur le développement de la tremblante expérimentale de la souris, in: "Virus Non-conventionnels et Affections du Système Nerveux Central", L. Court and F. Cathala, eds., Masson et Cie., Paris (1983).

97. Beck, E., I.J. Bak, J.F. Christ, D.C. Gajdusek, C.J. Gibbs, Jr., and R. Hassler, Experimental kuru in the spider monkey. Histopathological and ultrastructural studies of the brain during early stages of incubation, Brain 98:595 (1975).

98. Moreau-Dubois, M.C., P. Brown, R.G. Rohwer, C.L. Masters, M. Franko, and D.C. Gajdusek, Experimental scrapie in golden Syrian hamsters: temporal comparison of in vitro cell-fusing activity with brain infectivity and histopathological changes, Infect. Immun. 37:195 (1982).

99. Viret, J., D. Dormont, D. Molle, L. Court, F. Leterrier, F. Cathala, C.J. Gibbs, Jr., and D.C. Gajdusek, Structural modification of nerve membranes during experimental scrapie evolution in mouse, Biochem. Biophys. Res. Commun. 101:830 (1981).

100. Gourmelon, P., L. Court, M.H. Bassant, P. Breton, and D. Dormont, Modifications électroencéphalographiques dans la tremblante expérimentale de la souris et du hamster, in: "Virus Non-conventionnels et Affections du Système Nerveux Central", L. Court and F. Cathala, eds., Masson et Cie., Paris (1983).

101. Grégoire, N., J.M. Gorde, C. Rousseau, D. Gambarelli, and G. Salamon, Etude des altérations locales du métabolism énergétique cérébral après inoculation de l'agent de la scrapie chez le hamster, in: "Virus Non-conventionnels et Affections du Système Nerveux Central", L. Court and F. Cathala, eds., Masson et Cie., Paris (1983).

102. Dickinson, A.G., H. Fraser, and G.W. Outram, Scrapie incubation time can exceed natural lifespan, Nature 256:732 (1975).

103. Medoff, G., J. Brajtburg, and G.S. Kobayashi, Antifungal agents useful in therapy of systemic fungal infections, Ann. Rev. Pharmcol. Toxicol. 23:303 (1983).

104. Chandrabose, K.A., Personal communication. (Department of Molecular Biology, Wellcome Research Laboratories, Research Triangle Park, North Carolina, 27709).

ANTI INFLUENZA VIRUS ACTIVITY OF AMANTADINE,

RIMANTADINE AND ANALOGUES

J.S. Oxford

Viral Products Division
National Institute for Biological Standards and
Control, Holly Hill, London NW6

INTRODUCTION

Two decades have passed since the initial discovery of the antiviral effect of the primary amine amantadine[1]. During this period extensive controlled clinical trials in more than 20,000 volunteers have established an unequivocal prophylactic effect against influenza A virus infection approximately equivalent to that of inactivated influenza vaccines[2-6]. In addition a mild therapeutic effect has been demonstrated against H2N2, H3N2 and H1N1 viruses[5]. Earlier fears that new influenza virus antigenic variants which constantly emerge and assume epidemiological significance would be less inhibited by amantadine have not been fulfilled. Recent field viruses (at least strains isolated up to 1981) are as well inhibited in vitro and in the clinic as the first H2N2 viruses tested in the early 60's. Recently there has been more interest in finding ways of using the compound usefully on a large scale. This fresh approach has probably been stimulated by the continued absence of any more effective influenza inhibitors and also by the heightened interest in antiviral compounds in general, following the increasingly successful use in the clinic of three new anti-herpes compounds, acyclovir, phosphonoformate and bromovinyl deoxyuridine[7,8].

An important omission to date has been the unestablished mode of action of amantadine molecules. This is partly a result of a paucity of data and a poor understanding of early events following infection of the cell with influenza virus. Once the precise point of action of amantadine is understood, the way may be open to the production of a whole new generation of molecules acting at this stage in the life cycle. It is also of some importance

159

to extend the antiviral action of this group of molecules to
include influenza B viruses which can cause mortality and large
epidemics, although not so frequently or devastatingly as influenza
A viruses. Indeed the earlier studies with other primary amines
established quite clearly that certain of these compounds inhibit
influenza B viruses as well as influenza A viruses[9,10]. Finally,
the studies with amantadine have shown the difficulties of successfully using an anti-respiratory virus compound in the field.

   The present review will consider some of the above mentioned
problems in more detail and, in particular, will attempt to
summarise recent work on the search for new and more active amantadine derivatives, studies on the mode of action of the compound,
and current clinical trials. The recent advances in our understanding of the molecular biology of viruses has resulted partly
from the development of rapid and simplified techniques of virus
genome analysis. Some of these techniques such as oligonucleotide
mapping or RNA:RNA hybridisation can be applied quite easily to the
analysis of the genome of field viruses and so provide data on
genetic and phenotypic variation among RNA or DNA viruses. Such
variation might have a profound influence on the ability or otherwise of virion populations to adapt to the advent of new selective
 pressures, namely specific inhibitors of virus replication. In this
regard amantadine resistant strains of influenza have been isolated
both in the laboratory[11] and the field[12].

THE SEARCH FOR MORE ACTIVE ANTIVIRAL AMANTADINE ANALOGUES

   The cyclic primary amine 1-amino adamantane (Amantadine/
symmetral) is an active inhibitor in vitro of human influenza A
viruses of the three antigenic subtypes H1N1, H2N2 and H3N2, avian
influenza A viruses and equine influenza A viruses[1,2]. No anti
influenza B virus activity has been detected to date with amantadine
analogues although other amines and simple ammonium compounds
possess in vitro activity against both influenza virus[9,10] types
(Table 1). In addition, the antiviral spectrum of amantadine is not
restricted solely to influenza A viruses since the parent or
analogue molecules may exert in vitro antiviral effects against
rubella, parainfluenza viruses, arena viruses and dengue virus[1,2],
although, in general, the degree of antiviral activity detected is
rather small and at present limited to laboratory investigations.
In general terms most of the antiviral compounds in existence
including amantadine have a rather restricted antiviral
activity[3,7,8]. New amantadine molecules are still being synthesized and investigated for antiviral and other biological properties
such as anti-Parkinsonian activity. These analogues include
thioamides, thioacetamides, amidines, pyrimidines, ketones,
imidoacid esters, and carboxylic acid esters[13,14]. The biological
activity of some recently synthesized derivatives is summarised
in Fig 1 and Table 2. We can conclude however that, at present,

the two most useful, effective antiviral compounds in this group
remain amantadinc and rimantadine. Both compounds have a rather
similar degree and spectrum of antiviral activity in vitro, in vivo
and in the clinic although they differ in details. Extensive
clinical data exists for both compounds and more recently carefully
controlled comparative clinical trials have also been performed.
Rimantadine would appear to induce rather fewer side reactions in
man than amantadine, although a necessary reservation here is that
the tissue distribution of the two compounds may vary slightly and
before a final conclusion is reached the two amantadine molecules
should be administered to achieve pre-determined and equivalent
plasma levels. Only under these conditions can a precise comparison
of relative toxicities be determined.

The tissue distribution of the amantadine derivative
(1-amino-3, 5-dimethyl adamantane) was established in post mortem
tissues of a 77 year old woman with Parkinson's disease. The
patient had been treated with 2 x 10 mg of the amantadine molecule
daily for 53 days and levels of the compound in the tissues were
kidney (0.18 µg/g), lung (0.17), spleen (0.1), blood (0.07),
cerebellum (0.22). Also levels of amantadine were determined in
tissue specimens in a 5 month old girl with influenza A virus
pneumonia. 2.5 mg/kg of amantadine each 12 h had been administered
and tissue specimens were obtained 4½ hours after the final dose.
Serum concentrations of amantadine ranged from 0.8 to 1.64 µg/ml
whilst higher concentrations of amantadine were found in the lung
(21.4 µg/ml). These levels of compound would significantly
inhibit the replication of recent H3N2 and H1N1 influenza A viruses
for example based on in vitro data.

IN VITRO ANTIVIRAL ACTIVITY OF AMANTADINE MOLECULES

Early laboratory data appeared to show heterogeneity of
response of influenza A viruses of different subtypes to inhibition
by amantadine and gave rise to concern that new epidemic subtypes of
influenza A might not be inhibited by amantadine. As a result the
compound was only licensed initially in the USA for prophylaxis of
influenza A viruses of the H2N2 subtypes which were demonstrably
sensitive to the compound. When the new pandemic influenza A/Hong
Kong/68 (H3N2) virus appeared, an opportunity was lost to use and
test the compound on a large scale during the first wave of the
epidemic at a time when no influenza vaccine was available. However,
the compound is now licensed for the prophylaxis and therapy of all
influenza A viruses. More recent experience would suggest that
these biological differences could result from accumulated mutations
in laboratory strains. More recently an apparently unique epidemiological situation has arisen with the co-cirulation in the
community of influenza A viruses of two antigenic subtypes, namely
H3N2 and H1N1[15]. The latter viruses are related antigenically and
biochemically to H1N1 viruses which circulated and caused epidemics

between 1947-1956[16] . These recent viruses are sensitive to the
antiviral activity of amantadine. Another apparently unique epidemiological event occurred in 1976 with the spread, albeit limited
to a single army camp in the USA, of the swine influenza A virus
A/New Jersey/76 (H1N1). Again, this 'new' virus was shown to be
well inhibited by amantadine, as have all recent H3N2 viruses
throughout the period of antigenic drift from 1968 to the present
day (Table 3). It should be realised, however, that only a relatively small number of such viruses (compared with the number of
field isolates) have been analysed for sensitivity to amantadine
and it is possible and indeed probably that resistant and
partially-resistant strains exist already in the community. Also,
neither rimantadine or amantadine are completely effective in
tissue culture or animal studies in preventing viral replication
when cells are challenged with a high virus dose. This would
suggest that a low proportion of virions vary in their response
to these antiviral compounds.

PROPHYLACTIC AND THERAPEUTIC EFFICACY OF AMANTADINE AND RIMAN-
TADINE IN CLINICAL TRIALS

    Many, but not all, organised clinical trials conducted under
vigorous double-blind, placebo-controlled conditions have established a prophylactic and therapeutic effect of amantadine and
rimantadine against influenza A viruses[2]. There is general agree-
ment in the literature that the protective effect of amantadine
as a prophylactic against influenza A virus averages 70 to 80%
and ranges from 0 to 100%. To place this figure in perspective,
clinical trials with influenza vaccines have, over the same period
of time shown similar results with protective effects ranging from
0 to 100% with a mean approaching 70 to 80%. For both amantadine
and vaccine with 70% protective efficacy and a 10% influenza virus
attack rate, 14 persons would have to be immunized or given amantadine to prevent influenza in one person. Assuming a 0.1%
mortality rate than approximately 14000 persons would have to be
treated or immunized to prevent one death. An exhaustive comparison
of early clinical trials, both prophylactic and therapeutic has
been published and this data will not be presented again here.
Rather we will present data from a single recently conducted trial
to illustrate the degree of antiviral activity that is commonly
obtained. Amantadine or rimantadine have now been investigated as
an antiviral in many countries of the world and in many differing
clinical situations ranging from semi closed communities, such as
prisons, hospitals and boarding schools to larger scale groups such
as university students and also to families in the community. In
most countries controlled clinical trials are still being carried
out and the prospect of more widespread use of the compound is still
treated with caution. In contrast, in the USSR, rimantadine is
used on a wider scale in the community and may now be purchased
in pharmacies without prescription.

A number of excellent double-blind placebo-controlled trials
have been carried out recently using HlNl viruses and the data
obtained has confirmed the results from earlier trials with H2N2
and H3N2 viruses. The results of a recent controlled clinical
trial are presented in Table 4. A well-controlled comparison of
the prophylactic effects of rimantadine and amantadine in an
influenza A outbreak caused by influenza A H3N2 (20% of cases)
and HlNl viruses (80% of cases) has been reported[18]. A total of
450 volunteers enrolled with a mean age of 25.0 $\pm$ 0.5 years and
with no significant differences in age, race, male/female ratios
or level of pre-existing HI antibody. Throat swabs for virus
isolation were taken twice weekly and volunteers were assigned to
amantadine, rimantadine or placebo groups. A 100 mg tablet was
taken twice daily for seven days and any symptoms were recorded.
Each week the volunteers returned the symptom diary to the
co-ordinating centre and received a further seven days supply of
tablets. If any respiratory illness occurred volunteers were
asked to return at once to the centre, and were examined by a
physician. Influenza-like illness was defined as a cough and/or
fever greater than $37.7^{O}C$ and two or more of the following
symptoms: sore throat, headache and myalgia. The trial lasted six
weeks and a serum sample was obtained at the beginning and again
at the end of the study for serological analysis. Significantly
more placebo recipients (40.9%) developed influenza-like illness
compared with amantadine (8.9%) or rimantadine (14.3%) groups
giving a reduction in the rate of illness of 78.2% and 65.0%
respectively (Table 4). However, as noted in previous trials[2]
the rates of laboratory-confirmed influenza-like illness were
reduced by 85.4% by rimantadine and 91.2% by amantadine, suggesting
that a proportion of influenza-like illness observed during the
study was not caused by influenza A virus. Of a small group of
these amantadine patients sampled, 89% had drug detectable in the
urine (52-438μg/ml) suggesting a good compliance rate. Altogether
62 volunteers left the study during the six weeks because of
possible side-effects. The withdrawal rates were 10.8% for placebo,
9.5% for rimantadine and 22.1% for amantadine and the excess rate
in the latter group was mainly caused by CNS effects including
insomnia, jitteriness and difficulty in concentrating, although
symptoms generally cleared within 48 hours of the cessation of
medication. Both amantadine and rimantadine were thus highly
effective in preventing illness and/or infection with no statistically significant differences between the efficacy rates of the
two compounds. However, amantadine-treated persons had significantly more CNS side effects with an excess rate of 9.0% compared
to placebo patients. The authors concluded that rimantadine might
be the compound of choice for the chemoprophylaxis of influenza A
infection in young volunteers, but further trials should be
carried out in the elderly and high risk individuals.

In summary, therefore, amantadine or rimantadine in most
clinical trials with influenza A viruses have not resulted in 100%
protection against infection or clinical disease. Rather, the
prophylactic efficacy would approximate to that of influenza vaccine
but with the qualification that amantadine has no inhibitory effect
against influenza B viruses. The compound has been used successfully
to prevent the spread of influenza A viruses in families, hospitals,
factories, universities, open communities (cities) and closed
communities such as army camps and prisons and antiviral effects
are reproducible and consistent in persons infected with viruses
of the different subtypes H1N1, H2N2 and H3N2. A quite clear therapeutic effect is also obtained, resulting in a 1-2 day faster
recovery, lower temperatures and fewer clinical symptoms with these
same viruses, and, perhaps more significantly from the point of view
of virus spread in the community, reduced excretion of virus
(reviewed in 2).

THE POSSIBILITY OF USING AMANTADINE OR RIMANTADINE ON A WIDER SCALE

At present health authorities in most countries recommend the
routine use of influenza vaccine to prevent influenza infections in
persons designated at special risk of mortality e.g. older persons,
diabetics, asthmatics and persons with chronic obstructive
pulmonary disease and heart disease. Little attempt has been made
to provide protection for the community at large or to abort an
epidemic, in spite of the well documented economic and social disruption caused [19] [20] [21]. Certain groups could be protected [22].
Firstly, unvaccinated children and adults at high risk of serious
morbidity and mortality because of underlying diseases, which include
pulmonary, cardiovascular, metabolic, neuromuscular, or immunodeficiency diseases should be included. Adults whose activities
were vital to community function and who had not been vaccinated
with an appropriate contemporary influenza vaccine could be given
amantadine prophylactically: for example, policemen, firemen,
selected hospital personnel. (Such persons are in frequent contact
with others who may have influenza and should be considered at
higher risk of contracting influenza than the general population.)
A third group would be persons in semiclosed institutional envir-
onments, especially older persons, who had not received the
current influenza vaccine. The groups for which the panel
concluded the benefit-to-risk considerations were less clear
included all elderly patients (65 years or older) who had not
received influenza vaccine and household contacts of an index case.
(Studies in the UK and USSR would indicate families to be an ideal
additional target group.) Finally, an obvious possibility would be to
supplement vaccine-induced protection with amantadine prophylaxis.
Only a few studies have investigated the possible additive effects
of vaccine and amantadine in this approach and preliminary results
have been encouraging raising protective efficacies from 32-37% in

vaccine or rimantadine groups to 60% in the group given both
vaccine and rimantadine[5].

AMANTADINE-RESISTANT INFLUENZA VIRUSES

Earlier studies[11] showed that amantadine-resistant influenza A
viruses could be selected by passage of virus in mice treated with
very high (150 mg/kg/day) doses of the compound. Before passage
in vitro the influenza A virus was inhibited by 0.3 μg/ml amantadine
whereas after a single passage in the presence of the drug 6 μg/ml
of amantadine was required to inhibit replication of some virus
isolates. After six passages in vivo most influenza strains were
completely resistant to amantadine and the related drug rimantadine.
More recently a number of laboratories have investigated the genetic
basis of amantadine resistance by producing virus recombinants
between amantadine resistant and amantadine susceptible influenza A
viruses[25,26,27,28]. In this way transfer of drug resistance can be
correlated with transfer of a particular gene or group of genes.
At present the results from different laboratories are somewhat
conflicting, although several groups agree that gene 7, coding for
matrix protein, appears to co-segregate with amantadine resistance.
The interpretations , however, are complicated to some extent by
the observation that using different in vitro techniques the same
influenza A virus may appear inhibited or resistant to the drug[27].
In addition, data on determinants of influenza virulence[29,30] suggest
a multi-gene linkage with these biological properties. It is quite
possible however that several gene products are involved and that
the product of gene 7 may have 'helper' activity.

At present little extensive field work has been carried out to
search for rimantadine or amantadine resistant viruses in contacts
or in persons being treated for influenza. This is an important
aspect to investigate since it is quite likely that amantadine
resistance could spread between field viruses by genetic reassortment. Some influenza H3N2 or H1N1 viruses circulating at present
in the community are known to be recombinants containing genes of
both virus sub-types[31] and thus intra or inter-typic recombination
is probably occurring with a relatively high frequency. Heider
et al[12] have recently reported two relatively resistant influenza A
H3N2 virus isolates in Berlin where rimantadine has not been used
as a prophylactic. However, even the amantadine sensitivie viruses
in this work showed a rather poor dose response to amantadine and
therefore the study needs to be extended and confirmed.

INHIBITION OF VIRUS REPLICATION BY AMANTADINE; MODE OF ACTION AT THE MOLECULAR LEVEL

Biological studies demonstrated that amantadine acted at an
early stage in influenza A virus infection[32] and later, more
detailed studies established the point of action at approximately

the late stage of virus uncoating[33]. More recently, recognition
that the N terminus of the HA2 polypeptide of influenza haemagglutinin has an amino acid sequence similar to that of the fusion
sequence of the F protein of Sendai viruses[34,35] and also the
demonstration of fusion and haemolysis events between influenza
viruses and cells at low pH[36,37,38,39] have led to the hypothesis
of an important role of fusion during infection of cells with
influenza A virus. A possible re-interpretation of the above data
on the mode of action of amantadine at present would be that
influenza viruses penetrate susceptible cells by viropexis[40,41,42],
and thus enter the cell cytoplasm in coated vesicles. Comparable
pre-lysosomal cytoplasmic vacuoles have been shown to have a low pH
and under these conditions influenza HA mediated fusion could occur,
since a configuration change in the HA could result in the N
terminus of HA2, which is normally some distance from either end
of the molecule contacting the membrane of the vacuole[35]. Fusion
of the viral and vacuole membranes would then occur, resulting in
release of viral RNA and subsequent transport to the cell nucleus,
where initial viral RNA transcription is known to occur. Amantadine
and other amines are known to increase the pH of intracytoplasmic
vacuoles and so the drug could act by simply increasing the pH to
6.5 or 7.0 when fusion could not occur and viral infection would
be blockaded. The hypothesis is most attractive, but some important
observations remain contradicting at present. Thus, as we noted
above, amantadine resistance is known to be correlated with gene 7
(coding for M protein) and not gene 4 (coding for HA). In addition,
there is the problem also referred to above, of why some viruses
such as influenza B would be resistant to amantadine. Finally, in
a recent study Richman et al[43] showed that when amantadine-treated
cells were washed in compound free medium, though relatively high
concentrations of amantadine remained intracellularly, the cells
were now susceptible to infection. Yet immediately before, with
amantadine in the culture medium and equilibrated in the cytoplasm,
the cells resisted infection. This implies an antiviral role of
amantadine at the superficial external plasma membrane of the cell
rather than an intracellular action. Earlier biological experiments with amantadine-treated cells showed that after incubation
with trypsin (which removed surface adhering material and presumably
amantadine) cells become susceptible to viral infection.

        We have investigated the pH optimum of haemolysis of a number
of A and B viruses as part of a separate study on biological
characteristics of field isolates of influenza virus. If the pH
optimum of fusion as shown by haemolysis for influenza B viruses
and drug-resistant A viruses was higher than for amantadine
sensitive influenza A viruses, fusion and hence infection would
proceed even in the presence of amantadine at a pH 6.6, whereas
corresponding events would be blocked with most amantadine-
sensitive influenza A viruses. Although a small but reproducible
difference in pH optimum and pH maximum for haemolysis was noted

between an amantadine-resistant virus (maximum pH 6.2) and the
parental amantadine-sensitive virus (maximum pH 5.6) this was
nevertheless close to the range of pHs shown by other influenza A
viruses. In addition, recombinants such as X-49 and NIB-4, NIB-7
and NIB-8 which are known to have inherited gene 7 and other
genes from A/PR8/34 and are thus relatively resistant to amantadine
like the A/PR8/34 parent, show a similar pH optimum and maximum
for haemolysis to the second parental strains which are all inhibited
by amantadine. Influenza B/HK/73 virus showed a high pH optimum
and a high pH maximum of haemolysis, but B/Lee/40 was similar to
influenza A viruses in its pH profile although neither virus is
inhibited by amantadine. Thus, these experiments were inconclusive
and further work is required to investigate and establish any role
for a fusion event, to determine if low pH haemolysis of red
blood cells is caused by a fusion process and to look for a
correlation between pH optimum for this biological event and
amantadine resistance or susceptibility.

POSSIBLE INFLUENCE OF GENETIC AND PHENOTYPIC VARIATION AMONG
INFLUENZA VIRUSES ON AMANTADINE-INDUCED PROTECTION

        Genetic and phenotypic heterogeneity of viruses may be of
considerable practical importance in attempts to control certain
virus diseases by chemo- or immuno-prophylaxis. Thus, certain
naturally circulating influenza A viruses may be resistant to
antiviral drugs such as amantadine by virtue of mutations in gene 7
or may differ in virulence, antigenic or biological properties and
thus may be able to circumvent drug induced immunity. Studies of
influenza A and B viruses circulating in the community have shown
that quite extensive genetic and phenotypic variation occurs. Thus,
a number of recently isolated influenza viruses of both H3N2 and
H1N1 antigenic subtypes have a temperature sensitive (ts) phenotype
and differ in virulence for volunteers[44,45]. Laboratory studies
of artificially induced influenza ts mutants have demonstrated
clearly that such mutants are attenuated for man and that the shutoff temperature is related to attenuation. The occurrence of non-
ts and ts viruses in nature indicates that influenza A viruses of
varying virulence are circulating in the community. In our studies,
even viruses isolated from the same city varied considerably in
the phenotypic ts character[44]. Influenza A and B viruses
circulating in the community may also differ in respect to the
biological property of plaquing in MDCK cells and to the electrophoretic properties of structural and non-structural polypeptides
and RNA[46]. Finally, analysis of field isolates of influenza A and
B viruses using panels of monoclonal antibodies to virus HA indicates
a considerable degree of antigenic heterogeneity even among viruses
isolated in rather circumscribed outbreaks in single towns or
schools. It is quite possible that biological techniques currently
used to estimate the degree of inhibition of viruses by amantadine

are too insensitive and significant but small differences in drug
resistance between viruses may be missed. These differences might,
however, have a significant effect on the circulation of viruses
in the community.

    In conclusion, amantadine and rimantadine remain the most
effective inhibitors of influenza A virus discovered to date, with
demonstrable and reproducible protective efficacy against influenza A
virus infections equivalent to that of influenza vaccines. Laboratory and clinical studies have clearly established that all the
known subtypes of human influenza A viruses (H1N1), H2N2 and H3N2)
are inhibited by amantadine. A limitation of the compounds is the
lack of a broad spectrum antiviral effect (influenza B virus is
not inhibited) but nevertheless these compounds could be used with
much more vigour than at present to ameliorate influenza in
individuals and perhaps even control influenza A virus epidemics
in selected communities.

    The biological activity of these molecules is summarised
    in table 2

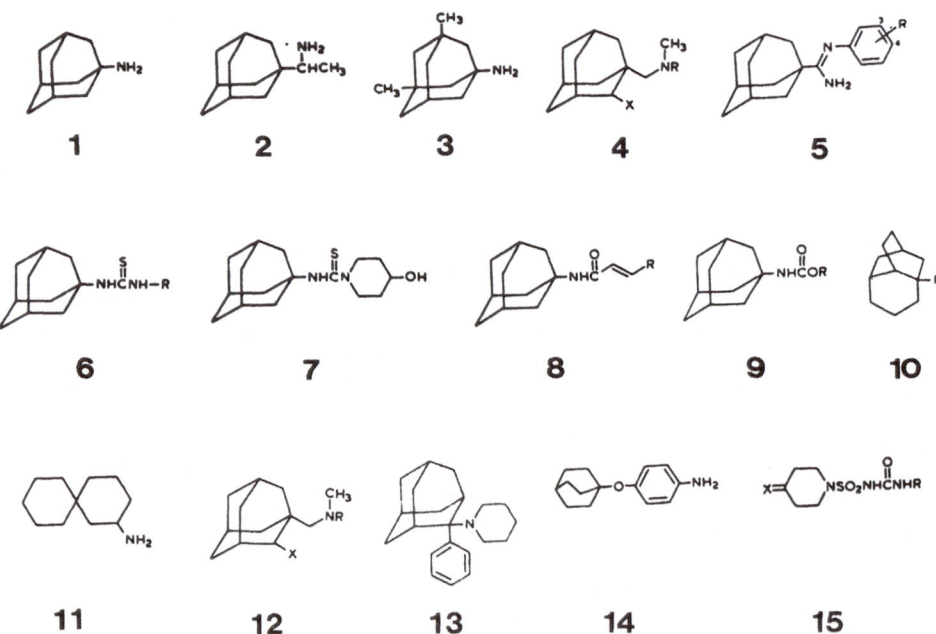

Fig 1.   Molecular structure of amino adamantane and
         some recently synthesised analogues

Table 1   Inhibition of influenza A and B viruses by
             amantadine and ammonium ions

| Virus | Reduction in virus end point titre ($\log_{10}$TCID$_{50}$/ml) | | |
	amantadine (25ug/ml)	ammonium acetate (100ug/ml)	rimantadine (25Ng/ml)
A/NWS(H1N1)	2.0	1.7	1.5
A/Singapore/1/57(H2N2)	2.5	3.5	not tested
A/Scotland/49/57(H2N2)	6.0	4.5	5.5
B/England/13/65	0.5	2.6	0.5

MK cell cultures were infected with influenza viruses and
incubated in the presence or absence of drugs for 3-4 days.
Virus infectivity endpoints were determined by haemadsorption.
A reduction of 1.0 $\log_{10}$TCID$_{50}$/ml or 90 per cent inhibition
of virus growth is considered significant in this test.

    The earlier concern about serious CNS toxicity of amantadine
has not been sustained and nor have earlier indications that the
compounds would not be effective against new antigenic variants
of influenza A virus been confirmed. Extensive use of amantadine
compounds would also generate data to provide a scientific basis
for the evaluation of more effective and broad spectrum antivirals
in the future. Finally, field studies in the search for amantadine
resistant viruses could usefully be extended because drug resistance is likely to become a practical problem of primary importance.

Table 2   Biological activities of amantadine and recently
           synthesised analogues

Molecule*	Chemical structure	Biological activity
1	amino adamantane (amantadine)	Influenza A virus inhibitor in vitro, in vivo and in man Effective in Parkinson's disease
2	α Methyl 1 adamantane methylamine (rimantadine)	Influenza A virus inhibitor in vitro, in vivo and in man (no activity in Parkinson's disease)
3	1-amino-3,5,dimethyl-adamantane (memantine, DMAA)	More potent than amantadine as stimulator of motor activity. Parkinson's drug
4	1 substituted adamantyl hydrazine derivative	Anti mycoplasma and fungal and herpes (HSV) activity
5	N substituted 1 adamantyl carboxamidines	Antiviral activity versus influenza A, vaccinia and herpes virus. The 4 amino derivative had anti polio activity whilst the adamantyl acetamidine inhibited NDV
6	acylated adamantyl thio urea derivatives	In vivo anti influenza A activity (not active versus HSV, SFV or Sendai)
7	alkyl substituted thio ureas	In vitro activity versus HSV, vaccinia and adenovirus
8	amide derivatives: N-(1-adamantyl) cinnamide	In vitro versus influenza, HSV, vaccinia. In vivo versus influenza A
9	propyl carbamate	antifungal
10	4 homotwistane derivatives	Amino and amino methyl analogues were active against NDV
11	aminospirane	In vitro against influenza A and B viruses. In vivo activity versus influenza A. The dialkyl aminoalkyl derivative inhibited rhino-virus type 14 in vitro.

Molecule	Chemical structure	Biological activity
12	2 substituted 1-(amino alkyl) adamantanes	Some molecules may have anti Parkinson's activity
13	phencyclidine analogues	Anti cholinergic activity
14	bicyclo-octyloxyaniline	Hypo beta lipoproteinaemic agent
15	4,4-disubstituted piper-idine	Hypoglycaemic activity

* see Fig 1 for molecular structure of
  the compounds

Table 3   Inhibition of polypeptide synthesis of influenza A and B
           viruses by amantadine

	% inhibition of virus induced polypeptide synthesis by 25 ug/ml amantadine		
Virus	NP	NS1	M
HswinelN1			
A/NJ/8/76	80.0	92.0	80.5
H1N1 subtype			
A/Jap/93NS/78	69.8 $\pm$ 15.3	91.7 $\pm$ 3.2	99.1 $\pm$ 1.3
A/Lackland/AFB/3/78	90.5	96.6	94.0
A/Brazil/11/78	96.5	98.4	ND
A/Fukushima/78	95.0	95.0	ND
H2N2 subtype			
A/Leningrad/549/80	77.4	91.9	86.8
H3N2 subtype			
A/England/641/78	87.4	86.8	86.5
A/England/939/78	83.9	93.8	93.1
A/England/938/78	60.2	87.5	69.0
A/Alaska/78	73.9	88.2	99.2
A/Bangkok/1/79	74.3	90.1	93.6
Influenza B/Singapore/79	0	0	0

Vero cell cultures were infected with 10 EID_{50}/cell of virus
and incubated in the presence or absence of amantadine overnight
when the cells were pulsed with S^{35} methionine for 30 min and
cell lysates analysed by electrophoresis in polyacrylamide gels[46].
Following autoradiography, the quantities of virus induced polypeptides were estimated by densitometer analysis of autoradiographs
and analysis of the tracings using a Kontron MOP digiplan apparatus.
Note that polypeptide synthesis of all the influenza A viruses is
well inhibited by amantadine but no antiviral effect is detected
against a representative influenza B virus.

Table 4   Effect of rimantadine and amantadine in preventing
            influenza-like illness and laboratory confirmed
            influenza among volunteers[18]

Group	Number	Number with influenza-like illness or influenza	Percent
Influenza like illness			
Placebo	132	54	40.9
Rimantadine	133	19	14.3
Amantadine	113	10	8.9
Laboratory confirmed influenza			
Placebo	132	27	20.5
Rimantadine	133	4	3.0
Amantadine	113	2	1.8

Illness was defined as cough and/or fever $37.7^{\circ}C$ p.o and two
or more of the following: sore throat, headache, myalgia.

Laboratory confirmed influenza was determined by virus isolation
and/or serum antibody rises.

REFERENCES

1.   W.L. Davies, R.R. Grunert, R.F. Haff, J.W. McGahen, E.M.
Neumayer, M. Paulshock, J.C. Watts, T.R. Wood, E.C. Hermann and
C.E. Hoffman, Antiviral activity of 1-adamantanamine Science
144:862 (1964)

2.   J.S. Oxford and A. Galbraith, Antiviral activity of amantadine:
a review of laboratory and clinical data. Pharmacology and
Therapeutics 11:181 (1980)

3.   C.H. Stuart Harris, J.S. Oxford, Eds "Problems in Antiviral
Therapy' Academic Press, London In press

4.   M.K. Indulen, V.A. Kalninya, Studies of the antiviral effect
and the mode of action of the anti-influenza compound rimantadine.
In: Recent Developments in Antiviral Chemotherapy L.Collier and
J.S. Oxford (Eds) Academic Press, London (1980)

5.   D.M. Zlydnikov, O.I. Kubar, T.P. Kovaleva and L.E. Kamforin,
Study of rimantadine in the USSR: a review of the literature.
Reviews of Infectious Diseases 3:408 (1981)

6.   J.R. La Montagne and G.J. Galasso Report of a workshop on
clinical studies of the efficacy of amantadine and rimantadine
against influenza virus J.Infect.Dis. 138:928 (1978)

7.   L. Collier and J.S. Oxford Eds. Developments in Antiviral
therapy. Academic Press, London (1980)

8.   G.J. Galasso, T.C. Merigan and R.A. Buchanan Eds. Antiviral
agents and viral disease of man. Raven Press, New York (1979)

9.   M.D. Eaton, I.E. Low, A.R. Scala and S. Uretsky Inhibition
by ammonium ion of the growth of influenza virus in chorioallantoic
tissue. Virology 18:102 (1962)

10. J.S. Oxford and G.C. Schild Immunofluorescent studies on the
inhibition of influenza A and B in mammalian cell cultures by
amantadine. J.Gen.Virol 2:377 (1968)

11. J.S. Oxford, I.S. Logan and C.W. Potter In vivo selection
of an influenza A2 strain resistant to amantadine Nature 226:82
(1970)

12. H. Heider, B. Adamczyk, H.W. Presber, C. Schroeder, R. Feldblum
and M.K. Indulen Occurrence of amantadine and rimantadine-resistant
influenza A virus strains during the 1980 epidemic Acta Viro 25:395
(1981)

13.   D.L. Swallow.   Antiviral agents.   Prog. Drug. Res. 22:267 (1978)

14.   J.W. Tilley and M.J. Kramer   Amino adamantane derivatives. Progress in Medicinal Chemistry 18:1 (1981)

15.   A.P. Kendal, J.M. Joseph, G. Kobayashi, D. Nelson, C.R. Reyes, M.R. Ross, J.L. Sarandria, R. White, D.F. Woodall, G.R. Noble and W.R. Dowdle.   Laboratory-based surveillance of influenza virus in the United States during the winter of 1977-78.   I.Periods of prevalence of H1N1 and H3N2 influenza strains, their relative rates of isolation in different age groups and detection of antigenic variants.   Amer.J.Epidemiol. 110:449 (1979)

16.   J.F. Young, U. Desselberger and P.Palese   Evolution of human influenza A viruses in nature: sequential mutations in the genomes of new H1N1 isolates   Cell 18:73 (1979)

17.   G. Douglas   Amantadine: Should it be used as an antiviral? Controversies in Therapeutics.   Publ. W.B. Saunders Co. p271 (1980)

18.   R. Dolin, R.C. Reichman, P. Madore, R. Maynard, P.M. Linton, and J. Webber-Jones   Controlled trial of amantadine and rimantadine in the prophylaxis of influenza A infection   New England J.Med. 307:580 (1982)

19.   C.H. Stuart-Harris   Clinical aspects of the respiratory tract In: Chemoprophylaxis and virus infections of the respiratory tract. Volume 1 p 26   Ed. J.S. Oxford   CRC Press Inc. Cleveland, Ohio (1977)

20.   C.H. Stuart-Harris   The epidemiology and prevention of influenza.   American Scientist 69:166 (1981)

21.   C.H. Stuart-Harris and G.C. Schild   In: Influenza, the viruses and the disease.   Edward Arnold, London .(1976)

22.   J. Elliot   Consensus on amantadine use in influenza A. J.Amer.Med.Assoc.   242:2383 (1979)

23.A.B. Sabin   Amantadine hydrochloride: Analysis of data related to its proposed use for prevention of A2 influenza virus disease in human beings.   J.Amer.Med.Assoc.   200:135 (1967)

24.   A.B. Sabin   Amantadine and influenza: evaluation of conflicting reports J.Infect.Dis. 138:557 (1978)

25.   M.D. Lubeck, J.L. Schulman and P. Palese   Susceptibility of influenza A viruses to amantadine is influenced by the gene coding for M protein.   J.Virol. 28:710 (1978)

26.   A.J. Hay, N.C.T. Kennedy, J.J. Skehel and G. Appleyard
The matrix protein gene determines amantadine sensitivity of
influenza viruses. J.Gen.Virol.   42:189 (1979)

27.   C. Scholtissek and G.P. Faulkner Amantadine resistant and
sensitive influenza A strains and recombinants.   J.Gen.Virol.
44:807 (1979)

28.   M. Hamzawi, R. Jennings, and C.W. Potter The amantadine
sensitivity of recombinant and parental influenza virus strains.
Med.Microbiol. and Immunology   169:259 (1981)

29.   A. Sugiura and M. Ueda Neurovirulence of influenza virus
in mice I. Neurovirulence of recombinants between virulent
and avirulent virus strains. Virology 101:440 (1980)

30.   T. Ogawa and M. Ueda Genes involved in the virulence of
an avian influenza virus. Virology 113:304 (1981)

31.   Y. Ghendon, A. Klimov, N. Gorodkova and L. Dohner Genome
analysis of influenza A virus strains isolated during an epidemic
of 1979-1980 J.Gen.Virol.   56:303 (1981)

31.   Y. Ghendon, A. Klimov, N. Gorodkova and L. Dohner Genome
analysis of influenza A virus strains isolated during an epidemic
of 1979-1980. J.Gen.Virol 56:303 (1981)

32.   C.E. Hoffman, E.M. Neumayer, R.F. Haff and R.A. Coldsby
Mode of action of the antiviral activity of amantadine in tissue
culture. J. Bacteriology 90:623 (1965)

33.   A.G.Bukrinskaya, N.K. Vorkunova and R.A. Narmanbetova
Rimantadine hydrochloride blocks the second step of influenza virus
uncoating Arch.Virol 66:275 (1980)

34.   I.A. Wilson, J.J. Skehel and D.C. Wiley Structure of the
haemagglutinin membrane glycoprotein of influenza virus at 3Å
resolution. Nature 289:366 (1981)

35.   J.J. Skehel, P.M. Bayley, E.B. Brown, S.R. Martin, M.D.
Waterfield, J.M. White, I.A. Wilson and D.C. Wiley Changes in the
conformation of influenza virus haemagglutinin at the pH optimum
of virus mediated membrane fusion Proc Nat.Acad.Sci.USA
79:968 (1982)

36.   J. White, K. Matlin and A. Helenius Cell fusion by
Semliki Forest, influenza and vesicular stomatitis viruses
J.Cell Biology 89:674 (1981)

ANTIVIRAL ACTION OF 2-(α-HYDROXYBENZYL)-BENZIMIDAZOLE (HBB)

Hans J. Eggers

Institut für Virologie der Universität zu
Köln, Fürst-Pückler-Str. 56, D-5000 Köln 41
FRG

INTRODUCTION

Since 1952 the antiviral activity of benzimi-dazoles has been studied extensively (1). A major stimulus during this work has been the central finding that virus-inhibiting activity and cell toxicity of chemical derivatives of benzimidazoles may vary inde-pendently (2). The introduction of 2-(α-hydroxybenzyl)-benzimidazole (HBB) as an antiviral agent marked a most important step of the investigations on selective inhibition of viral replication (3,4). It was found that HBB inhibits multiplication of poliovirus 2 in monkey kidney cell culture at concentrations which caused no microscopic changes in the cells. On the other hand, influenza virus multiplication remained unaffected (4).

During the last two decades a wealth of information has been accumulated on the antiviral action of HBB and some of its chemical derivatives. It is now well established that, contrary to common belief some 20 years ago, virus multiplication can be selctively inhibited by low molecular weight compounds. In the present discussion I shall focus on investigations on the mechanism of action of HBB.

ANTIVIRAL ACTIVITY OF HBB IN CELL CULTURE

Virus-inhibitory spectrum and effects of HBB on uninfected cells

The antiviral spectrum of HBB exhibits a remarkable pattern: at concentrations of HBB nontoxic to cells, viz. 220 µM, the multiplication of many members of the genus Enterovirus is being inhibited, whereas members of other virus families are HBB-insusceptible (5,6). Details concerning the antiviral spectrum of HBB can be found in previous reviews (7,8). Suffice it to point out here that the pattern of HBB susceptibility or insusceptibility even within the picornavirus family follows almost exclusively taxonomic characteristics (8,9).

As to be expected for any inhibitor, the susceptibility of various picornaviruses to HBB varies considerably (5,6). However, these differences appear to be dependent on the virus. So far, no example of varying susceptibility of picornaviruses due to the host cell is known.

As indicated above, concentrations of HBB highly inhibitory to susceptible enteroviruses (220 µM) did not affect the virus yields of insusceptible viruses. This finding strongly suggested that essential metabolic pathways of cells might not be affected by HBB. This conclusion has been confirmed in further investigations on the effects of HBB on metabolic reactions of uninfected primary monkey kidney cells. Concentrations of up to 493 µM HBB caused no effects on the following metabolic activities (up to 3 hr): cumulative O_2 uptake, glucose utilization, lactic acid production, adenosine ^{14}C uptake into RNA, and L-alanine ^{14}C uptake into proteins (5). At 219 µM HBB, the uridine ^3H uptake into RNA was also unaffected during the duration of the experiment (6.5 hr) (10). In addition, extensive morphological observations of HBB-treated cells (219 µM) demonstrated no significant differences from untreated cultures during the duration of the experiment (5 to 7 days). Minimum morphological changes were seen at 493 µM concentrations, but even at 876 µM concentrations morphological changes were only moderate (5,11,12,13). Cell multiplication - a broad and sensitive indicator of cellular metabolism - was also found to be unaffected in the presence of 219 µM HBB. This has been demonstrated for primary monkey kidney cells as well as HeLa cells (5,10).

All of these data substantiate the thesis that HBB
<u>selectively</u> inhibits members of the picornavirus
family.

Effects of HBB on the viral replication cycle

The time course of the HBB-sensitive process in
the viral multiplication cycle has been determined by
removing or adding the compound at various times after
virus infection. Detailed studies with echovirus 12 and
coxsackievirus A9 replicating in monkey kidney cell
cultures under single cycle conditions have revealed
that no HBB-inhibitable process takes place during the
first 2 1/2 hr of the latent period. The HBB-inhibitable period begins during the last 1 1/2 hr of the
latent phase and extends until the latest rapid
increase phase of the virus (10,14,15). Once the
HBB-inhibitable period has begun, addition of the compound has an immediate inhibitory effect on the
production of infectious virus: latest by 15 min,
infectious virus production has come to a complete
stop. These observations made in mass culture hold true
also for individual cells: HBB stops ongoing virus
multiplication (16).

The described kinetics of the antiviral action of
HBB indicated that early steps of virus-cell interaction might not be affected by the compound. In subsequent experiments early virus-cell interactions were
measured directly in the presence of HBB, and, in fact,
virus adsorption, penetration, and uncoating were found
unaffected by the compound (10,17,18).

Since picornaviral RNA and protein synthesis are
continuing in the untreated, infected cells throughout
the rapid increase phase of virus production (19), the
kinetics of the HBB-inhibitable process suggested an
effect on picornavirus macromolecular synthesis. This
turned out to be the case: synthesis of infectious
viral RNA was demonstrated to be inhibited by HBB (10).
Inhibition of synthesis of viral RNA by HBB has also
been demonstrated through measurements of the rates of
incorporation of uridine ^{3}H into virus-specific RNA in
actinomycin-treated cells (20). This finding excludes
the theoretical possibility that HBB might in some way
inhibit only the acquisition of infectivity of viral
RNA which has already been synthesized.

It should be pointed out here that HBB could never
be shown to have any direct inactivating effect on the
infectivity of HBB-susceptible virions or their RNA
(5,10).

As indicated above, HBB exhibited an immediate
effect on ongoing synthesis of infectious virus.
Similar observations were made concerning synthesis of
viral RNA. Within 30 min after addition of 200 µM HBB
to cells infected with echovirus 12 during various
times of the rapid increase phase of virus, no viral
RNA synthesis was detectable in uridine ^{3}H pulses (8).

Stop of viral RNA synthesis eventually stops
synthesis of viral polypeptides. It was possible,
however, to demonstrate a differential effect of HBB on
viral protein and virion synthesis by use of the
hemagglutinating echovirus 12. It could be shown that
the increase in virus hemagglutinin for a period of ca.
1 to 1 1/2 hr after addition of HBB was exclusively due
to formation of empty capsids, whereas synthesis of the
RNA-containing infectious virus had come to the above-
mentioned rapid stop (10,15). Furthermore, it could be
shown that the empty capsids produced after addition of
HBB were comprised of polypeptides which had been
synthesized after addition of the compound (15). This
experiment clearly proves that HBB does not directly
inibit synthesis of viral capsid proteins.

The evidence so far presented suggests that at
least some aspects of the virus-inhibitory activity of
HBB can be explained by its inhibiting effects on viral
RNA synthesis. The fact that viral RNA synthesis was
strongly inhibited by HBB, whereas cellular RNA
synthesis remained unimpaired by the compound - the
first demonstration of specific chemical inhibition of
the synthesis of viral nucleic acid - led, besides
other considerations, to a search for a virus-induced
RNA polymerase. Such a virus-induced RNA polymerase
activity was, in fact, detected in picorna-
virus-infected cells (21,22,23). This finding prompted
a study on the effects of HBB on the appearance in
enterovirus-infected cells of the virus RNA polymerase
activity. Addition of HBB to poliovirus-infected cells
during the early rapid increase phase of virus
multiplication caused a decrease in viral polymerase
activity to almost background level, as demonstrated in
an assay in cell-free extracts. On the other hand, the

compound appeared to have no effect on the activity of the viral enzyme preparation when added to the cell-free RNA polymerase assay (22).

At the time, these findings have been interpreted as follows: HBB inhibits the synthesis of a virus-directed RNA polymerase, the enzyme responsible for the replication of viral RNA. With inhibition of synthesis of the RNA polymerase the replication of viral RNA also comes to a stop which, in turn, finally leads to inhibition of synthesis of viral capsid polypeptides.

This interpretation, however, in the light of our present knowledge, raises a number of difficulties. E.g., it would imply a differential effect on viral polypeptide synthesis, since capsid protein synthesis continues for at least 1 hr after addition of HBB to infected cells. After addition of the compound to enterovirus-infected cells, a central finding is the rapid halt not only of multiplication of infectious virus, but also of viral RNA synthesis. This observation makes it unlikely that inhibition of viral RNA synthesis is mediated through inhibition of synthesis of viral RNA polymerase. Granted, enterovirus RNA polymerase activity appears unstable during the viral replication cycle (23). However, the rapid and complete stop of viral RNA synthesis after addition of HBB cannot be explained by this mechanism, viz., inhibition of synthesis of viral RNA polymerase with subsequent disappearance of enzyme activity by metabolic turnover. Rather, a direct effect of HBB on viral RNA synthesis appears much more plausible. In analogy to the results obtained by Tershak in the poliovirus system with guanidine (26,27) and considering the presently known data on the antiviral action of HBB we propose the initiation step of RNA synthesis to be a most probable site of the antiviral action of HBB.

This interpretation appears to be in conflict with the observation that the in vitro enzyme activity is apparently unimpaired by HBB (22). However, the following considerations must be kept in mind. The preparations originally used were very crude, reacting probably under nonoptimal conditions. The maximum reaction time was 15 min. It has been suggested, therefore, that incorporation of precursors into RNA in this system may, in a large part, be due to a completion of nascent RNA chains rather than initiation of new chains

(24,25). As to the antiviral action of guanidine, this
thesis appears to be supported by the detailed
investigations of Tershak (26,27).

In conclusion, all experimental data available so
far are compatible with the hypothesis that HBB
inhibits initiation of viral RNA synthesis. In molecular terms, however, the precise site of action of HBB
is not yet known. The complexity of the replication
process of picornaviruses makes the analysis rather
difficult (27,28,29,30,31). We consider a study of
conformational modifications of viral proteins by
various effectors a fruitful approach (32).

HBB resistance and dependence: Rescue

Passage of drug-sensitive virus in the presence of
HBB readily yields HBB-resistant virus (5). Clones of
varying degrees of HBB resistance are being obtained.
Sensitivity and resistance to HBB are genetic properties of the virion (33).

In addition to HBB-resistant mutants,
HBB-dependent mutants of various enteroviruses have
been isolated (34,35,36). This represented the first
demonstration of drug dependence in viruses. Like HBB
resistance also HBB dependence is genetically
determined (36). HBB is not required for early interactions of dependent virus with the cells but for RNA
synthesis of dependent virus (36). During the
exponential increase period of dependent virus, the
continued presence of compound is required (14).
Present data available suggest that the drug-sensitive
and drug-dependent processes may be biochemically
analogous, but further experiments on this point are
required.

HBB-dependent mutants give rise among their
progeny to HBB-independent virus particles with
considerable frequency. In a special system, apparently
free of selective pressure, it was determined that the
revertants from HBB dependence comprised mutants of a
wide spectrum of HBB sensitivity and resistence (37).
An estimate of the mutation frequency from HBB
dependence to HBB independence revealed a - at the time
- surprisingly high mutation frequency of about 10^{-4}
mutations per replication (37), an order of magnitude,
now well accepted for picornaviruses (38).

Rescue phenomena have been demonstrated with guanidine-sensitive, guanidine-resistant, and guanidine-dependent polioviruses (39,40,41,42,43). The results obtained with these systems strongly suggest that the assisting virus supplies a function for the rescued one, the viral RNA polymerase being a likely carrier of such a function.

Analogous experiments have been done with an HBB system. It has been demonstrated that HBB-dependent coxsackievirus A9 can be rescued in monkey kidney cells by HBB-sensitive echovirus 7 in the absence of the compound. Conversely, however, no significant rescue of HBB-sensitive echovirus 7 by HBB-dependent coxsackie-virus A9 could be demonstrated in cultures treated with HBB (43).

As indicated above, a direct effect of HBB on viral RNA synthesis, not mediated by the effects on synthesis of viral RNA polymerase, appears plausible. Thus, rescue of an HBB-sensitive virus in the presence of HBB may not be possible. Experiments to clarify this problem are under way in our laboratory.

Effects of HBB on picornavirus-infected cells

HBB significantly inhibits the development of picornavirus-induced morphological changes in monkey kidney cells, but the ultimate degeneration of the infected cells cannot be prevented, even in the absence of detectable virus replication (44). These virus-induced morphological changes in the presence of HBB are quite different from those observed in infected, untreated controls. The most plausible hypo-thesis to explain the ultimate degeneration of infected, HBB-treated cells is derived from the experi-mental finding that HBB does not prevent the early virus-induced shutoff of cellular protein and RNA synthesis. The shutoff in the presence of 100 μM HBB - a concentration inhibiting completely echovirus 12 replication - occurs with kinetics indistinguishable from that of infected, untreated cells (8). It is obvious that under such conditions the cell cannot survive any length of time, though cytopathic effects are not visible before 24 hr post infection. It should be stressed, however, that the cytopathic effects in untreated, infected cells, occurring within 6 hr after infection, are morphologically different (47,48,49,50).

All of these findings are fully compatible with the kinetics of the antiviral action of HBB described above: HBB leaves unaffected early virus-cell inter- actions including virus uncoating and virus-induced cellular shutoff. It interferes with viral RNA synthesis thereby ultimately inhibiting viral protein (capsid) synthesis which may be a most important factor causing common picornavirus-induced morphological alterations (48).

CONCLUSIONS

2-(α-Hydroxybenzyl)-benzimidazole (HBB) selectively inhibits many members of the picornavirus group. HBB does not interfere with early virus-cell interactions including virus adsorption, penetration, uncoating and virus-induced shutoff of cellular protein and RNA synthesis. The compound inhibits viral RNA syn- thesis, probably by interfering with initiation of RNA synthesis at the virus-induced RNA polymerase. Inhibition of viral protein synthesis, in all probability, is a secondary effect due to inhibition of viral RNA synthesis by HBB.

The virus selectivity of the compound is evident from the fact that - at virus-inhibitory concentrations - it leaves vital functions of the cell unaffected including cell multiplication. The virus selectivity of HBB is also manifest in that the compound does not affect members of families other than that of the picornavirus family.

Though it has been thought for a long time that HBB might not be protective in enterovirus-infected animals, possibly due to the emergence of HBB-resistant or HBB-dependent mutants, recent investigations on prophylactic and therapeutic effects in infected animals have been encouraging (8,51,52). It could be shown that the less dramatic antiviral effects of HBB (and its derivatives) in animals - as compared to those in cell culture - are due to metabolic inactivation of the compounds rather than due to the emergence of drug-resistant virus mutants (8). Therapeutic trials with benzimidazoles in virus-infected human volunteers are ongoing (53).

REFERENCES

1. I. Tamm, K. Folkers, and F. L. Horsfall, Jr., Inhibition of influenza virus multiplication by 2,5-dimethylbenzimidazole. Yale J. Biol. Med. 24:559 (1952).

2. I. Tamm, Selective inhibition of virus multiplication by synthetic chemicals. Bull. N.Y. Acad. Med.,31:537 (1955).

3. A. C. Hollinshead, and P. K. Smith, Effects of certain purines and related compounds on virus propagation. J. Pharmacol. Exp. Ther. 123:54 (1958).

4. I. Tamm, and M. M. Nemes, Selective inhibition of poliovirus multiplication. J. Clin. Invest. 38:1047 (1959).

5. H. J. Eggers, and I. Tamm, Spectrum and characteristics of the virus inhibitory action of 2-(α-hydroxybenzyl)-benzimidazole. J. Exp. Med. 113:657 (1961).

6. H. J. Eggers, and I. Tamm, 2-(α-Hydroxybenzyl)-benzimidazole (HBB) as an aid in virus classification. Virology 13:545 (1961).

7. I. Tamm, and L. A. Caliguiri, 2-(α-Hydroxybenzyl)-benzimidazole and related compounds, in: The International Encyclopedia of Pharmacology and Therapeutics, vol. 1, D. J. Bauer, ed., Pergamon, Oxford, New York, pp. 115 (1972).

8. H. J. Eggers, Benzimidazoles. Selective inhibitors of picornavirus replication in cell culture and in the organism, in: Chemotherapy of Viral Infections, P. E. Came, and L. A. Caliguiri, eds., Handb. Exp. Pharm., Vol. 61. Springer, Berlin, Heidelberg, New York, pp. 377 (1982).

9. G. Siegl, and H. J. Eggers, Failure of guanidine and 2-(α-hydroxybenzyl)-benzimidazole to inhibit replication of hepatitis A virus in vitro, J. Gen. Virol. 61:111 (1982).

10. H. J. Eggers, and I. Tamm, On the mechanism of selective inhibition of enterovirus multiplication by 2-(α-hydroxybenzyl)-benzimidazole. Virology 18:426 (1962).

11. I. Tamm, R. Bablanian, M. M. Nemes, C. H. Shunk, F. M. Robinson, and K. Folkers, Relationship between structure of benzimidazole derivatives and selective virus inhibitory activity, J. Exp. Med. 113:625 (1961).

12. I. Tamm, H. J. Eggers, R. Bablanian, A. F. Wagner, and K. Folkers, Structural requirements of selective inhibition of enteroviruses by 2-(α-hydroxybenzyl)-benzimidazole and related compounds. Nature 223:785 (1969).

13. I. Tamm, and H. J. Eggers, Differences in the selective virus inhibitory action of 2-(α-hydroxy-benzyl)-benzimidazole and guanidine HCl, _Virology_ 18:439 (1962).

14. H. J. Eggers, E. Reich, and I. Tamm, The drug-requiring phase in the growth of drug-dependent enteroviruses, _Proc. Natl. Acad. Sci. USA_ 50:183 (1963).

15. S. Halperen, H. J. Eggers, and I. Tamm, Evidence for uncoupled synthesis of viral RNA and viral capsids, _Virology_ 24:36 (1964).

16. L. A. Caliguiri, H. J. Eggers, N. Ikegami, and I. Tamm, A single-cell study of chemial inhibition of enterovirus multiplicaton, _Virology_ 27:551 (1965).

17. H. J. Eggers, and E. Waidner, Effect of 2-(α-hydroxy-benzyl)-benzimidazole and guanidine on the uncoating of echovirus 12, _Nature_ 227:952 (1970).

18. H. J. Eggers, Selective inhibition of uncoating of echovirus 12 by rhodanine, _Virology_ 78:241 (1977).

19. S. E. Luria, J. E. Darnell, Jr., D. Baltimore, and A. Campbell, _in_: General Virology, 3rd ed., J. Wiley and Sons, New York, Santa Barbara, Chichester, Brisbane, Toronto (1978).

20. H. J. Eggers, and I. Tamm, Inhibition of enterovirus ribonucleic acid synthesis by 2-(α-hydroxybenzyl)-benzimidazole, _Nature_ 197:1327 (1963).

21. D. Baltimore, and R. M. Franklin, Preliminary data on a virus-specific enzyme system responsible for the synthesis of viral RNA, _Biochem. Biophys. Res. Commun._ 9:388 (1962).

22. D. Baltimore, H. J. Eggers, R. M. Franklin, and I. Tamm, Poliovirus-induced RNA polymerase and the effects of virus-specific inhibitors on its production, _Proc. Natl. Acad. Sci. USA_ 49:843 (1963).

23. H. J. Eggers, D. Baltimore, and I. Tamm, The relation of protein synthesis to formation of polio-virus RNA polymerase, _Virology_ 21:281 (1963).

24. L. A. Caliguiri, and I. Tamm, Action of guanidine on the replication of poliovirus RNA, _Virology_ 35:408 (1968).

25. L. A. Caliguiri, and I. Tamm, Distribution and translation of poliovirus RNA in guanidine-treated cells, _Virology_ 36:223 (1968).

26. D. R. Tershak, Inhibition of poliovirus polymerase by guanidine in vitro, _J. Virol._ 41:313 (1982).

27. D. R. Tershak, F. H. Yin, and B. D. Korant, Guanidine, _in_: Chemotherapy of Viral Infections, P.E. Came, and L.A. Caliguiri, eds., Handb. Exp. Pharm., Vol. 61, Springer, Berlin, Heidelberg, New

York, pp. 343 (1982).

28. R. R. Rueckert, T. J. Matthews, O. M. Kew, M. Pallansch, C. McLean, and D. Omilianowski, Synthesis and processing of picornaviral polyprotein, in: Molecular Biology of Picornaviruses, R. Perez-Bercoff, ed., Plenum, New York, pp. 113 (1979).

29. P. D. Cooper, B.'B. Wentworth, and D. McCahon, Guanidine inhibition of poliovirus: a dependence of viral RNA synthesis on the configuration of structural protein, Virology 40:486 (1970).

30. B. D. Korant, Poliovirus coat protein as the site of guanidine action, Virology 81:17 (1977).

31. Th. Adrian, B. Rosenwirth, and H. J. Eggers, Isolation and characterization of temperature-sensitive mutants of echovirus 12, Virology 99:329 (1979).

32. H. J. Eggers, and I. Tamm, Antiviral chemotherapy, Annu. Rev. Pharmacol. 6:231 (1966).

33. I. Tamm, and H. J. Eggers, Unique susceptibility of enteroviruses to inhibition by 2-(α-hydroxybenzyl)-benzimidazole and derivatives, in: 2nd International Symposium of Chemotherapy, Part. II, H. P. Kuemmerle, P. Preziosi, and P. Rentchnick, eds., Karger, Basel New York, pp. 88 (1963).

34. H. J. Eggers, Discussion to: Genetic recombination with Newcastle disease virus, poliovirus, and influenza (by G. K. Hirst). Cold Spring Harbor Symp. Quant. Biol. 27:309 (1962).

35. H. J. Eggers, and I. Tamm, A variant of coxsackie A9 virus which requires 2-(α-hydroxybenzyl)-benzimida-zole (HBB) for optimal growth (Abstr.), in: VIIIth Int. Congr. Microbiol. p. 85 (1962).

36. H. J. Eggers, and I. Tamm, Drug dependence of entero-viruses: variants of coxsackie A9 and ECHO 13 viruses that require 2-(α-hydroxybenzyl)-benzimida-zole for growth, Virology 20:62 (1963).

37. H. J. Eggers, and I. Tamm, Coxsackie A9 virus: mutation from drug dependence to drug independence, Science 148:97 (1965).

38. B. K. Nottay, O. M. Kew, M. H. Hatch, J. T. Heyward, and J. F. Obijeski, Molecular variation of type 1 vaccine-related and wild polioviruses during repli-cation in humans, Virology 108:405 (1981).

39. C. E. Cords, and J. J. Holland, Replication of poliovirus RNA induced by heterologous virus, Proc. Natl. Acad. Sci. USA 51:1080 (1964).

40. J. J. Holland, and C. E. Cords, Maturation of polio-virus RNA with capsid protein coded by heterologous enteroviruses, Proc. Natl. Acad. Sci. USA 51:1082 (1964).

41. E. Wecker, and G. Lederhilger, Curtailment of the latent period by double-infection with polioviruses, Proc. Natl. Acad. Sci. USA 52:246 (1964).

42. V. I. Agol, and G. A. Shirman, Interaction of guanidine-sensitive and guanidine-dependent variants of poliovirus in mixedly infected cells, Biochem. Biophys. Res. Commun. 17:28 (1964).

43. N. Ikegami, H. J. Eggers, and I. Tamm, Rescue of drug-requiring and drug-inhibited enteroviruses, Proc. Natl. Acad. Sci. USA 52:1419 (1964).

44. R. Bablanian, H. J. Eggers, and I. Tamm, Inhibition of enterovirus cytopathic effects by 2-(α-hydroxy-benzyl)-benzimidazole, J. Bacteriol. 91:1289 (1966).

45. B. Rosenwirth, and H. J. Eggers, Echovirus 12-induced host cell shutoff is prevented by rhoda-nine, Nature 267:370 (1977).

46. B. Rosenwirth, and H. J. Eggers, Structure and replication of echovirus type 12. 2. Viral polypep-tides synthesized in the infected cell, Eur. J. Biochem. 92:61 (1978).

47. R. Bablanian, H. J. Eggers, and I. Tamm, Studies on the mechanism of poliovirus-induced cell damage. I. The relation between poliovirus-induced metabolic and morphological alterations in cultured cells, Virology 26:100 (1965).

48. R. Bablanian, H. J. Eggers, and I. Tamm, Studies on the mechanism of poliovirus-induced cell damge. II. The relation between poliovirus growth and virus-induced morphological changes in the cells, Virology 26:114 (1965).

49. S. Dales, H. J. Eggers, I. Tamm, and G. E. Palade, Electron microscopic study of the formation of poliovirus, Virology 26:379 (1965).

50. K. Bienz, D. Egger, Y. Rasser, and W. Bossart, Kinetics and location of poliovirus macromolecular synthesis in correlation to virus-induced cytopathology, Virology 100:390 (1980).

51. H. J. Eggers, Successful treatment of enterovirus-infected mice by 2-(α-hydroxybenzyl)-benzimidazole and guanidine, J. Exp. Med. 143:1367 (1976).

52. E. C. Herrmann, Jr., J. A. Herrmann, and D. C. DeLong, Comparision of the antiviral effects of substituted benzimidazoles and guanidine in vitro and in vivo, Antiviral Research 1:301 (1981).

53. R. J. Phillpotts, D. C. DeLong, J. Wallace, R. W. Jones, S. E. Reed, and D. A. J. Tyrrell, The activity of enviroxime against rhinovirus infection in man, Lancet I:1342 (1981).

ANTIRHINOVIRUS DRUGS

D. A. J. Tyrrell

Director
MRC Common Cold Unit
Harvard Hospital
Coombe Road
Salisbury SP2 8BW
Wiltshire England

INTRODUCTION

Common colds are among the mildest of diseases affecting man, but they are also very frequent. Indeed, adults are said to have from 2 to 5 episodes a year on the average and certain predisposed children may have 12 attacks a year. The disease is often sufficiently severe to cause absence from work or education. In predisposed patients the colds may go on to bronchitis or may precipitate heart failure or upset the control of diabetes.

It is now over 20 years since rhinoviruses were first cultivated and they were found in cases of typical common colds in adults. It was also found that they caused typical colds when administered intranasally to human volunteers. Thus it was proved that they cause the disease. Later studies showed that they could be cultured from children as well as adults and also from relapses in chronic bronchitis and from acute exacerbations of the wheezy bronchitis syndrome in children.[1] We are still not clear how large their contribution is to the whole clinical problem. Accurate virus diagnosis is not easy. Using a sensitive tissue culture system a virus may be cultured from nasal secretions in about $\frac{1}{4}$ of cases of colds. However some viruses can be cultured in organ cultures of nasal or tracheal epithelium, or in other specially sensitive cell cultures although they do not grow in the usual tissue cultures. There are so many serotypes that serological diagnosis is impractical. Therefore the best estimate of the proportion of colds caused by rhinoviruses is that obtained

from studies in which ample specimens were collected from cases
and a range of highly sensitive test cultures were used. We
recently reported such a study from which we conclude that approaching half of typical common colds are caused by rhinoviruses.[2]
Although other viruses are involved, particularly in children, the
results of tissue culture based studies suggest that a similar
proportion of colds in children are also due to rhinoviruses.
Since there are over one hundred serotypes of rhinoviruses and all
seem to cause colds to a similar extent it is not likely that a
useful antirhinovirus vaccine will be developed, but it is plausible
that an antirhinovirus drug could be effective although, as we shall
see later, rhinoviruses may differ greatly in their response to
antivirals as well as to specific antibodies.

Pathology and pathogenesis

Full pathological studies on subjects infected with rhinoviruses have not been done but some general facts have been established. The viruses are specially well adapted to growth in the
nasal mucosa, in that they grow best in cultures of the nasal and
tracheal epithelial cells and do not attack other cells such as
squamous epithelia. They also replicate best at the reduced
temperatures of the nose, about 33°. Thus virus replication is
apparently confined to the superficial cells of respiratory mucosa –
there is no evidence that virus spreads into submucosal tissue or
the bloodstream, though it can spread along the mucosal surface
and affect the trachea and bronchi.

Virus is not usually detected in the nose on the day after
infection, but appears on the second day and it is late on that
date or some time on the next that symptoms develop, such as
sneezing or a running nose. As we shall see later all these facts
are important when deciding how to use an antiviral drug to
influence the course of the disease.

The structure and replication of rhinoviruses

Rhinoviruses are in most respects typical members of the
picornavirus family[3] and the main features are set out in Table 1.
The characteristic 27nm particle has icosahedral symmetry and contains no lipid. The capsid is formed of 4 structural proteins VP1,
VP2, VP3 and VP4 and contains a single strand of positive sense
RNA. The RNA is translated to form a polymerase as the first step
in the replication of the RNA molecule. The whole molecule is
however translated and subsequently cleaved by a protease and gives
rise to the capsid proteins also. This process is best understood
in poliovirus and foot-and-mouth disease virus but something similar
certainly happens with rhinoviruses. The synthesis of RNA and
protein, cleaving of the peptides and the formation of the particle
are all highly integrated and occur at the internal membranes of

Table 1. Some features of the biology of rhinovirus

Particle   — diameter 27nm. Lipid absent. Cubic
                symmetry. Peptides VP1, VP2, VP3, VP4.
Nucleic acid — RNA single stranded, positive sense
Replication   — cytoplasmic single translation product,
                cleaved to provide polymerase and capsid
                protein

the cell. Virus is then released and passes through the medium to
attach to enter another cell through interactions with cell receptors.
All these virus specific effects are potential points in replication
which an antiviral might block. In addition it is known that the
initial attachment and uncoating step is quite specific — at least
it appears to involve sites which are quite distinct from those
used by other picornaviruses and might in theory be a site of antiviral action.

The search for antirhinovirus drugs

Most of the work in developing antirhinovirus drugs has begun
by screening substantial numbers of organic compounds. This was
started soon after it became possible to grow the viruses in readily
available cells, such as sensitive strains of human fibroblasts and
Hela cells. It is still not realised that antirhinovirus tests can
be done as easily as, say, antiherpesvirus tests.

A fairly standard array of tests can be used as shown in
Table 2. Inhibition of CPE is convenient and can be made relatively
quantitative by using serial dilutions of drug in the medium of
replicate cultures challenged with a standard amount of virus. Drug
sensitivity can also be assessed by adding serial dilutions of virus
to cultures maintained in medium alone or in medium containing a
dilution of drug; the effect detected is the reduction in the end
point of virus titration and the test is particularly useful for
comparing the susceptibility of different viruses. It is also valuable to supplement these tests by yield experiments in which virus
is grown in drug treated cultures and the amount produced is estimated by titrating in a second system, using end point or plaque
assays.

Since some antiviral substances, like interferon, have different
effects in different cells it has been found to be important to test
at least selected antivirals in cultures of several different susceptible cells, e.g. human lung fibroblasts, and Hela cells. It would
be desirable to test them in the cells in which we desire them to
act in man, namely respiratory epithelium. This can be done by using
organ cultures of human foetal, nasal or tracheal epithelium, and
enviroxime was shown to be highly active in such cultures before
it was tested in human volunteers.

Table 2. Tests used to search for antirhinovirus activity

A. Inhibition of cytopathic effect a) Roller tubes
                                    b) Microtiter plates
                                    c) Plaque reduction
    e.g. 1. titration in presence or absence of drug –
            single concentration.
         2. titration of dilutions of drug with single
            dose of virus.
B. Reduction of yield
    a) Tissue culture
    b) Organ cultures of respiratory epithelium

It is also important to test any candidate compounds for
activity against a substantial number of different rhinoviruses.
This is needed for two reasons. Firstly to find if possible a drug
which has high activity against a wide range of virus isolates and
serotypes: unless it is effective against most viruses it will have
no impact on the clinical problem. Secondly, in planning the first
volunteer trials it is important to select a virus with maximum
sensitivity to the substance – the tests are often negative and are
cumbersome and expensive to repeat so when investigating a new
group of compounds it is desirable to do the test with the most
favourable virus available.

Mode of action and pharmacokinetics

Once a substance which is potent in vitro has been found it
is desirable to acquire a general picture of what effects it may have
before giving it to man. Of course appropriate toxicological studies
in animals are needed but they will not be discussed in detail here.

It is sometimes thought that the mode of action of the drug
should be worked out before it is used. However it is my view that
this is not essential, at least not in basic or molecular detail.
On the other hand, some general aspects of its effect at the cellular
level may be useful in planning its evaluation. For instance there
are likely to be advantages in a drug which stops virus replication
when added to a cell in which virus infection is already under way
if one wishes to try to treat a cold rather than to prevent one. On
the other hand the effects of many drugs are readily reversed by
removing them from the culture medium. Interferon has the important
property of inducing virus resistance that lasts for hours after it
has been in contact with the cell and so makes it possible to produce
an effect with long intervals between successive intranasal admini-
strations. We therefore would like to have synthetic antivirals
whose effect is not annulled by simply removing them from the tissue
culture medium.

Finally it can be crucially important to consider factors such as solubility, metabolism and drug distribution. Some drugs e.g. a chalcone, may not be absorbed when given by mouth, and so a prodrug was designed which is absorbed and then hydrolysed to provide the active antiviral in the circulation. Other highly insoluble drugs may not be absorbed in an aqueous formulation but can be absorbed in a non aqueous solution e.g. dichloroflavan in maize oil. However even if it is absorbed tissue distribution may not be satisfactory - it is remarkable that amantadine seems to be concentrated in lung and probably other respiratory tissues. However the dichloroflavan just mentioned is a recent example of a drug which is found in inhibitory concentration in the circulation but at very low levels in the secretions and presumably at very low levels in the nasal epithelial cells, since it is not effective in volunteers against rhinovirus infection. On the other hand if a drug is to be given as a local spray or aerosol its solubility may not matter so much, though insoluble substances may be difficult to formulate.

Background and general considerations

There has been interest in discovering drugs against rhinoviruses ever since it became clear that the organisms were responsible for a substantial proportion of colds about 20 years ago. However the first synthetic drug that was shown to be effective against a respiratory virus was amantadine. It will be recalled that it was shown that oral administration prevented influenza virus infection, first in mice and then in men. Later it was shown to be effective if given therapeutically after the disease had commenced. These observations were important because there was a good deal of pessimism in some quarters about whether antiviral drugs could be effective in any respiratory diseases. It was postulated that the drug might not reach the respiratory epithelium or that once the disease was under way virus replication would already be largely over and so antiviral treatment would have no effect. Yet the fact that amantadine prevented and treated infection in mice and man indicated that these problems could be solved. On the other hand work on rhinoviruses was more difficult as there are many more virus types, no satisfactory animal models have been found and, as we shall see later, antiviral substances are often not effective given by mouth.

There have been various approaches to finding antirhinovirus drugs. Interferon turned out to be a powerful inhibitor of rhinovirus replication in vitro and indeed when, after many failures, it was found over 10 years ago that sufficient doses of human leucocyte interferon, given as repeated nasal sprays, prevented rhinovirus colds. (see 4) This encouraged investigators to look for synthetic antirhinovirus substances and to use them by intranasal administration. However for editorial reasons interferon is excluded from this chapter. Nevertheless we should mention that a good deal of

work was done on interferon inducers during the 1970's as a way of
using the interferon mechanism even though interferon itself was
in short supply. The idea was to overcome the difficulty of making
exogenous infection by stimulating the respiratory tract to make its
own. Polyinosinic polycyidylic acid (poly I.C.) is well known and
both it and natural double stranded RNA, such as that extracted from
the mycophage of Penicillium inhibits virus replication _in vitro_
at very low concentrations. Both were given by intranasal spray to
volunteers who were challenged with rhinoviruses.(5, 6) Poly I.C.
produced a slight clinical effect but ds RNA not only produced little
effect but caused nasal irritation. Understandably little further
work was done. However low molecular weight molecules can also act
as inducers and were therefore studied. Tilorone was unacceptably
toxic but substituted propanediamines were more promising. Indeed
two were sufficiently active _in vitro_ and non toxic to be evaluated
in volunteers. Little interferon was apparently produced when
CP20961 was adminstered intranasally and although some effects
against rhinovirus infections were reported this could not be
replicated elsewhere and there seem to have been no recent studies.
(7, 8)

    We should also mention one early study done with a strain of
coxsackievirus type A21. This organism is a definite enterovirus by
laboratory tests but behaves like a rhinovirus as it multiplies
preferentially in the respiratory tract and produces common colds.
It was found that it could be inhibited _in vitro_ by low concentrations of fusidic acid and some other related 'steroidal' antibiotics.
Sodium fusidate by mouth was therefore used in a trial to prevent
infection and disease with coxsackie A21.(9) Although virus inhibitory concentrations were reached in the volunteers' serum no antiviral effect or clinical benefit was seen. The reason appeared to
be that the drug was excluded from the respiratory tract; at least
it could not be found in nasal secretions collected from treated
volunteers who developed colds.

    It was also possible to test compounds which were known to be
effective against influenza. One case was that of spiramantadine,
a derivative of amantadine, which was effective in preventing
experimental influenza in volunteers. It was found to be active
against rhinoviruses _in vitro_ and so it was then tested as an oral
drug in volunteers.(10) However it had no effect – presumably it
reached the respiratory epithelium but not in sufficient concentrations to prevent the growth of rhinoviruses which were less sensitive _in vitro_ than influenza A viruses.

    A number of isoquinoline compounds have been studied as anti-
virals and it was shown that two UK2054 and 2371 were active against
influenza virus _in vitro_ and subsequently when given orally to
volunteers. The fact that they were effective suggested that they
were distributed to respiratory epithelium. It was found that they

also had _in vitro_ activity against rhinoviruses so it was logical to
test them against these organisms in volunteers. However no effect
was detected and again it was thought that this was probably because
these viruses were significantly less sensitive to inhibition by the
compounds than were the influenza viruses.[11] Another isoquinoline
(DIQA) obviously was active against a range of viruses in mice
although it had no such activity in tissue cultures. It was nevertheless tested in volunteers using rhinovirus 24 as a challenge.
There was apparently a reduction of symptoms and virus excretion
that was just statistically significant,[12] but I know of no follow
up work.

Isoprinosine has been extensively studied as an antiviral - it
is a p-aminobenzoic acid salt of inosinedimethyl aminoisopropanol.
It has been tested in a number of animal and _in vitro_ models. There
is little evidence that it is effective though it does seem to be
remarkably non toxic. Nevertheless two groups have tested it as an
antirhinovirus drug in volunteers, giving it by mouth. RV9 and RV31
were used as challenge in one study[13] and RV44 or RV32 in the
other but there was no evidence of significant protection. There
was slight enhancement of the antibody response but this did not
reach statistical significance either, though the substance is now
being promoted under another name as an immune enhancer; and if it
does have an effect of this sort it might provide benefit to patients.
[14] Indeed Professor S. M. Chu of Beijing tells me that this group
is investigating the possibility that some traditional Chinese herbal
remedies, (such as Radix Astragalus membranacea, which is taken regularly as an aqueous extract to reduce the incidence of colds, and
may indeed be effective) may work rather by enhancing immune responses than by directly impairing virus replication.

In contrast there have been extensive studies on the extent and
mechanism of the antiviral effects of a number of substituted benzimidazoles. These are active against a number of RNA viruses, including influenza, entero- and rhino- viruses.[15] They have been valuable tools in understanding the mechanisms of virus replication.
One of them α hydroxy benzyl, ribofuranosyl benzimidazole (HBB) is
active against rhinoviruses _in vitro_ and not cytotoxic. However
substances arising from this work have not been brought to trial
for the prevention of colds. Likewise although zinc ions are
powerful inhibitors of rhinoviruses no one seems to have applied
these to preventing or treating disease.[16]

This chapter does not give details of all the compounds that
have been found at one time or another to have antirhinovirus
activity. Instead it concentrates on substances that have, in
addition, been found to be sufficiently non toxic, well tolerated
and metabolically stable to be used for attempts to prevent infection in volunteers, the side of the work with which I have been
particularly associated.

The methods of recruiting, housing and handling the volunteers
have already been described in some detail[17] so it is unnecessary
to repeat them here. One point is worth making however. In spite
of all the efforts in finding potent antivirals it is still true that
trials in man usually show no effect, so it has been our policy when
investigating a new drug to confirm first that it is well tolerated
and then to aim to do an experiment that will give an acceptable
negative result, which will not need to be followed by further
more stringent trials at the cost of more time and effort. We
therefore aim to give the drug by as intensive a regime as seems
possible and reasonable. We always give it for about a day before
we give virus and continue it for 4 to 6 days. We always use a
virus which has been shown _in vitro_ to be the most sensitive to
the drug that is available and we always give a small inoculum.
The hypothesis we are testing is that with everything in its favour,
the drug might reduce the frequency of colds substantially, e.g.
by 50% and we want to exclude that possibility at conventional
levels of probability, though calculating the power of our trials
is not all that easy.

Some specific antirhinovirus compounds

Several substituted isoquinoline compounds were found to be
active _in vitro_ against rhinoviruses at concentrations lower than
those of substances mentioned above. They are however absorbed and
non toxic. A test was done in gibbons which can be infected with
certain human rhinoviruses, although they do not show symptoms of
respiratory disease, and some activity was detected.[18] Tests in
volunteers given drug by mouth (SKF21687 and SKF30097) and challenged
with rhinoviruses showed no significant effect.[19] Later a drug
(SKF40491) was given by the intranasal route and again no statistically significant activity was found.[20]

At about the same time tests were done with a compound produced
by Glaxo laboratories GL R9-338 - it again produced little effect
but a reduction in virus shedding that was not statistically significant. Another isoquinoline derivative RP 19326 was given intranasally and produced benefits in clinical symptoms that were just
significant. It is important that all these experiments were done
using viruses which were found by _in vitro_ experiments to have
maximal sensitivity to the drugs. The drugs were also given frequently and the studies were designed to give as clear a negative
result as the resources available would allow. However the numbers
of volunteers used were quite small and only a strong effect could
be excluded with any great certainty. Concentrations of about
4 µg were required for marked antiviral activity _in vitro_ so the
results implied that more potent compounds were needed.

Table 3. Trials of prophylaxis at Common Cold Unit

Compound	Route	In vitro inhibitory conc. µg/ml.	Serotype of challenge virus	Number of colds Treated	Placebo
SKF 40491	i.n.	4	3	7/10	4/9
GL R9-338	i.n.	4	9	4/11	5/13
RP 19326	i.n.	4	9	5/11	4/8
Enviroxime LY 122772	oral & i.n.	0.2	9	4/18	6/18*
4'6 Dichloroflavan 683.C.	oral	0.003	9	9/26	10/23
Chalcone Ro 09-0415	oral	0.01+	9	8/28	10/29

*statistically significant reduction in clinical scores and nasal secretion
+of active metabolite

RECENT STUDIES

Recently drugs have become available that are active at concentrations of </μg, a good deal lower than those just mentioned. example is enviroxime, which happens to be a benzimidazole derivative and was selected from a long series of compounds synthesized at the Eli Lilly laboratories. It was not tested for antirhinovirus effects in animals, but because we have experience of antirhinovirus drugs which are active in one virus-susceptible cell and not in another it was tested for its antiviral activity in organ cultures of human respiratory epithelium.[21] It was shown that it inhibited virus growth but that the effect was readily reversed on withdrawing it from the medium. It may act by inhibiting polymerase activity but whatever the mechanism it has similar potency when tested against a wide range of serotypes of virus.

In the trials it was given by both oral and intranasal routes[22] on the hypothesis that if it were active we would be willing to do further experiments to decide which was the effective route. It reduced the symptoms produced by rhinoviruses but it was less effective than interferon; it also had some irritant effects on the nasal mucosa, at least in the original formulation and induced nausea in some subjects. Later experiments indicated that if only the intranasal route was used some effect could be produced though not on every occasion[23, 24] Drug was given intranasally by an intensified regime which was delayed until after virus had been given.[25] Colds were apparently modified if treatment started before symptoms appeared. The final conclusion had to be that although enviroxime is active against a wide range of serotypes it is of limited efficacy in man. It is probably not active if given by mouth since it does not enter secretions after oral administration. It would be reasonable to survey further members of the series to look for some that are secreted after oral dosage and do not produce central nausea or vomiting.

Another compound has been developed in Wellcome Research laboratories, namely a dichloroflavan. This is believed to be the most active antirhinovirus substance yet described, although the exact degree of activity varies from serotype to serotype.[26] It is unusual in that it can inactivate the infectivity of virus particles, probably by interacting with the viral peptides - indeed virus can be reactivated by extraction with chloroform. It is lipophilic but well absorbed as a solution in oil. The blood levels are substantial but when given by the oral route it did not prevent infection with or symptoms of rhinovirus 9 which was particularly sensitive to it.[27] However if an intranasal formulation can be developed it would be worthwhile reinvestigating such a strikingly active substance.

An antiviral chalcone has been developed by the study of an oriental herbal remedy for colds which was found to contain an anti-viral activity. Chemically it is not too far from the structure of dichloroflavan and seems to also have the effect of inactivating virus particles.[28] It was too irritant to give by nose and was not well absorbed. However a prodrug, which was inactive but absorbed and then converted to the active form by loss of a phosphate group has been developed. It apparently has no effect in our volunteers when given by mouth. We suspect that this is because it does not reach the respiratory mucosa.[29]

It thus seems that though these substances are more active they are still not potent prophylactics in man. The reasons are complex but it is worth remembering that they are still not as active as interferon and so we are not able to introduce as much antiviral activity into the nose with a synthetic antiviral as is possible when using interferons.

Comment

In spite of the title of this meeting I have referred mainly to drugs found by screening and not by design and I do not think we know enough about virus structure and replication to define our target in molecular terms, as may perhaps be possible with herpes-viruses. I have emphasized rather that the drug must be sufficiently active, non toxic and appropriately distributed if it is to show any effect in human volunteers. There are hopes that the faults of enviroxime may be rectified by searching among related compounds for one which, in addition to being active, does not induce nausea and is excreted via the respiratory mucosa. Such a molecule might well be a useful orally active antirhinovirus drug. If a drug is to be useful it must act on many serotypes and it is likely to do this if it acts on the well conserved parts of the particle. It may be significant that enviroxime apparently acts on the polymerase while dichloroflavan apparently affects capsid peptides which are clearly variable, and affects different serotypes to a very different extent.

We can look further ahead and ask whether we might find an anti-rhinovirus drug that could be used therapeutically, i.e. to modify a cold that has already shown at least some initial symptoms. Pre-liminary unpublished experiments using interferons suggest that this is not easy to do. It has been suggested by pessimists that this will always be impossible, but amantadine apparently modifies clinically apparent influenza and enviroxime has a slight effect on colds in the incubation period. Thus to affect a cold it might be necessary to have a highly active substance or mixture of substances, which could interupt a virus infection which was already underway. It might be easier to produce an effect in a longer lasting infection like an attack of bronchitis. This would be a worthwhile target for

later work and we should continue to keep it in mind but meantime
we should confirm and then apply in the community the successful
prophylaxis of colds in inoculated volunteers. For instance, if
enviroxime prophylaxis can be shown to work with normal subjects it
might be acceptable to patients, such as those with chronic cardio-
respiratory disease, to take it regularly, at least for part of the
year, in order to reduce the chance of a relapse induced by a common
cold. However I suspect that this stage will be reached with interferon before it becomes practical with any of the low molecular
weight synthetic antivirals discussed here.

It may also be important to explore better ways of delivering drugs
to the affected mucosa for instance by fine particle aerosols.
Amantadine and ribavirin are being studied in this way and a recent
paper suggests that aerosolized ribavirin can improve the course of
bronchitis of infants due to respiratory syncytial virus.(30)

REFERENCES

1. GWALTNEY, J. M. (1957) Rhinoviruses. Yale J. Biol. Med. 108,
    444-453
2. LARSON, H. E., REED, S. E., and TYRRELL, D. A. J. (1980)
    Isolation of rhinoviruses and coronaviruses from 38 colds in
    adults. J. med. Virol. 5, 221-229
3. MACNAUGHTON, M. R. (1982) The structure and replication of
    rhinoviruses. Current Topics in Microbiology and Immunology
    97, 1-26
4. GREENBERG, S. B., HARMON, M. W., TYRRELL, D. A. J. and SCOTT,
    G. M. (1981-82) Trials of interferon in respiratory infections
    in man. Texas Rep. Biol. Med. 41, 549-553
5. HILL, D. A., BARON, S., PERKINS, J. C., WORTHINGTON, M.,
    VAN KIRK, J. E., MILLS, J., KAPIKIAN, A. Z., CHANOCK, R. M. (1972)
    Evaluation of an interferon inducer in viral respiratory disease.
    J. Amer. Med. Assoc. 219, 1179-1184
6. AOKI, F. Y., REED, S. E., CRAIG, J. W., TYRRELL, D. A. J. and
    LEES, L. J. (1958) Effect of a polynucleotide inducer of fungal
    origin on experimental rhinovirus infection in humans.
    J. Infect. Dis. 137, 82-86
7. DOUGLAS, R. G. and BETTS, R. F. (1974) Effect of induced inter-
    feron in experimental rhinovirus infections in volunteers.
    Infect. Immun. 9, 506-510
8. STANLEY, E. D., JACKSON, G. G., DIRDA, V. A. and RUBENIS, M.
    (1976) Effect of a topical interferon inducer on rhinovirus
    infections in volunteers. J. Infect. Dis. (Supp/June 1976)
    121-127
9. ACORNLEY, J. E., BESSELL, C. J., BYNOE, M. L., GOTFREDSEN, W. O.
    and KNOYLE, J. M. (1967) Antiviral activity of sodium fusidate
    and related compounds. Brit. J. Pharmacol. Chemother. 31,
    210-220

10. MATHUR, A., BEARE, A. S. and REED, S. E. (1973) *In vitro*
     activity and preliminary clinical trials of a new adamantane
     compound. Antimicrob. Agents Chemother. 4, 421-426
11. REED, S. E. and BYNOE, M. L. (1970) The antiviral activity
     of isoquinoline drugs for rhinoviruses *in vitro* and *in vivo*.
     J. med. Microbiol. 3, 346-352
12. TOGO, Y., SCHWARTZ, A. R. and HORNICK, R. B. (1973) Antiviral
     effect of 3,4-dihydro-1-isoquinoline acetamide hydrochloride
     in experimental human rhinovirus infection. Chemotherapy. 18,
     17-
13. SOTO, A. J., HALL, T. S. and REED, S. E. (1973) Trial of the
     antiviral action of isoprinosine against rhinovirus infection
     of volunteers. Antimicrobial Agents Chemother. 5, 332-334
14. ANON (1982) Inosiplex: Antiviral, immunomodulator, or neither?
     Lancet i, 1052-1054
15. TAMM, I., CALIGUIRI, L. A. (1972) 2-($\propto$-hydroxybenzyl) benzi-
     midazole and related compounds *in: Chemotherapy of virus
     diseases I.* Ed. D. J. BAUER 115-180
16. KORANT, B. D., BAUER, J. C., BUTTERWORTH, B. E. (1974) Zinc
     ions inhibit replication of rhinoviruses. Nature 248, 558-560
17. BEARE, A. S., and REED, S. E. (1977) The study of antiviral
     compounds in volunteers *in: Chemoprophylaxis and virus infection
     of the respiratory tract.* Vol II Ed J. Oxford C.R.C. Press,
     Cleveland, Ohio p27-55
18. PINTO, C. A., BAHN SEN, H. P., RAVIN, L. J., HAFF, R. F. and
     PAGANO, J. E. (1972) The antiviral effect of a triazinoindole
     (SKand F 40491) in rhinovirus infected gibbons. Proc. Soc.
     exp. Biol. Med. 141, 467-74
19. TOGO, Y., SCHWARTZ, A. R. and HORNICK, R. B. (1973) Failure of
     a 3-substituted triazoinoinsole in the prevention of experi-
     mental human rhinovirus infection. Chemotherapy, 18, 17-26
20. REED, S. E., CRAIG, J. W. and TYRRELL, D. A. J. (1976) Four
     compounds active against rhinovirus: comparison *in vitro* and
     in volunteers. J. Infect. Dis. (Suppl) 138 A128-A135
21. DELONG, D. C. and REED, S. E. (1980) Inhibition of rhinovirus
     replication in organ culture by a potential antiviral drug.
     J. infect. Dis. 141, 87-91
22. PHILLPOTTS, R. J., JONES, R. W., DELONG, D. C., REED, S. E.,
     WALLACE, J. and TYRRELL, D. A. J. (1981) The activity of
     enviroxime against rhinovirus infection in man. Lancet ii,
     1342-1344
23. HAYDEN, F. G. and GWALTNEY, J. M. Jr (1982) Prophylactic
     activity of intranasal enviroxime against experimentally induced
     rhinovirus type 39 infection. Antimicrob. Agents Chemother.
     21, 892-897
24. LEVANDOWSKI, R. A., PACHUCKI, C. T., RUBENIS, M. and JACKSON,
     G. G. (1983) Topical enviroxime against rhinovirus infection.
     Antimicrob. Agents Chemother. 22, 1004-7
25. PHILLPOTTS, R. J., WALLACE, J., TYRRELL, D. A. J. and TAGART,
     V. B. (1983) A study of the therapeutic activity of enviroxime

against rhinovirus infection in volunteers. <u>Antimicrob. Agents Chemother.</u> <u>23</u>, 671-675

26. BAUER, D. J., SELWAY, J. W. T., BATCHELOR, J. F., TISDALE, MARGARET, CALDWELL, IAN C. and YOUNG, D. A. B. (1981) 4', 6-Dichloroflavan (BW683c) a new antirhinovirus compound. <u>Nature</u> <u>292</u> (5821) 369-70

27. PHILLPOTTS, R. J., WALLACE, J. W., TYRRELL, D. A. J., FREESTONE, D. and SHEPHERD, W. N. (1983) Failure of oral 4, 6, dichloroflavan to protect against rhinovirus infection in man. <u>Arch. Virol.</u> <u>75</u>, 115-21

28. ITHITSUKA, H., NINOMIYA, Y. T., OHSANA, C., FIJIU, M. and SUHARA, Y. (1982) Direct and specific inactivation of rhinovirus by chalcone Ro 09-0410 <u>Antimicrob. Agents Chemother.</u> <u>22</u>, 617-21

29. PHILLPOTTS, R. J., HIGGINS, P. G., WILLMAN, J., TYRRELL, D. A. J., LENNOX-SMITH, I. (1983) Evaluation of the antirhino-virus chalcone Ro 09-0410 in volunteers. Manuscript in preparation.

30. HALL, C. B., McBRIDE, J. T., WALSH, E. E., BELL, D. M., GALA, C. L., HILDRETH, S., TEN EYCK. L. G. and HALL, W. J. (1983) Aerosolized ribavirin treatment with respiratory syncytial viral infection. <u>New Eng. J. Med.</u> <u>308</u>, 1443-1447

PYRIMIDINE NUCLEOSIDE ANALOGUES AS ANTIVIRAL AGENTS

Erik De Clercq

Rega Institute for Medical Research
Katholieke Universiteit Leuven
B-3000 Leuven, Belgium

INTRODUCTION

The majority of the antiviral drugs that have been licensed for clinical use or are being considered for clinical use are pyrimidine nucleoside analogues and these compounds are primarily directed towards the treatment of herpesvirus (HSV-1, HSV-2 and VZV) infections. Thus, idoxuridine (IDU, 5-iodo-2'-deoxyuridine) and trifluridine (TFT, 5-trifluoromethyl-2'-deoxyuridine) are widely used as eye drops, at 0.1 % and 1 % respectively, in the topical treatment of herpetic keratitis; 5-ethyl-2'-deoxyuridine is used in W.-Germany as a 0.3 % gel, 0.15 % eye drops or 0.5 % solution for subconjunctival injection, in the treatment of herpetic keratitis; and 5-iodo-2'-deoxycytidine is used in France as 0.15 % eye drops or 1 % eye ointment for the topical treatment of herpetic keratitis, and as 1 % ointment for the topical treatment of herpetic skin lesions. IDU can also be used for the topical treatment of cutaneous HSV and VZV infections, i.e. when applied at 10 % in DMSO (dimethylsulfoxide). Finally, cytarabine (Ara-C, cytosine arabinoside, 1-β-D-arabinofuranosylcytosine) has been administered occasionally in the systemic treatment of herpetic encephalitis and herpes zoster. The recommended daily dose was 40 mg/m^2/day for 5 days. At present, ara-C is not longer used as an antiviral drug. Instead, it is commonly used in combination with other drugs such as thioguanine or daunomycin in the treatment of acute lymphoblastic leukemia and acute myeloblastic leukemia.

Abbreviations. ID_{50}, 50 % inhibitory dose; HSV-1, herpes simplex virus type 1; HSV-2, herpes simplex virus type 2; VZV, varicella-zoster virus; CMV, cytomegalovirus; EBV, Epstein-Barr virus.

In recent years several new pyrimidine nucleoside analogues
have been developed which offer great promise for both topical and
systemic therapy of HSV and VZV infections. Foremost among these new
promising antiviral drugs are bromovinyldeoxyuridine (BVDU, (E)-5-
(2-bromovinyl)-2'-deoxyuridine) and fluoroiodoaracytosine (FIAC,
1-(2-deoxy-2-fluoro-β-D-arabinofuranosyl)-5-iodocytosine). Their antiviral potentials will be discussed within the context of the other
pyrimidine nucleoside analogues. The pyrimidine nucleosides are divided into ten classes : 5-substituted 2'-deoxyuridines, 5-substitu-
ted 2'-deoxycytidines, 5-substituted arabinofuranosyluracil derivatives, 5-substituted arabinofuranosylcytosine derivatives, 3'- and
5'-amino derivatives of 5-substituted 2'-deoxyuridines, 2'-halogeno-
arabinofuranosylpyrimidines, pyrimidine carbocyclic nucleoside analogues, azapyrimidine nucleoside analogues, deazapyrimidine nucleoside analogues and pyrimidine acyclic nucleoside analogues. Each
class of compounds will be discussed from the following viewpoints :
(i) structure-function relationship, (ii) spectrum of antiviral activity, (iii) mechanism of action, and (iv), where appropriate, efficacy in animal models and humans.

5-SUBSTITUTED 2'-DEOXYURIDINES

Although a few 5-substituted 2'-deoxyuridines, i.e. 5-iodo-
dUrd, 5-trifluoromethyl-dUrd, 5-methylamino-dUrd and 5-ethyl-dUrd,
were described about 15 - 20 years ago, most dUrd analogues date from
the last 5 years. The list of 5-substituted dUrd derivatives has become so long that for an adequate description of their properties
the compounds have to be divided in several classes.

A first class would then comprise those compounds that are virtually inactive as antiviral agents (ID_{50} for HSV-1 > 100 µg/ml)
(Fig. 1) : i.e. 5-mercaptomethyl-dUrd,[1] 5-thiocyanomethyl-dUrd,[1]
5-methylsulfinylmethyl-dUrd,[2] 5-dimethylaminomethyl-dUrd,[3] 5-iodo-
acetamidomethyl-dUrd,[3] 5-propyloxymethyl-dUrd,[4] 5-benzyloxymethyl-
dUrd,[4] 5-hexyl-dUrd,[5] 5-paranitrobenzyloxy-dUrd[6] and many others.
Apparently, the 5-substituents of these dUrd analogues are too bulky
to impart antiviral activity.

The second class of dUrd analogues is composed of those compounds that are rather selective but not very potent as antiviral
agents (ID_{50} for HSV-1 : ∿ 1 - 10 µg/ml) (Fig. 2) : i.e. 5-ethynyl-
dUrd,[7,8] 5-propyl-dUrd,[9] 5-isopropyl-dUrd (E. De Clercq and D. Shugar : unpublished observations, 1980), 5-methoxymethyl-dUrd,[10] 5-
methylthiomethyl-dUrd,[3] 5-methylsulfonylmethyl-dUrd,[2] 5-azidomethyl-
dUrd,[3,11] 5-methylamino-dUrd,[12] 5-propynyloxy-dUrd[6] and 5-cyanome-
thoxy-dUrd.[13] The selectivity of these compounds is primarily based
upon a specific phosphorylation by the virus-induced dThd kinase.
How they finally act is not fully understood. After they have been
converted to the 5'-mono-, 5'-di- and 5'-triphosphate, the dUrd
analogues may interfere with a variety of enzymes, i.e. dThd kinase,

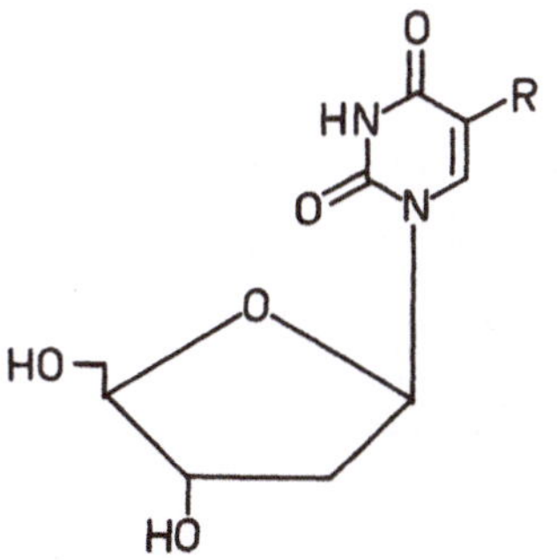

R = −CH$_2$SH   5-mercaptomethyl-dUrd
     −CH$_2$CN   5-cyanomethyl-dUrd
     −CH$_2$SCN   5-thiocyanomethyl-dUrd
     −CH$_2$SOCH$_3$ 5-methylsulfinylmethyl-dUrd
     −CH$_2$SCH$_2$CH$_3$ 5-ethylmercaptomethyl-dUrd
     −CH$_2$N(CH$_3$)$_2$ 5-dimethylaminomethyl-dUrd
     −CH$_2$NHCH$_2$C$_6$H$_5$ 5-benzylaminomethyl-dUrd
     −CH$_2$NHCOCH$_2$I   5-iodoacetamidomethyl-dUrd
     −CH$_2$OCH$_2$CH$_2$CH$_3$ 5-propyloxymethyl-dUrd
     −CH$_2$OCH$_2$C$_6$H$_5$ 5-benzyloxymethyl-dUrd
     −CH$_2$CH$_2$CH$_2$CH$_2$CH$_2$CH$_3$ 5-hexyl-dUrd
     −OCH$_2$C$_6$H$_4$(pNO$_2$) 5-paranitrobenzyloxy-dUrd
     −CH=N-NHCSNH$_2$ 5-thiosemicarbazone of 5-formyl-dUrd

Fig. 1. 5-Substituted 2'-deoxyuridines.
        Structural formulae of those 5-substituted 2'-deoxyuridine
        (dUrd) derivatives that are virtually inactive as antiviral
        agents (ID$_{50}$ for HSV-1 > 100 µg/ml).

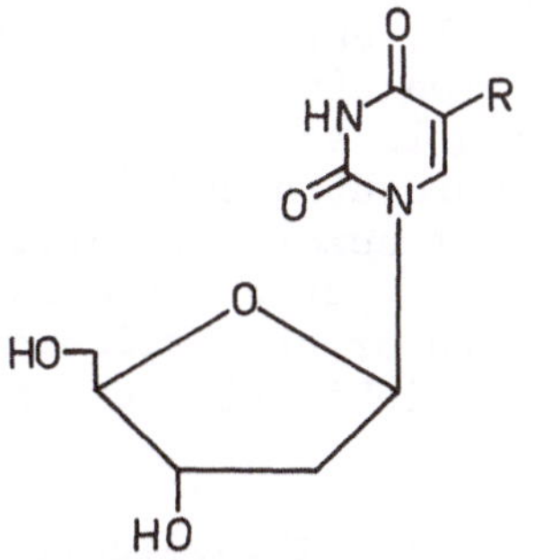

R = −CH$_2$CH$_3$ 5-ethyl-dUrd
     −CH$_2$CH$_2$CH$_3$ 5-propyl-dUrd
     −CH(CH$_3$)$_2$ 5-isopropyl-dUrd
     −CH$_2$-O-CH$_3$ 5-methoxymethyl-dUrd
     −CH$_2$-S-CH$_3$ 5-methylthiomethyl-dUrd
     −CH$_2$-SO$_2$-CH$_3$ 5-methylsulfonylmethyl-dUrd
     −CH$_2$N$_3$ 5-azidomethyl-dUrd
     −NH-CH$_3$ 5-methylamino-dUrd
     −O-CH$_2$C≡CH   5-propynyloxy-dUrd
     −O-CH$_2$C≡N   5-cyanomethoxy-dUrd

Fig. 2. 5-Substituted 2'-deoxyuridines.
        Structural formulae of those 5-substituted 2'-deoxyuridine
        (dUrd) derivatives that are selective but not very potent
        in their antiviral activity (ID$_{50}$ for HSV-1 : ∿ 1 - 10 µg/ml).

dTMP kinase, dTDP kinase, dTMP synthetase, dCMP deaminase, CDP reductase, and DNA polymerase. Some of the dUrd analogues, i.e. 5-
ethyl-dUrd and 5-propyl-dUrd, may also be incorporated into DNA, but
it is not evident that this incorporation is actually required for
their antiviral activity.[14]

   The third class of dUrd analogues includes those compounds that
contain a small, electron-withdrawing C-5 substituent (Fig. 3): i.e.
5-fluoro-dUrd, 5-chloro-dUrd, 5-bromo-dUrd, 5-iodo-dUrd, 5-trifluo-
romethyl-dUrd, 5-ethynyl-dUrd,[15] 5-vinyl-dUrd,[15] 5-(1-chlorovinyl)-
dUrd,[15] 5-formyl-dUrd,[3,16,17] 5-nitro-dUrd[18] and 5-hydroxymethyl-

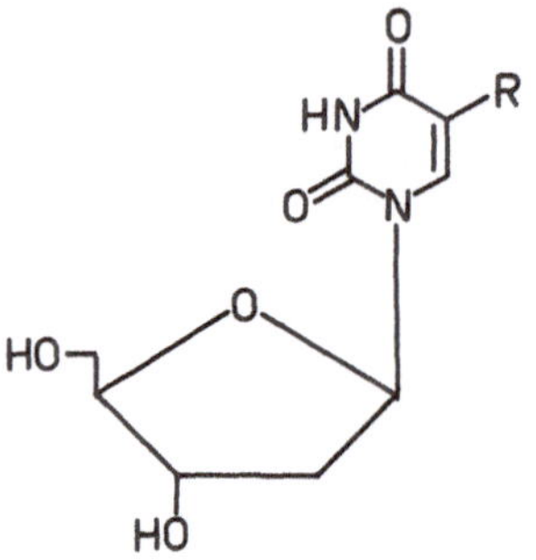

R =	
-F	5-fluoro-dUrd
-Cl	5-chloro-dUrd
-Br	5-bromo-dUrd
-I	5-iodo-dUrd (IDU)
$-CF_3$	5-trifluoromethyl-dUrd (TFT)
-C≡CH	5-ethynyl-dUrd
$-CH{=}CH_2$	5-vinyl-dUrd
$-CCl{=}CH_2$	5-(1-chlorovinyl)-dUrd
-CHO	5-formyl-dUrd
$-NO_2$	5-nitro-dUrd
$-CH_2OH$	5-hydroxymethyl-dUrd

Fig. 3. 5-Substituted 2'-deoxyuridines.
Structural formulae of those 5-substituted 2'-deoxyuridine (dUrd) derivatives that are potent but not very selective in their antiviral activity (ID_{50} for HSV-1 : ∿ 0.1 – 1 µg/ml).

dUrd.[1,11] These compounds are rather potent inhibitors of HSV-1 and HSV-2 replication (ID_{50} : ∿ 0.1 – 1 µg/ml), but they do not display any selectivity in their antiherpes action since they inhibit normal cell metabolism at a concentration which is equal to or even lower than the antiviral concentration. 5-Fluoro-dUrd and its congeners (Fig. 3) are also efficient inhibitors of tumor cell proliferation,[19] and the target for their antitumor action appears to be dTMP synthetase. Both the antiviral and antitumor properties of this class of compounds may be mediated by an inhibition of the dTMP synthetase reaction.

From 5-ethynyl-dUrd several other 5-(1-alkynyl)derivatives have been prepared and these compounds represent the fourth class of the dUrd analogues (Fig. 4), i.e. 5-propynyl-dUrd, 5-butynyl-dUrd, 5-pentynyl-dUrd, 5-hexynyl-dUrd and 5-heptynyl-dUrd.[5] The most potent of this series is 5-ethynyl-dUrd with an ID_{50} of 0.5 and 0.1 µg/ml for HSV-1 and HSV-2, respectively. 5-Propynyl-dUrd is almost as active as 5-ethynyl-dUrd, but, as the alkynyl side chain further increases in length, the antiviral activity wanes. As soon as the side chain exceeds 5 carbon atoms in line, the activity is lost completely. Substitution of an hydroxyl group at the terminal carbon of the side chain also weakens the antiviral potency. All 5-alkynyl-dUrd analogues resemble 5-ethynyl-dUrd in that they inhibit normal cell metabolism at roughly the same concentration as HSV-1 replication. It is likely, therefore, that all 5-alkynyl-dUrd analogues are targeted at dTMP synthetase.

The fifth class of dUrd analogues is composed of the derivatives of 5-vinyl-dUrd (Fig. 5), i.e. (*E*)-5-(2-fluorovinyl)-dUrd,[20] (*E*)-5-(2-chlorovinyl)-dUrd,[21] (*E*)-5-(2-bromovinyl)-dUrd (BVDU),[1,15] (*E*)-5-(2-iodovinyl)-dUrd (IVDU),[15] (*E*)-5-(2-cyanovinyl)-dUrd,[22,23]

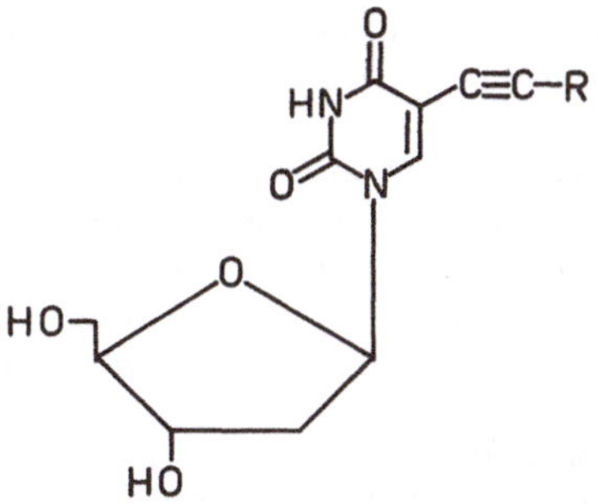

R =	
-H	5-ethynyl-dUrd
$-CH_3$	5-propynyl-dUrd
$-CH_2OH$	5-(3-hydroxypropynyl)-dUrd
$-CH=CH_2$	5-(but-3-ene-1-ynyl)-dUrd
$-CH_2CH_3$	5-butynyl-dUrd
$-CH_2CH_2OH$	5-(4-hydroxybutynyl)-dUrd
$-CH_2CH_2CH_3$	5-pentynyl-dUrd
$-CH_2CH_2CH_2OH$	5-(5-hydroxypentynyl)-dUrd
$-CH_2CH_2CH_2CH_3$	5-hexynyl-dUrd
$-CH_2CH_2CH_2CH_2CH_3$	5-heptynyl-dUrd
$-C_6H_5$	5-phenylethynyl-dUrd

Fig. 4. 5-Substituted 2'-deoxyuridines.
Structural formulae of the 5-(1-alkynyl)-2'-deoxyuridine (dUrd) derivatives.

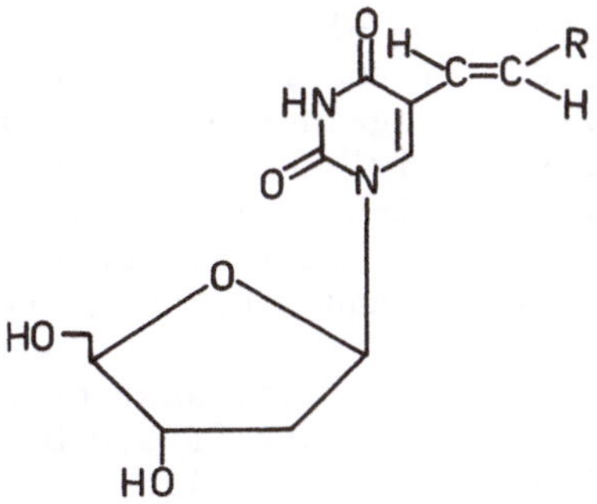

R =	
-H	5-vinyl-dUrd
-F	(E)-5-(2-fluorovinyl)-dUrd
-Cl	(E)-5-(2-chlorovinyl)-dUrd
-Br	(E)-5-(2-bromovinyl)-dUrd (BVDU)
-I	(E)-5-(2-iodovinyl)-dUrd (IVDU)
-CN	(E)-5-(2-cyanovinyl)-dUrd
$-CH_3$	(E)-5-propenyl-dUrd
$-CF_3$	(E)-5-(3,3,3-trifluoropropenyl)-dUrd
$-SCH_3$	(E)-5-(2-methylthiovinyl)-dUrd
$-OCH_2CH_3$	(E)-5-(2-ethoxyvinyl)-dUrd
$-CH_2CH_3$	(E)-5-butenyl-dUrd
$-CH_2CH_2CH_3$	(E)-5-pentenyl-dUrd
$-CH_2CH_2CH_2CN$	(E)-5-(5-cyanopentenyl)-dUrd
$-CH_2CH_2CH_2Cl$	(E)-5-(5-chloropentenyl)-dUrd
$-CH_2CH_2CH_2CH_3$	(E)-5-hexenyl-dUrd
$-C_6H_5$	(E)-5-styryl-dUrd

Fig. 5. 5-Substituted 2'-deoxyuridines.
Structural formulae of the 5-(1-alkenyl)-2'-deoxyuridine (dUrd) derivatives.

(E)-5-propenyl-dUrd,[24] (E)-5-(3,3,3-trifluoropropenyl)-dUrd,[25] (E)-5-(2-methylthiovinyl)-dUrd,[23] (E)-5-(2-ethoxyvinyl)-dUrd,[20] (E)-5-butenyl-dUrd,[23] (E)-5-pentenyl-dUrd,[23] (E)-5-(5-cyanopentenyl)-dUrd, (E)-5-(5-chloropentenyl)-dUrd (E. De Clercq, P. Vincent and L. Pichat: unpublished observations, 1982), (E)-5-hexenyl-dUrd[23] and (E)-5-styryl-dUrd.[26] The most potent of this series are BVDU and IVDU : they inhibit HSV-1 replication at an ID_{50} of 0.007-0.01 µg/ml. Less potent are the propenyl and 3,3,3-trifluoropropenyl derivatives with an ID_{50} of 0.07 µg/ml, the cyanovinyl derivative with an ID_{50} of 4 µg/ml, and the styryl derivative with an ID_{50} of 40 µg/ml, whereas the pentenyl and hexenyl derivatives are devoid of antiviral activity. As has been ascertained for a number of 5-(2-X-vinyl)-dUrd analogues, i.e. bromovinyl-dUrd[27] and 5-propenyl-dUrd,[23] the *E*

("Entgegen") isomer is more active as an inhibitor of HSV-1 than the
Z ("Zusammen") isomer.

From the structure-function analysis of the 5-substituted 2'-
deoxyuridines it appears that the optimum inhibition of HSV-1 replication in cell culture is acquired when the 5-substituent is unsaturated and conjugated with the pyrimidine ring, is not longer than 4
carbon atoms in length, has E stereochemistry, includes a hydrophobic, electronegative function, but does not contain a branching
point.[23] All these structural features are combined in BVDU which is
the most potent of all anti-HSV-1 agents that have been described so
far.

BVDU has already been the subject of several review articles.
[28-31] Its spectrum of activity is confined to herpesviruses : it inhibits the replication of HSV-1, VZV, pseudorabies virus, bovine
herpesvirus and simian varicella virus at a concentration of 0.001 -
0.01 µg/ml. HSV-2 is 100- to 200-fold less susceptible to inhibition
by BVDU, and the differential sensitivity of HSV-1 and HSV-2 to BVDU
could be advocated as a useful marker test for the identification of
clinical HSV isolates. BVDU is also effective against baculovirus
(nuclear polyhedrosis virus), herpesvirus saimiri and EBV, albeit at
concentrations higher than those required for inhibition of HSV-1 or
VZV. BVDU is only slightly active against CMV;[32] it is inactive
against adenovirus, vesicular stomatitis virus and other RNA viruses.

BVDU does not affect normal cell metabolism, i.e. host cell DNA
synthesis, unless it is added at a concentration that is 3,000- to
10,000-fold higher than the concentration required for inhibition of
HSV-1 DNA synthesis.[33,34] The selectivity of BVDU as an antiherpes
agent depends primarily on its phosphorylation by the virus-induced
dThd kinase. This enzyme converts BVDU successively to its 5'-mono-
phosphate (BVDUMP) and 5'-diphosphate (BVDUDP).[35,36] Such phosphorylation would not occur in uninfected cells. Indeed, BVDU has a
much greater affinity for the HSV-1 (and VZV)-induced dThd kinase
than for the cytosol dThd kinase.[37] Moreover, BVDU has a greater affinity for the HSV-1 dThd kinase than for the HSV-2 dThd kinase. In
contrast with HSV-1 dThd kinase, HSV-2 dThd kinase would lack dTMP
kinase activity.[36] This may explain why BVDU is stuck at the 5'-
monophosphate stage in HSV-2-infected cells, whereas it is further
phosphorylated to the 5'-di- and 5'-triphosphate stage in HSV-1-
infected cells. Since BVDUMP is further processed by dTMP synthetase
to a series of putatively inert metabolites,[38] its trapping in
HSV-2-infected cells would result in a more rapid elimination of the
drug, hence contribute to its lower activity against HSV-2.

Upon conversion of BVDU to the 5'-triphosphate, BVDUTP would
interact with the DNA polymerization reaction. BVDUTP inhibits HSV-1
DNA polymerase to a significantly greater extent than the cellular
DNA polymerase α, β and γ, and this preferential inhibition of viral

DNA polymerase further contributes to its selective antiviral action.[39] In addition, BVDUTP can serve as an alternate substrate for both viral and cellular DNA polymerases and hence be incorporated into DNA.[40] Since the initial phosphorylation of BVDU to BVDUMP and BVDUDP is restricted to the virus-infected cell, the eventual incorporation of BVDU into DNA will also be confined to the virus-infected cell. When added at a supraoptimal concentration (30 µM or ∿ 10 µg/ml), BVDU is incorporated into both viral DNA and cellular DNA of the virus-infected cell,[40] but, if added within the optimal concentration range (0.1–0.5 µM or 0.03–0.15 µg/ml), BVDU is incorporated only into viral DNA.[41] The incorporation of BVDU into viral DNA would render the DNA more vulnerable to degradation, and also suppress its template activity for RNA synthesis.[42] Whatever consequences the incorporation of BVDU into viral DNA may have, the extent of this incorporation is closely correlated with the degree of virus yield reduction.[41]

BVDU has proven highly efficacious in the topical and systemic (i.e. subcutaneous, intraperitoneal or oral) treatment of various herpesvirus infections in animal models, i.e. cutaneous and orofacial HSV-1 infections in mice, HSV-1 encephalitis in mice, HSV-1 keratitis (including stromal keratitis) and HSV-1 uveitis (iritis) in rabbits, simian varicella virus infection in monkeys and pseudorabies virus infection in pigs (as reviewed by De Clercq[28-31]). There are also a number of clinical conditions in which BVDU has shown efficacy: i.e. in the topical treatment (as 0.1 % eye drops) of dendritic corneal ulcers, geographic corneal ulcers and deep stromal keratitis, and in the systemic treatment (orally at 7.5–15 mg/kg/day for 5 days) of ophthalmic zoster, disseminated zoster in cancer patients, varicella-zoster in leukemic children and mucocutaneous HSV-1 infections in immunocompromised adults. So far, BVDU has only been submitted to open clinical trials and it is obvious that the promising results obtained in these trials should be confirmed by double-blind controlled studies before the therapeutic efficacy of BVDU could be established unequivocally.

5-SUBSTITUTED 2'-DEOXYCYTIDINES

5-Substituted 2'-deoxycytidines (dCyd) may be expected to exhibit a greater selectivity as antiherpes agents than 5-substituted 2'-deoxyuridines since their processing in the virus-infected cell requires the assistance of at least two herpesvirus-induced enzymes, dThd (dCyd) kinase and dCMP deaminase, as compared with only one virus-induced enzyme, dThd (dCyd) kinase for the dUrd analogues.[43] The dCyd analogues may finally be incorporated into DNA as the dUrd analogue, thereby following the pathway 5-X-dCyd → 5-X-dCMP → 5-X-dUMP → 5-X-dUDP → 5-X-dUTP → DNA. Alternatively, the dCyd analogues may follow the pathway 5-X-dCyd → 5-X-dUrd → 5-X-dUMP → 5-X-dUDP → 5-X-dUTP → DNA.

If, however, special measures are taken to prevent deamination
of 5-X-dCyd or 5-X-dCMP, i.e. by addition of 2'-deoxytetrahydrouri-
dine or alkylation at the N-4 position of the pyrimidine moiety, the
5-X-dCyd analogues may as such be incorporated into DNA,[44] thus following the pathway 5-X-dCyd → 5-X-dCMP → 5-X-dCDP → 5-X-dCTP → DNA.
From an antiviral viewpoint there are some advantages associated
with the latter pathway : (<u>i</u>) 5-X-dCyd is not catabolyzed by dUrd
(dThd) phosphorylase, as are the dUrd analogues; (<u>ii</u>) 5-X-dCMP is
not dehalogenated by thymidylate synthetase, as is, for example,
IDUMP; (<u>iii</u>) 5-X-dCMP does not inhibit thymidylate synthetase, hence
cause cytotoxicity, as do several dUMP analogues, i.e. 5-nitro-dUMP
and 5-ethynyl-dUMP; (<u>iv</u>) 5-X-dCTP does not inhibit host nucleoside
diphosphate reductase, as do some dUTP analogues, i.e. (<u>E</u>)-5-prope-
nyl-dUTP; and finally (<u>v</u>), once incorporated, 5-X-dCyd may affect
the methylation of DNA which appears to be an important part of the
maturation process of herpesviruses.[44]

Several 5-substituted dCyd analogues (Fig. 6) have been evaluated for their antiviral effects : i.e. 5-bromo-dCyd and 5-iodo-
dCyd,[45,46] 5-vinyl-dCyd,[47] 5-nitro-dCyd,[43] 5-ethynyl-dCyd,[43] (<u>E</u>)-5-
(2-bromovinyl)-dCyd,[43] (<u>E</u>)-5-(2-iodovinyl)-dCyd,[43] 5-hydroxymethyl-
dCyd[1] and (<u>E</u>)-5-propenyl-dCyd.[23] Of these dCyd analogues, (<u>E</u>)-5-(2-
bromovinyl)-dCyd and (<u>E</u>)-5-(2-iodovinyl)-dCyd have emerged as the
most potent and most selective inhibitors of HSV-1 replication
(ID$_{50}$: 0.07 µg/ml). As a rule, 5-substituted dCyd analogues are
equally potent, or only slightly less potent, but markedly more selective in their antiherpes activity than the corresponding 5-sub-
stituted dUrd analogues. This is exemplified by 5-ethynyl-dCyd <u>vs</u>
5-ethynyl-dUrd, 5-nitro-dCyd <u>vs</u> 5-nitro-dUrd, 5-vinyl-dCyd <u>vs</u> 5-
vinyl-dUrd, and 5-hydroxymethyl-dCyd <u>vs</u> 5-hydroxymethyl-dUrd.

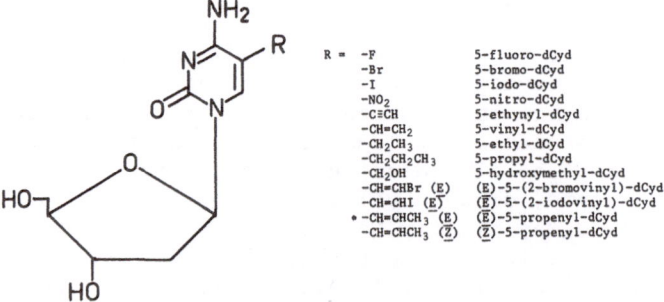

Fig. 6. 5-Substituted 2'-deoxycytidines.
Structural formulae.

None of the dCyd analogues has been the subject of thorough in
vivo studies. Since there are several reasons to believe that they
may have an heightened efficacy over that of the dUrd analogues
(see supra), the dCyd analogues, and in particular the (E)-5-(2-
halogenovinyl)-dCyd derivatives, should be further pursued for their
therapeutic potentials in the treatment of HSV and VZV infections.

5-SUBSTITUTED ARABINOFURANOSYLURACIL DERIVATIVES

The antiviral activity spectrum of the 5-substituted arabinofuranosyluracil (ara-U) derivatives is, like that of their dUrd
counterparts, primarily confined to herpesviruses, i.e. HSV-1, HSV-2
and VZV. Several ara-U analogues, i.e. 5-nitro-ara-U,[48] 5-cyano-
ara-U,[48] 5-hydroxy-ara-U,[48] 5-butyl-ara-U,[49] 5-methoxymethyl-ara-U,
[49] 5-butynyl-ara-U[5] and 5-(3,3,-dimethylbutynyl)-ara-U,[5] are virtually devoid of antiviral activity, whereas other ara-U analogues,
i.e. 5-fluoro-ara-U,[50] 5-iodo-ara-U,[50] 5-formyl-ara-U,[50] 5-propynyl-
ara-U,[5] 5-propenyl-ara-U,[51,52] 5-butenyl-ara-U[52] and propynyloxy-
ara-U[48] inhibit HSV-1 replication at a concentration of 10 - 30 µg/
ml (Fig. 7).

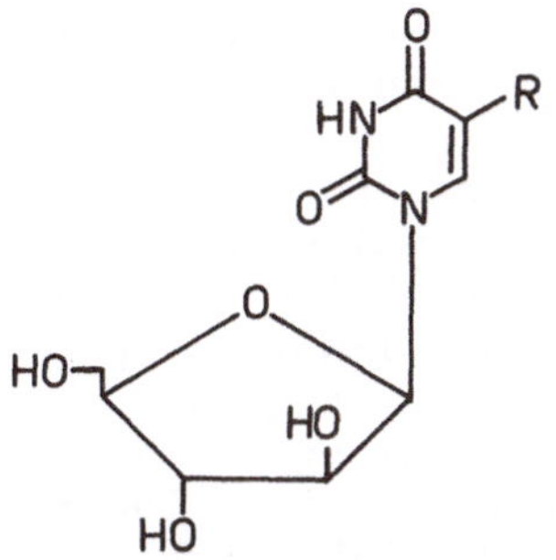

R =	-H	ara-U
	-F	5-fluoro-ara-U
	-I	5-iodo-ara-U
	-OH	5-hydroxy-ara-U
	-CN	5-cyano-ara-U
	-NO$_2$	5-nitro-ara-U
	-CHO	5-formyl-ara-U
	-CH$_2$OCH$_3$	5-methoxymethyl-ara-U
	-CH$_2$CH$_2$CH$_3$	5-propyl-ara-U
	-CH$_2$CH$_2$CH$_2$CH$_3$	5-butyl-ara-U
	-CH=CHCH$_3$	5-propenyl-ara-U
	-CH=CHCH$_2$CH$_3$	5-butenyl-ara-U
	-C≡CCH$_3$	5-propynyl-ara-U
	-C≡CCH$_2$CH$_3$	5-butynyl-ara-U
	-C≡CC(CH$_3$)$_3$	5-(3,3-dimethylbutynyl)-ara-U
	-OCH$_2$C≡CH	5-propynyloxy-ara-U

Fig. 7. 5-Substituted arabinofuranosyluracil derivatives.
Structural formulae of those 5-substituted 1-β-D-arabinofu-
ranosyluracil (Ara-U) derivatives that are either inactive
or weakly active as antiviral agents (ID$_{50}$ for HSV-1 : > 10
µg/ml).

There are, however, a number of ara-U analogues which inhibit
HSV-1 replication at a concentration significantly lower than 10 µg/
ml (Fig. 8). The most potent of this series are (E)-5-(2-bromovinyl)-
ara-U (BVaraU)[53-55] and (E)-5-(2-chlorovinyl)-ara-U[53] with an ID$_{50}$
for HSV-1 of 0.1 µg/ml. Then follow 5-vinyl-ara-U,[52] 5-methyl-ara-U
(ara-T),[56,57] 5-trifluoromethyl-ara-U,[58] 5-ethyl-ara-U[49,59] and 5-
ethenyl-ara-U.[5]

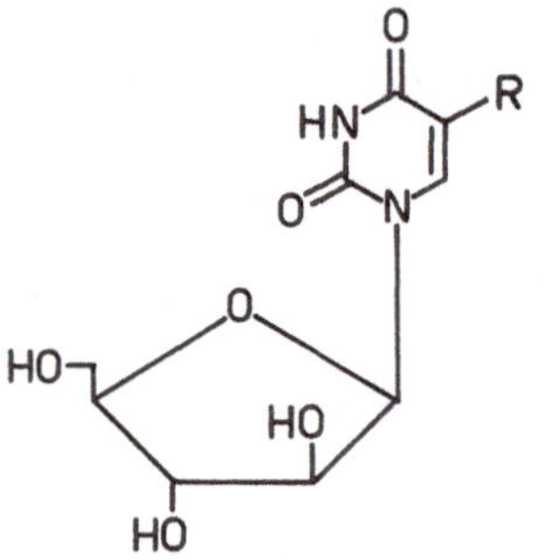

R = $-CH_3$ 5-methyl-ara-U (ara-T)
     $-CF_3$ 5-trifluoromethyl-ara-U
     $-CH_2CH_3$ 5-ethyl-ara-U
     $-CH=CH_2$ 5-vinyl-ara-U
     $-C\equiv CH$ 5-ethynyl-ara-U
     $-CH=CHCl$ (E)   (E)-5-(2-chlorovinyl)-ara-U
     $-CH=CHBr$ (E)   (E)-5-(2-bromovinyl)-ara-U (BVaraU)

Fig. 8. 5-Substituted arabinofuranosyluracil derivatives.
Structural formulae of those 5-substituted 1-β-D-arabinofu-
ranosyluracil (ara-U) derivatives that show rather potent
antiviral activity (ID_{50} for HSV-1 : < 10 μg/ml).

As has been noted for the dUrd analogues, the antiviral potency
of the 5-substituted ara-U analogues critically depends on the
length of the C-5 side chain. Irrespective of the nature of the C-5
side chain (alkyl, alkenyl or alkynyl), the antiviral activity decreases with increasing number of carbon atoms; thus, in order of
(decreasing) activity : ara-T > 5-ethyl-ara-U > 5-propyl-ara-U > 5-
butyl-ara-U; 5-vinyl-ara-U > 5-propenyl-ara-U > 5-butenyl-ara-U;
and 5-ethynyl-ara-U > 5-propynyl-ara-U > 5-butynyl-ara-U.

As a rule, the 5-substituted ara-U analogues are less active
than the corresponding 5-substituted dUrd analogues, i.e. 5-ethyl-
ara-U < 5-ethyl-dUrd, 5-propyl-ara-U < 5-propyl-dUrd, 5-nitro-ara-U
< 5-nitro-dUrd, 5-propynyl-ara-U < 5-propynyl-dUrd, 5-propenyl-ara-U
< 5-propenyl-dUrd, etc. The only exception to the rule is ara-T
which is markedly more active as an antiviral agent than its dUrd
counterpart, dThd. Also, BVaraU is slightly more active against VZV
than the corresponding dUrd analogue, BVDU.[60,61] With an ID_{50} of
0.001 μg/ml, BVaraU is the most potent anti-VZV agent described so
far.[61] On the other hand, BVaraU is not inhibitory for HSV-2 up to
a concentration of approximately 20 μg/ml, that is 200-fold higher
than its ID_{50} for HSV-1.

In contrast with BVaraU, ara-T is equally effective against
HSV-1, HSV-2 and VZV (ID_{50} : ∿ 0.25-0.5 μg/ml).[57,62] Ara-T is also
effective against EBV,[63] but inactive against CMV.[62]

Like many other antiherpes agents, ara-T would owe its selective antiviral activity to a specific phosphorylation by the virus-
induced dThd kinase.[57] Ara-T may also be phosphorylated in uninfected cells, albeit less efficiently than in HSV-1 infected cells.[64]
After it has been phosphorylated to the 5'-triphosphate (ara-TTP),
ara-T may interfere with the DNA polymerization reaction. Indeed,

ara-TTP is a competitive inhibitor of dTTP for several DNA polymerases.[65,66] In its inhibitory effect on DNA polymerases, ara-TTP
shows no preference for HSV DNA polymerase. Hence, its selectivity
as an antiherpes agent would reside solely at the dThd kinase level.

Other ara-U analogues, i.e. BVaraU, may act in a similar fashion
as ara-T. Like ara-T, BVaraU is a preferential substrate for the viral dThd kinase.[67] In its 5'-triphosphate form, BVaraU competes with
dTTP for both viral and cellular DNA polymerases, and, unlike BVDU,
it is not internally incorporated into DNA of virus-infected cells.[55]

There is only scarce information on the efficacy of the ara-U
analogues in animal models. Ara-T has proven to be effective in the
systemic treatment of HSV-1 encephalitis in mice[68] but other evidence of its in vivo efficacy in the therapy of HSV infections has not
been provided.

5-SUBSTITUTED ARABINOFURANOSYLCYTOSINE DERIVATIVES

Ara-C (cytosine arabinoside) is a potent antiherpes agent
(ID$_{50}$ for HSV-1 : 0.04 µg/ml). However, ara-C is far from specific
in its antiviral action, since it inhibits normal cell metabolism at
a concentration slightly lower than the minimum antiviral concentration. The mechanism of action of ara-C is based on an inhibition of
the DNA polymerase reaction by ara-CTP (ara-C 5'-triphosphate), which
is generated from ara-C in both virus-infected and uninfected cells.
In its inhibitory effects on DNA polymerase, ara-CTP does not dis-
criminate between viral and cellular DNA polymerases. It simply inhibits viral and cellular DNA synthesis to the same extent.

Possibly, ara-C may acquire a greater selectivity in its antiviral action upon introduction of the appropriate substituent in
the C-5 position. (E)-5-(2-halogenovinyl)-derivatives of ara-C have
not been synthesized yet, but some other 5-substituted ara-C analogues have been prepared (Fig. 9) and evaluated for their antiviral

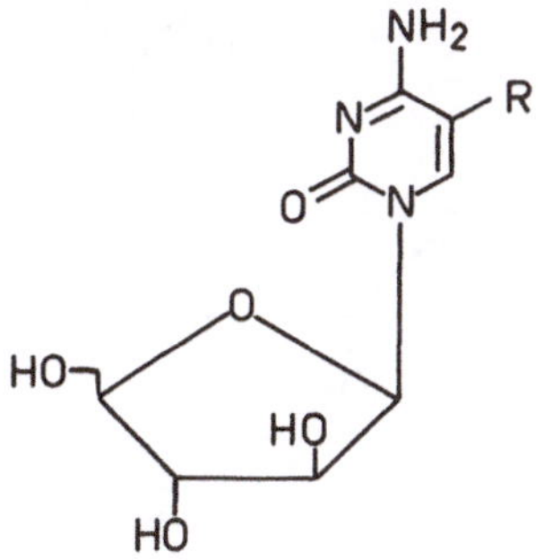

R =		
-H		ara-C
-F		5-fluoro-ara-C
-CH$_3$		5-methyl-ara-C
-CH$_2$CH$_3$		5-ethyl-ara-C
-CH$_2$CH$_2$CH$_3$		5-propyl-ara-C
-CH(CH$_3$)$_2$		5-isopropyl-ara-C

Fig. 9. 5-Substituted arabinofuranosylcytosine derivatives.
Structural formulae of 1-β-D-arabinofuranosylcytosine
(Ara-C) and 5-substituted ara-C analogues.

properties, i.e. 5-fluoro-ara-C,[69] 5-methyl-ara-C,[70] 5-ethyl-
ara-C,[59] 5-propyl-ara-C and 5-isopropyl-ara-C (E. De Clercq, J. Descamps, J. Balzarini and D. Shugar: unpublished observations, 1980).
5-Fluoro-ara-C is just as toxic as ara-C and the 5-alkyl-ara-C analogues are either inactive (5-propyl-ara-C, 5-isopropyl-ara-C) or
only slightly active (5-methyl-ara-C, 5-ethyl-ara-C). Yet, 5-methyl-
ara-C and 5-ethyl-ara-C can be considered as "prodrugs" of ara-T
and 5-ethyl-ara-U, respectively. To the extent that they act as
substrates for dCyd deaminase, 5-methyl-ara-C and 5-ethyl-ara-C
should be readily converted to the corresponding ara-U analogues,
at least in those cells and tissues which contain high levels of
dCyd deaminase.

3'- AND 5'-AMINO DERIVATIVES OF 5-SUBSTITUTED 2'-DEOXYURIDINES

In attempts to increase the potency or selectivity of the 5-
substituted 2'-deoxyuridines, or to enlarge their activity spectrum,
several 3'- and 5'-amino derivatives have been prepared : i.e.
5'-amino-5-iodo-2',5'-dideoxyuridine (AIU) and 3'-amino-(E)-5-(2-
bromovinyl)-2',3'-dideoxyuridine (BVADU) (Fig. 10).

Upon replacement of the 5'-hydroxyl moiety of IDU by an amino
group, both the antiviral potency and cytotoxicity are decreased, but
the toxicity is decreased to a greater extent than the antiviral potency, so that the therapeutic index is actually improved.[71,72]
Similarly, 5-bromo-dUrd, 5-chloro-dUrd, 5-trifluoromethyl-dUrd and
dThd itself gain in selectivity but lose in potency upon substitution of 5'-amino for 5'-hydroxyl.[72,73] If, in addition to the 5'-
hydroxyl, the 3'-hydroxyl group is also replaced by an amino group,
as in 3',5'-diamino-3',5'-dideoxythymidine, complete lack of antiviral activity ensues.[74]

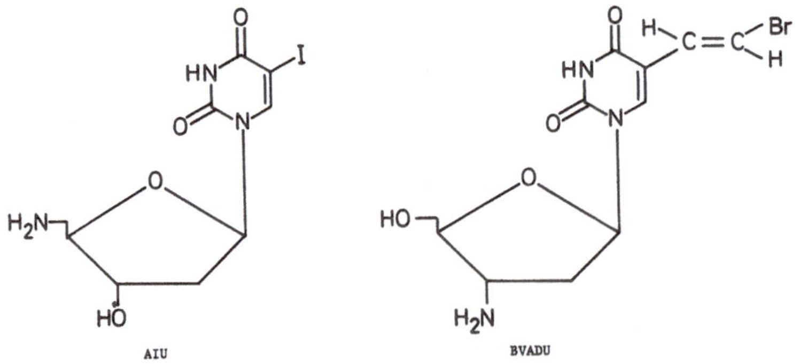

Fig. 10. 3'- and 5'-amino derivatives of 5-substituted 2'-deoxy-
uridines.
Structural formulae of 5'-amino-5-iodo-2',5'-dideoxyuridine
(AIU) and 3'-amino-(E)-5-(2-bromovinyl)-2',3'-dideoxyuri-
dine (BVADU).

With BVDU as the parent compound, little, if any, antiviral activity is retained upon substitution of an amino group at the C'-5 position.[75] However, the 3'-amino derivative of BVDU retains about 1/10 of both the antiherpes potency and cytotoxicity of the parent compound, thus achieving a similar therapeutic index.[53,76]

AIU and BVADU exhibit the same antiviral activity spectrum as IDU and BVDU. This includes HSV-1, HSV-2 and VZV.[53,71,76,77] Yet, BVADU is about 200-fold more potent than AIU (ID_{50} for HSV-1 : 0.1 and 20 μg/ml, respectively) and both BVADU and AIU are markedly less inhibitory for HSV-2 than for HSV-1.

The selectivity of AIU and BVADU as antiviral agents primarily depends on their phosphorylation by the virus-induced dThd kinase. This phosphorylation would restrict the further action of the compounds to the virus-infected cell. As has been demonstrated with AIU,[78] the virus-induced dThd kinase would convert its substrate successively to the 5'-mono- and 5'-diphosphate. Upon further conversion to the 5'-triphosphate by a cellular kinase, AIU may be incorporated internally into DNA. This incorporation would obviously be restricted to DNA of the virus-infected cell,[79] but, within the infected cell, AIU is incorporated into both viral and cellular DNA. This incorporation leads to extensive DNA breakage because of the formation of the labile phosphoramidate (P-N) bonds.[80] Although the incorporation of AIU into viral DNA may be held responsible for its antiviral activity, the precise mechanism by which this antiviral activity is achieved, has not been elucidated. Also subject to further study is the question whether BVADU, like AIU, is incorporated into DNA.

AIU has proven to be as effective as IDU in the treatment of herpetic keratitis, but only if applied at a 20- or 60-fold higher dose (as a 10 % or 30 % eye ointment).[81] In other animal models,[82] no antiviral activity could be demonstrated with either topical or systemic AIU, while IDU was clearly effective when used at similar treatment regimens. AIU is apparently too weak an antiviral agent (minimum inhibitory concentration : ∿ 20 μg/ml)[69] to be useful as an antiviral drug. Whether BVADU offers greater promise for the treatment of herpesvirus infections remains to be evaluated.

2'-HALOGENO-ARABINOFURANOSYLPYRIMIDINES

Several 5-substituted 2'-halogeno-arabinofuranosylcytosines and -uracils exhibit a highly potent and selective antiherpes activity, the most potent being FIAC (2'-fluoro-5-iodo-ara-C) and FMAU (2'-fluoro-5-methyl-ara-U) (Fig. 11). Their ID_{50} for HSV-1, HSV-2 and VZV is about 0.01 μM (∿ 0.003 μg/ml),[83-85] and for CMV it is 0.1-0.3 μM (∿ 0.03-0.1 μg/ml).[32] The inhibitory activity of FIAC and FMAU against CMV is the more interesting since no other compounds have been found effective against CMV at such low concentrations.

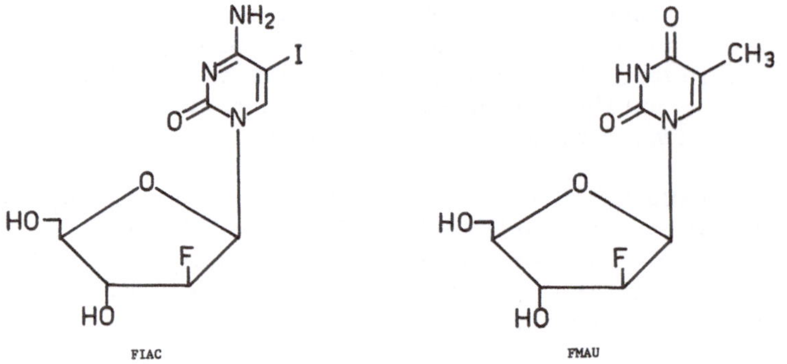

Fig. 11. 2'-Halogeno-arabinofuranosylpyrimidines.
Structural formulae of 1-(2-deoxy-2-fluoro-β-D-arabinofura-
nosyl)-5-iodocytosine (2'-fluoro-5-iodo-ara-C, FIAC) and
1-(2-deoxy-2-fluoro-β-D-arabinofuranosyl)-5-methyluracil
(2'-fluoro-5-methyl-ara-U, FMAU).

Of equal interest is the inhibitory effect of FMAU on the growth of
murine and human leukemia cells, including leukemic cell lines resistant to ara-C (1-β-D-arabinofuranosylcytosine).[86,87]

The fact that FIAC is at least 1000 times more effective in inhibiting HSV replication than its isomer with the fluorine in the
"down" (ribo) configuration demonstrates clearly that the 2'-fluoro
substituent in the "up" (arabino) configuration is essential for the
antiherpes activity of FIAC.[85] Furthermore, those arabinonucleosides
that contain a 2'-fluoro substituent exhibit, in general, more potent antiherpes activity than do the corresponding 2'-chloro or 2'-
bromo congeners. However, two 2'-chloro derivatives, i.e. 2'-chloro-
5-iodo-ara-C and 2'-chloro-5-methyl-ara-C, are significantly more
effective against HSV-2 than against HSV-1, and in this sense they
may serve as useful probes to distinguish between HSV types 1 and
2 in clinical isolates.[85]

Akin to several other pyrimidine nucleoside analogues, FIAC
would owe its selective antiherpes activity to a specific phosphorylation by the virus-induced dThd (dCyd) kinase.[84,88] Indeed, FIAC
and the other 2'-fluoro-arabinosides have a greater affinity for the
viral dThd kinase than for the cellular (cytosol or mitochondrial)
dThd kinase.[67] Once it has been converted to its 5'-triphosphate,
FIAC may interfere with viral DNA polymerase,[89] and also be incorporated into DNA, either internally or at the 3'-end.[90] At the nucleoside or nucleotide level, FIAC may be subject to deamination, de-
iodination and methylation, and thereby give rise to a number of
metabolites, i.e. FIAU (2'-fluoro-5-iodo-ara-U), FAU (2'-fluoro-
ara-U) and FMAU.[91] The 5'-triphosphates of these metabolites may

also interfere with DNA polymerase and eventually be incorporated
into DNA. One of the consequences of the incorporation of FIAC (or
its metabolites) into DNA would be an increased susceptibility of
the DNA to degradation by DNases.[90] Since the initial phosphorylation of FIAC is confined essentially to the herpesvirus-infected
cell, it will affect primarily viral DNA synthesis, and, as has been
noted with other selective antiherpes agents (such as BVDU), FIAC
inhibits viral DNA synthesis at a concentration that is lower by 3-4
orders of magnitude than the concentration required to suppress cellular DNA synthesis in uninfected cells.[33]

FIAC and FMAU (even more so than FIAC) have proven highly efficacious in the systemic treatment of experimental HSV-1 and HSV-2
infections, i.e. encephalitis, in mice (J.J. Fox : personal communication, 1983). FIAC would also be superior to ara-A in the systemic
(intravenously at 400 mg/m^2/day for 5 days) therapy of severe VZV
infections in immunocompromised patients.

PYRIMIDINE CARBOCYCLIC NUCLEOSIDE ANALOGUES

Within the class of the pyrimidine carbocyclic nucleoside analogues, distinction has to be made between the carbocyclic analogues
of the ribonucleosides such as C-cytidine (C-Cyd) (Fig. 12) and
C-uridine, and the carbocyclic analogues of deoxyribonucleosides,
such as C-2'-deoxyuridine, C-3'-deoxyuridine and 5-substituted C-2'-
deoxyuridines, i.e. 5-iodo-C-2'-deoxyuridine (C-IDU) (Fig. 12).

C-cytidine, also referred to as carbodine, is as potent an inhibitor of influenza virus in vitro as ribavirin; however, it is
only marginally active against HSV-1, vaccinia virus and rhinovirus

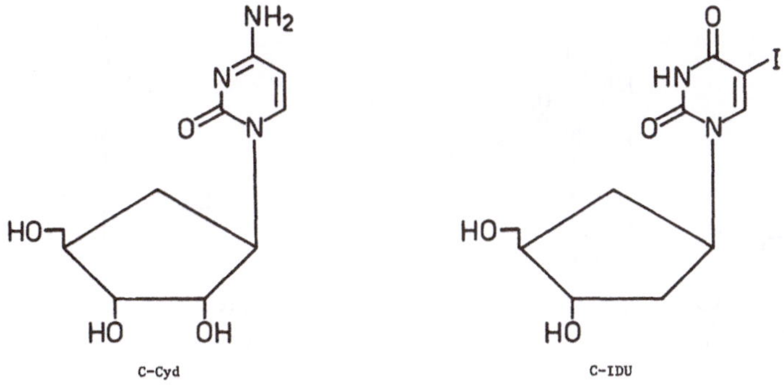

Fig. 12. Pyrimidine carbocyclic nucleoside analogues.
         Structural formulae of carbocyclic cytidine (C-Cyd) and
         carbocyclic 5-iodo-2'-deoxyuridine (C-IDU).

1A.[92] In vivo, C-Cyd does not exhibit any therapeutic activity in the treatment of lethal influenza virus infections in mice, probably because it is rapidly deaminated to its inactive metabolite C-uridine.

The mechanism of antiviral action of C-Cyd has not been established. Inside the cell, C-Cyd is phosphorylated to the 5'-triphosphate (C-CMP), and, in this form, it might interfere with the viral RNA transcription process. C-CMP may also inhibit the conversion of UTP to CTP by CTP synthetase, and, although such inhibitory effect would not produce any selective antiviral activity, it may be responsible for the cytotoxic and antitumor action of C-Cyd.

The 5-substituted C-2'-deoxyuridines (i.e. C-IDU) specifically inhibit HSV-1 and, to a lesser extent, HSV-2 replication.[93] The potency of C-IDU against HSV-1 is comparable to that of IDU itself. The 5-substituted C-2'-deoxyuridines are inactive against dThd kinase-deficient HSV-1, which indicates that they need to be activated by the virus-induced dThd kinase just as the true 5-substituted 2'-deoxyuridines are known to be. A plausible advantage of the carbocyclic dUrd derivatives may lie in their resistance to phosphorolytic cleavage by dUrd (dThd) phosphorylase. As a consequence, they may persist for a much longer time in the body, hence achieve a more pronounced or prolonged antiviral activity.

AZAPYRIMIDINE NUCLEOSIDE ANALOGUES

There are only two azapyrimidine nucleoside analogues which are accredited with antiviral properties. These are 6-azauridine (Aza[6]Urd) and 5,6-dihydro-5-azathymidine (Aza[5]h$_2$dThd, DHAdT) (Fig. 13). Aza[6]Urd has a broad-spectrum antiviral activity encompassing both DNA and RNA viruses, whereas the activity spectrum of DHAdT is confined to herpesviruses.

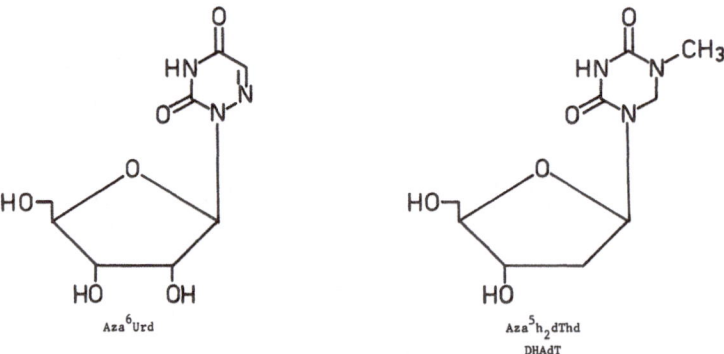

Fig. 13. Azapyrimidine nucleoside analogues.
         Structural formulae of 6-azauridine (Aza[6]Urd) and 5,6-dihydro-5-azathymidine (Aza[5]h2dThd, DHAdT).

The mechanism of action of aza⁶Urd would be based upon an inhibition of orotidylic acid decarboxylase, the enzyme that converts OMP to UMP, an important step in the de novo biosynthesis of pyrimidine nucleotides. To act as an inhibitor of OMP decarboxylase, aza⁶Urd must first be phosphorylated to its 5'-monophosphate by Urd kinase. This phosphorylation would occur readily in both virus-infected and uninfected cells. However, to the extent that virus-infected cells show an increased Urd kinase activity as compared to normal uninfected cells,[94] one may expect some selectivity in the antiviral action of aza⁶Urd.

Whether aza⁶Urd holds any promise as an antiviral agent is not clear. Its antiviral spectrum[95] is rather bizarre : it comprises both (−)RNA viruses (i.e. parainfluenza) and (+)RNA viruses (i.e. encephalomyocarditis, Venezuelan equine encephalitis), whereas related (−)RNA viruses (i.e. Newcastle disease) and (+)RNA viruses (i.e. mengo, Western equine encephalitis) are not included. Further work seems to be required to assess the true antiviral potentials of aza⁶Urd. Attempts should also be directed at the development of aza⁶Urd analogues, i.e. 5-substituted 6-azauridines.

As could be expected from a deoxynucleoside analogue, DHAdT is only effective against DNA viruses. It is particularly active against HSV-1, HSV-2 and VZV. It is only slightly inhibitory for vaccinia virus and not inhibitory for pseudorabies virus.[96] The antiherpes activity of DHAdT also extends to experimental HSV-1 infections in mice.[97,98] In its antiviral potency DHAdT is comparable to vidarabine (Ara-A, adenine arabinoside, 9-β-D-arabinofuranosyladenine). The structural features that are important for the antiviral activity of DHAdT have not been determined. Nor has its mechanism of action been resolved.

DEAZAPYRIMIDINE NUCLEOSIDE ANALOGUES

The pyrimidine (3-deazapyrimidine) nucleoside analogues, 3-deazauridine (c³Urd) and 3-deazacytidine (c³Cyd) (Fig. 14) have been pursued primarily as antitumor agents. However, these compounds possess significant antiviral activity against several RNA viruses, including rhino 1A, 13 and 56, coxsackie A21, influenza APR8, parainfluenza 1 and vesicular stomatitis virus.[99,100] 3-Deazauridine is also active against Gross murine leukemia virus in vitro but does not show marked activity against Rauscher murine leukemia virus in vivo.

The degree of selectivity and the mode of action of c³Urd (and c³Cyd) remain to be elucidated. Inside the cell c³Urd is readily metabolized to its 5'-triphosphate, and in this form it interacts with CTP synthetase, the enzyme that converts UTP to CTP.[103] While CTP synthetase may represent an attractive target in the design of antitumor agents, it seems less useful from an antiviral viewpoint,

since it does not guarantee any selectivity in the antiviral activity.

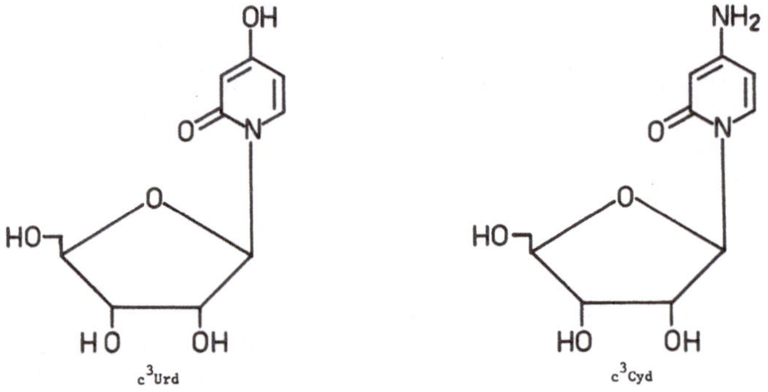

Fig. 14. Deazapyrimidine nucleoside analogues.
Structural formulae of 3-deazauridine (c^3Urd) and 3-deaza-
cytidine (c^3Cyd).

PYRIMIDINE ACYCLIC NUCLEOSIDE ANALOGUES

Based on the broad-spectrum antiviral activity of (S)-DHPA
(9-((S)-2,3-dihydroxypropyl)adenine),[104] the selective and potent
antiherpes activity of BVDU and ACV (acyclovir, 9-(2-(hydroxyethoxy)
methyl)guanine),[105,106] and the selective antiherpes activity of
AIU, a series of pyrimidine acyclic nucleosides has been synthesized, all sharing common structural features with either (S)-DHPA,
BVDU, ACV or AIU.

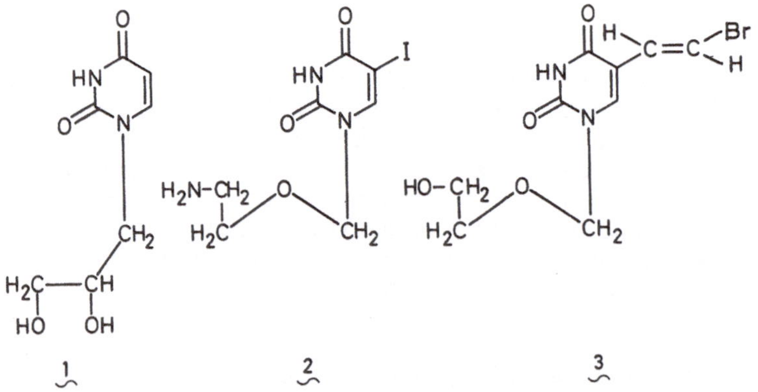

Fig. 15. Pyrimidine acyclic nucleoside analogues.
Structural formulae of 1-((S)-2,3-dihydroxypropyl)uracil
(1), 1-((2-aminoethoxy)methyl)-5-iodouracil (2) and 1-((2-
hydroxyethoxy)methyl)-(E)-5-(2-bromovinyl)uracil (3).

The pyrimidine acyclic nucleosides are either 2,3-dihydroxypro-pyl derivatives,[107] (2-hydroxyethoxy)methyl derivatives[108,109] or (2-aminoethoxy)methyl derivatives[110] of uracil, thymine, cytosine and 5-substituted uracils, including (E)-5-(2-bromovinyl)uracil (E. De Clercq and M.J. Robins: unpublished observations, 1983) (Fig. 15).

None of these acyclic nucleosides exhibits significant antivi-ral activity against HSV or any other DNA virus or RNA virus. In contrast with BVDU, ACV and AIU, which are excellent substrates for the HSV-1 encoded dThd kinase, the 1-(2-hydroxyethoxy)methyl- and 1-(2-aminoethoxy)methyl derivatives have little, if any, affinity for the viral dThd kinase,[109,110] and this lack of substrate acti-vity suffices to explain the absence of antiherpes activity with the pyrimidine acyclic nucleosides.

CONCLUSION

The pyrimidine nucleoside analogues which have been evaluated for their antiviral properties belong to the following classes : 5-substituted dUrd, dCyd, ara-U or ara-C analogues, 3'- or 5'-amino-5-substituted dUrd analogues, 2'-halogeno-5-substituted ara-U and ara-C analogues, pyrimidine carbocyclic and acyclic nucleoside ana-logues, and azapyrimidine and deazapyrimidine (pyridine) nucleoside analogues.

From these pyrimidine nucleoside analogues, several compounds emerged as potent and selective inhibitors of HSV-1 and VZV, viz. (E)-5-(2-bromovinyl)-dUrd (BVDU), (E)-5-(2-iodovinyl)-dUrd (IVDU), and their dCyd counterparts, (E)-5-(2-bromovinyl)-ara-U (BVaraU), the 3'-amino derivative of (E)-5-(2-bromovinyl)-dUrd (BVADU), 2'-fluoro-5-methyl-ara-U (FMAU) and 2'-fluoro-5-iodo-ara-C (FIAC). FIAC and FMAU are also effective inhibitors of HSV-2 and CMV. The prime reason for the selectivity (against HSV and VZV) of all these com-pounds reside in a specific phosphorylation by the virus-induced dThd kinase. The target for their antiviral action would be viral DNA synthesis. After they have been converted to the 5'-triphosphate, BVDU, IVDU, BVaraU, BVADU, FMAU and FIAC either interfere with viral DNA polymerase or are incorporated into viral DNA, thereby pertur-bing the normal functions of the viral DNA.

The carbocyclic and acyclic nucleoside analogues, and the aza- and deazapyrimidine nucleoside analogues, have not been pursued to the same extent as the dUrd, dCyd, ara-U and ara-C analogues. Yet, these compounds may have interesting potentials as antiviral agents, i.e. the carbocyclic nucleosides because they are less apt to degra-dation, and the aza- and deazapyrimidine nucleosides because they may act at targets other than viral DNA, hence interfere with a different range of viruses (other than the herpesviruses).

REFERENCES

1. J. Reefschläger, D. Bärwolff, P. Engelmann, P. Langen and H.A.
   Rosenthal, Efficiency and selectivity of (E)-5-(2-bromovinyl)-
   2'-deoxyuridine and some other 5-substituted 2'-deoxypyrimi-
   dine nucleosides as anti-herpes agents, Antiviral Research
   2:41 (1982).
2. C.L. Schmidt, C.T.-C. Chang, E. De Clercq, J. Descamps and M.P.
   Mertes, Synthesis of 5-[(methylthio)methyl]-2'-deoxyuridine,
   the corresponding sulfoxide and sulfone, and their 5'-phos-
   phates: antiviral effects and thymidylate synthetase inhibition, J. Med. Chem. 23:252 (1980).
3. E. De Clercq, J. Descamps, C.L. Schmidt and M.P. Mertes, Antiviral activity of 5-methylthiomethyl-2'-deoxyuridine and other
   5-substituted 2'-deoxyuridines, Biochem. Pharmacol. 28:3249
   (1979).
4. J.B. Meldrum, V.S. Gupta and J.R. Saunders, Cell culture studies
   on the antiviral activity of ether derivatives of 5-hydroxy-
   methyldeoxyuridine, Antimicrob. Agents Chemother. 6:393 (1974).
5. E. De Clercq, J. Descamps, J. Balzarini, J. Giziewicz, P.J. Barr
   and M.J. Robins, Nucleic acid related compounds. 42. Synthesis
   and biological activities of 5-alkynyluracil nucleosides,
   J. Med. Chem. 26:661 (1983).
6. P.F. Torrence, J.W. Spencer, A.M. Bobst, J. Descamps and E. De
   Clercq, 5-O-alkylated derivatives of 5-hydroxy-2'-deoxyuridine
   as potential antiviral agents. Anti-herpes activity of 5-pro-
   pynyloxy-2'-deoxyuridine, J. Med. Chem. 21:228 (1978).
7. E. De Clercq and D. Shugar, Antiviral activity of 5-ethyl pyrimidine deoxynucleosides, Biochem. Pharmacol. 24:1073 (1975).
8. Y.-C. Cheng, B.A. Domin, R.A. Sharma and M. Bobek, Antiviral action and cellular toxicity of four thymidine analogues : 5-
   ethyl-, 5-vinyl-, 5-propyl-, and 5-allyl-2'-deoxyuridine,
   Antimicrob. Agents Chemother. 10:119 (1976).
9. E. De Clercq, J. Descamps and D. Shugar, 5-Propyl-2'-deoxyuri-
   dine : a specific anti-herpes agent, Antimicrob. Agents Chemother. 13:545 (1978).
10. L.A. Babiuk, B. Meldrum, V.S. Gupta and B.T. Rouse, Comparison
    of the antiviral effects of 5-methoxymethyldeoxyuridine with
    5-iododeoxyuridine, cytosine arabinoside, and adenine arabi-
    noside, Antimicrob. Agents Chemother. 8:643 (1975).
11. G.T. Shiau, R.F. Schinazi, M.S. Chen and W.H. Prusoff, Synthesis
    and biological activities of 5-(hydroxymethyl, azidomethyl, or
    aminomethyl)-2'-deoxyuridine and related 5'-substituted analo-
    gues, J. Med. Chem. 23:127 (1980).
12. T.Y. Shen, J.F. McPherson and B.O. Linn, Nucleosides. III. Stu-
    dies on 5-methylamino-2'-deoxyuridine as a specific antiherpes
    agent, J. Med. Chem. 9:366 (1966).
13. G.-F. Huang, M. Okada, E. De Clercq and P.F. Torrence, Synthesis

and antiviral activity of 5-[(cyanomethylene)oxy]-2'-deoxy-
uridine, J. Med. Chem. 24:390 (1981).

14. Y.-C. Cheng, S. Grill and G. Dutschman, Time-dependent action of
    5-propyl deoxyuridine as antiherpes simplex virus type 1 and
    type 2 agents, Biochem. Pharmacol. 28:3529 (1979).

15. E. De Clercq, J. Descamps, P. De Somer, P.J. Barr, A.S. Jones
    and R.T. Walker, (E)-5-(2-bromovinyl)-2'-deoxyuridine : a po-
    tent and selective anti-herpes agent, Proc. Natl. Acad. Sci.
    USA 76:2947 (1979).

16. P. Langen, S.R. Waschke, K. Waschke, D. Bärwolff, J. Reefschlä-
    ger, P. Schulz, B. Preussel and C. Lehmann, 5-Formyl-2'-deoxy-
    uridine : cytostatic and antiviral properties and possible
    modes of action, Acta biol. med. germ. 35:1625 (1976).

17. J.S. Park, C.T.-C. Chang, C.L. Schmidt, Y. Golander, E. De
    Clercq, J. Descamps and M.P. Mertes, Oxime and dithiolane de-
    rivatives of 5-formyl-2'-deoxyuridine and their 5'-phospha-
    tes : antiviral effects and thymidylate synthetase inhibition,
    J. Med. Chem. 23:661 (1980).

18. E. De Clercq, J. Descamps, G.-F. Huang and P.F. Torrence, 5-
    Nitro-2'-deoxyuridine and 5-nitro-2'-deoxyuridine 5'-monophos-
    phate : antiviral activity and inhibition of thymidylate syn-
    thetase in vivo, Mol. Pharmacol. 14:422 (1978).

19. E. De Clercq, J. Balzarini, P.F. Torrence, M.P. Mertes, C.L.
    Schmidt, D. Shugar, P.J. Barr, A.S. Jones, G. Verhelst and
    R.T. Walker, Thymidylate synthetase as target enzyme for the
    inhibitory activity of 5-substituted 2'-deoxyuridines on mouse
    leukemia L1210 cell growth, Mol. Pharmacol. 19:321 (1981).

20. J. Reefschläger, D. Bärwolff, K. Dressler and P. Langen, Diffe-
    rential antiherpes activity of the (E)- and (Z)-isomer of 5-
    (2-fluorovinyl)-2'-deoxyuridine (FVUdR), Antiviral Research:
    submitted (1983).

21. E. De Clercq, J. Descamps, G. Verhelst, A.S. Jones and R.T.
    Walker, Antiviral activity of 5-(2-halogenovinyl)-2'-deoxyuri-
    dines, in: "Current Chemotherapy and Infectious Disease",
    J.D. Nelson and C. Grassi, eds., American Society of Micro-
    biology, Washington, D.C., p. 1372 (1980).

22. E. De Clercq, G. Verhelst, J. Descamps and D.E. Bergstrom, Dif-
    ferential inhibition of herpes simplex viruses, type 1 (HSV-1)
    and type 2 (HSV-2), by (E)-5-(2-X-vinyl)-2'-deoxyuridines,
    Acta microbiol. Acad. Sci. hung. 28:307 (1981).

23. J. Goodchild, R.A. Porter, R.H. Raper, I.S. Sim, R.M. Upton, J.
    Viney and H.J. Wadsworth, Structural requirements of olefinic
    5-substituted deoxyuridines for anti-herpes activity, J. Med.
    Chem.: in press (1983).

24. Y.-C. Cheng, S. Grill, J. Ruth and D.E. Bergstrom, Anti-herpes
    simplex virus and anti-human cell growth activity of E-5-pro-
    penyl-2'-deoxyuridine and the concept of selective protection
    in antivirus chemotherapy, Antimicrob. Agents Chemother. 18:
    957 (1980).

25. D.E. Bergstrom, J.L. Ruth, P.A. Reddy and E. De Clercq, Synthesis of (E)-5-(3,3,3-trifluoro-1-propenyl)-2'-deoxyuridine (TFPe-dUrd), and related analogs: potent and unusually selective antiviral activity of TFPe-dUrd against HSV-1, J. Med. Chem.: submitted (1983).

26. E. De Clercq, J. Balzarini, J. Descamps, C.F. Bigge, C.T.-C. Chang, P. Kalaritis and M.P. Mertes, Antiviral, antitumor, and thymidylate synthetase inhibition studies of 5-substituted styryl derivatives of 2'-deoxyuridine and their 5'-phosphates, Biochem. Pharmacol. 30:495 (1981).

27. A.S. Jones, S.G. Rahim, R.T. Walker and E. De Clercq, Synthesis and antiviral properties of (Z)-5-(2-bromovinyl)-2'-deoxyuridine, J. Med. Chem. 24:759 (1981).

28. E. De Clercq, Antiviral activity of 5-substituted pyrimidine nucleoside analogues, Pure & Appl. Chem. 55:623 (1983).

29. E. De Clercq, BVDU ((E)-5-(2-bromovinyl)-2'-deoxyuridine), in: "Antiviral Drugs and Interferon : The Molecular Basis of their Activity", Y. Becker, ed., Martinus Nijhoff Publishers, The Hague, in press (1983).

30. E. De Clercq, Selective anti-herpes drugs, in: "Proceedings of the International Symposium on Medical Virology", L.M. de La Maza, ed., Elsevier/North Holland, Amsterdam, in press (1983).

31. E. De Clercq, BVDU (bromovinyldeoxyuridine) : current status in antiviral therapy, in: "Control of Viral Diseases", E. Kurstak, ed., Marcel Dekker, Inc., New York, in press (1983).

32. E.-C. Mar, P.C. Patel, Y.-C. Cheng, J.J. Fox and E.-S. Huang, Effect of a series of nucleoside analogs on human cytomegalovirus replication in vitro, J. Gen. Virol.: submitted (1983).

33. A. Larsson and B. Öberg, Selective inhibition of herpesvirus deoxyribonucleic acid synthesis by acycloguanosine, 2'-fluoro-5-iodo-aracytosine, and (E)-5-(2-bromovinyl)-2'-deoxyuridine, Antimicrob. Agents Chemother. 19:927 (1981).

34. E. De Clercq, On the mechanism of anti-herpes action of E-5-(2-bromovinyl)-2'-deoxyuridine, in: "Herpetische Augenerkrankungen", R. Sundmacher, ed., J.F. Bergmann Verlag, München, p. 329 (1981).

35. J. Descamps and E. De Clercq, Specific phosphorylation of E-5-(2-iodovinyl)-2'-deoxyuridine by herpes simplex virus-infected cells, J. Biol. Chem. 256:5973 (1981).

36. J.A. Fyfe, Differential phosphorylation of (E)-5-(2-bromovinyl)-2'-deoxyuridine monophosphate by thymidylate kinases from herpes simplex viruses types 1 and 2 and varicella zoster virus, Mol. Pharmacol. 21:432 (1982).

37. Y.-C. Cheng, G. Dutschman, E. De Clercq, A.S. Jones, S.G. Rahim, G. Verhelst and R.T. Walker, Differential affinities of 5-(2-halogenovinyl)-2'-deoxyuridines for deoxythymidine kinases of various origins, Mol. Pharmacol. 20:230 (1981).

38. P.J. Barr, N.J. Oppenheimer and D.V. Santi, Thymidylate synthetase catalyzed conversions of E-5-(2-bromovinyl)-2'-deoxyuridylate, J. Biol. Chem.: submitted (1983).

39. H.S. Allaudeen, J.W. Kozarich, J.R. Bertino and E. De Clercq,
    On the mechanism of selective inhibition of herpesvirus repli-
    cation by (E)-5-(2-bromovinyl)-2'-deoxyuridine, Proc. Natl.
    Acad. Sci. USA 78:2698 (1981).
40. H.S. Allaudeen, M.S. Chen, J.J. Lee, E. De Clercq and W.H. Pru-
    soff, Incorporation of E-5-(2-halovinyl)-2'-deoxyuridines in-
    to deoxyribonucleic acids of herpes simplex virus type 1-in-
    fected cells, J. Biol. Chem. 257:603 (1982).
41. W.R. Mancini, E. De Clercq and W.H. Prusoff, The relationship
    between incorporation of E-5-(2-bromovinyl)-2'-deoxyuridine
    into herpes simplex virus type 1 DNA with virus infectivity
    and DNA integrity, J. Biol. Chem. 258:792 (1983).
42. J. Sagi, A. Czuppon, M. Kajtar, A. Szabolcs, A. Szemző and L.
    Ötvös, Modified polynucleotides. VI. Properties of a synthetic
    DNA containing the anti-herpes agent (E)-5-(2-bromovinyl)-2'-
    deoxyuridine, Nucleic Acids Res. 10:6051 (1982).
43. E. De Clercq, J. Balzarini, J. Descamps, G.-F. Huang, P.F. Tor-
    rence, D.E. Bergstrom, A.S. Jones, P. Serafinowski, G. Ver-
    helst and R.T. Walker, Antiviral, antimetabolic, and cytotoxic
    activities of 5-substituted 2'-deoxycytidines, Mol. Pharmacol.
    21:217 (1982).
44. L. Fox, M.J. Dobersen and S. Greer, Incorporation of 5-substitu-
    ted analogs of deoxycytidine into DNA of herpes simplex virus-
    infected or -transformed cells without deamination to the thy-
    midine analog, Antimicrob. Agents Chemother. 23:465 (1983).
45. I. Schildkraut, G.M. Cooper and S. Greer, Selective inhibition
    of the replication of herpes simplex virus by 5-halogenated
    analogues of deoxycytidine, Mol. Pharmacol. 11:153 (1975).
46. M.J. Dobersen, M. Jerkofsky and S. Greer, Enzymatic basis for
    the selective inhibition of varicella-zoster virus by 5-halo-
    genated analogues of deoxycytidine, J. Virol. 20:478 (1976).
47. S.G. Rahim, M.J.H. Duggan, R.T. Walker, A.S. Jones, R.L. Dyer,
    J. Balzarini and E. De Clercq, Synthesis and biological pro-
    perties of 2'-deoxy-5-vinyluridine and 2'-deoxy-5-vinylcyti-
    dine, Nucleic Acids Res. 10:5285 (1982).
48. P.F. Torrence, G.-F. Huang, M.W. Edwards, B. Bhooshan, J. Des-
    camps and E. De Clercq, 5-Substituted uracil arabinonucleosi-
    des as potential antiviral agents, J. Med. Chem. 22:316 (1979).
49. H. Machida, S. Sakata, A. Kuninaka, H. Yoshino, C. Nakayama
    and M. Saneyoshi, In vitro antiherpesviral activity of 5-al-
    kyl derivatives of 1-β-D-arabinofuranosyluracil, Antimicrob.
    Agents Chemother. 16:158 (1979).
50. J. Reefschläger, G. Herrmann, D. Bärwolff, B. Schwarz, D. Cech
    and P. Langen, Antiherpesviral potential of (E)-5-(2-bromo-
    vinyl)- and 5-vinyl-1-β-D-arabinofuranosyluracil and some
    other 5-substituted uracil arabinosyl nucleosides in two dif-
    ferent cell lines, Antiviral Research: in press (1983).
51. G. Stening, B. Gotthammar, A. Larsson, S. Alenius, N.G. Johans-
    son and B. Öberg, Antiherpes activity of (E)-5-(1-propenyl)-
    2'-deoxyuridine and 5-(1-propenyl)-1-β-D-arabinofuranosyl-

uracil, Antiviral Research 1:213 (1981).

52. H. Machida, A. Kuninaka, H. Yoshino, K. Ikeda and Y. Mizuno,
    Antiherpesvirus activity and inhibitory action on cell growth
    of 5-alkenyl derivatives of 1-β-D-arabinofuranosyluracil,
    Antimicrob. Agents Chemother. 17:1030 (1980).

53. H. Machida, S. Sakata, A. Kuninaka and H. Yoshino, Antiherpes-
    viral and anticellular effects of 1-β-D-arabinofuranosyl-E-5-
    (2-halogenovinyl)uracils, Antimicrob. Agents Chemother. 20:
    47 (1981).

54. E. De Clercq, R. Busson, L. Colla, J. Descamps, J. Balzarini and
    H. Vanderhaeghe, Antiviral activity of sugar-modified deriva-
    tives of (E)-5-(2-bromovinyl)-2'-deoxyuridine, in: "Current
    Chemotherapy and Immunotherapy", P. Periti and G.G. Grassi,
    eds., American Society for Microbiology, Washington, D.C.,
    p. 1062 (1982).

55. J. Descamps, R.K. Sehgal, E. De Clercq and H.S. Allaudeen, Inhi-
    bitory effect of E-5-(2-bromovinyl)-1-β-D-arabinofuranosylura-
    cil on herpes simplex virus replication and DNA synthesis,
    J. Virol. 43:332 (1982).

56. G.A. Gentry and J.F. Aswell, Inhibition of herpes simplex virus
    replication by araT, Virology 65:294 (1975).

57. J.F. Aswell, G.P. Allen, A.T. Jamieson, D.E. Campbell and G.A.
    Gentry, Antiviral activity of arabinosylthymine in herpesviral
    replication : mechanism of action in vivo and in vitro,
    Antimicrob. Agents Chemother. 12:243 (1977).

58. T.-S. Lin and Y.-S. Gao, Synthesis and biological activity of
    5-(trifluoromethyl)- and 5-(pentafluoroethyl)pyrimidine nu-
    cleoside analogues, J. Med. Chem. 26:598 (1983).

59. T. Kulikowski, Z. Zawadzki, D. Shugar, J. Descamps and E. De
    Clercq, Synthesis and antiviral activities of arabinofurano-
    syl-5-ethylpyrimidine nucleosides. Selective antiherpes acti-
    vity of 1-(β-D-arabinofuranosyl)-5-ethyluracil, J. Med. Chem.
    22:647 (1979).

60. H. Machida, A. Kuninaka and H. Yoshino, Inhibitory effects of
    antiherpesviral thymidine analogs against varicella-zoster
    virus, Antimicrob. Agents Chemother. 21:358 (1982).

61. S. Shigeta, T. Yokota, T. Iwabuchi, M. Baba, K. Konno, M. Oga-
    ta and E. De Clercq, Comparative efficacy of antiherpes drugs
    against various strains of varicella-zoster virus, J. Infect.
    Dis. 147:576 (1983).

62. R.L. Miller, J.P. Iltis and F. Rapp, Differential effect of
    arabinofuranosylthymine on the replication of human herpesvi-
    ruses, J. Virol. 23:679 (1977).

63. T. Ooka and A. Calender, Effects of arabinofuranosylthymine on
    Epstein-Barr virus replication, Virology 104: 219 (1980).

64. W.E.G. Müller, R.K. Zahn, J. Arendes and D. Falke, Phosphoryla-
    tion of arabinofuranosylthymine in non-infected and herpes-
    virus (TK$^+$ and TK$^-$)-infected cells, J. gen. Virol. 43:261
    (1979).

65. A. Matsukage, K. Ono, A. Ohashi, T. Takahashi, C. Nakayama and M. Saneyoshi, Inhibitory effect of 1-β-D-arabinofuranosylthymine 5'-triphosphate and 1-β-D-arabinofuranosylcytosine 5'-triphosphate on DNA polymerases from murine cells and oncornavirus, Cancer Res. 38:3076 (1978).

66. K. Ono, A. Ohashi, M. Ogasawara, A. Matsukage, T. Takahashi, C. Nakayama and M. Saneyoshi, Inhibition of deoxyribonucleic acid polymerases from murine cells and oncornavirus by 5-alkylated derivatives of 1-β-D-arabinofuranosyluracil 5'-triphosphate : substituent effects on inhibitory action, Biochemistry 20:5088 (1981).

67. Y.-C. Cheng, G. Dutschman, J.J. Fox, K.A. Watanabe and H. Machida, Differential activity of potential antiviral nucleoside analogs on herpes simplex virus-induced and human cellular thymidine kinases, Antimicrob. Agents Chemother. 20:420 (1981).

68. H. Machida, M. Ichikawa, A. Kuninaka, M. Saneyoshi and H. Yoshino, Effect of treatment with 1-β-D-arabinofuranosylthymine of experimental encephalitis induced by herpes simplex virus in mice, Antimicrob. Agents Chemother. 17:109 (1980).

69. E. De Clercq, J. Descamps, G. Verhelst, R.T. Walker, A.S. Jones, P.F. Torrence and D. Shugar, Comparative efficacy of antiherpes drugs against different strains of herpes simplex virus, J. Infect. Dis. 141:563 (1980).

70. J.F. Aswell and G.A. Gentry, Cell-dependent antiherpesviral activity of 5-methylarabinosylcytosine, an intracellular ara-T donor, Ann. N.Y. Acad. Sci. 284:342 (1977).

71. Y.C. Cheng, B. Goz, J.P. Neenan, D.C. Ward and W.H. Prusoff, Selective inhibition of herpes simplex virus by 5'-amino-2',5'-dideoxy-5-iodouridine, J. Virol. 15:1284 (1975).

72. T.-S. Lin, J.P. Neenan, Y.-C. Cheng, W.H. Prusoff and D.C. Ward, Synthesis and antiviral activity of 5- and 5'-substituted thymidine analogs, J. Med. Chem. 19:495 (1976).

73. T.-S. Lin, C. Chai and W.H. Prusoff, Synthesis and biological activities of 5-trifluoromethyl-5'-azido-2',5'-dideoxyuridine and 5-trifluoromethyl-5'-amino-2',5'-dideoxyuridine, J. Med. Chem. 19:915 (1976).

74. T.-S. Lin and W.H. Prusoff, Synthesis and biological activity of several amino analogues of thymidine, J. Med. Chem. 21:109 (1978).

75. R. Busson, L. Colla, H. Vanderhaeghe and E. De Clercq, Synthesis and antiviral activity of some sugar-modified derivatives of (E)-5-(2-bromovinyl)-2'-deoxyuridine, Nucleic Acids Res. Symposium Series no. 9:49 (1981).

76. E. De Clercq, J. Descamps, J. Balzarini, T. Fukui and H.S. Allaudeen, Antiviral activity of the 3'-amino derivative of (E)-5-(2-bromovinyl)-2'-deoxyuridine, Biochem. J. 211:439 (1983).

77. J.P. Iltis, T.-S. Lin, W.H. Prusoff and F. Rapp, Effect of 5-iodo-5'-amino-2',5'-dideoxyuridine on varicella-zoster virus in vitro, Antimicrob. Agents Chemother. 16:92 (1979).

78. M.S. Chen and W.H. Prusoff, Phosphorylation of 5-iodo-5'-amino-2',5',dideoxyuridine by herpes simplex virus type 1 encoded thymidine kinase, J. Biol. Chem. 254:10449 (1979).

79. M.S. Chen, D.C. Ward and W.H. Prusoff, Specific herpes simplex virus-induced incorporation of 5-iodo-5'-amino-2',5'-dideoxy-uridine into deoxyribonucleic acid, J. Biol. Chem. 251:4833 (1976).

80. P.H. Fischer, M.S. Chen and W.H. Prusoff, The incorporation of 5-iodo-5'-amino-2',5'-dideoxyuridine and 5-iodo-2'-deoxyuri-dine into herpes simplex virus DNA, Biochim. Biophys. Acta 606:236 (1980).

81. C.A. Puliafito, N.L. Robinson, D.M. Albert, D. Pavan-Langston, T.-S. Lin, D.C. Ward and W.H. Prusoff, Therapy of experimental herpes simplex keratitis in rabbits with 5-iodo-5'-amino-2', 5'-dideoxyuridine, Proc. Soc. Exp. Biol. Med. 156:92 (1977).

82. I.S. Sim, N. Stebbing and N.H. Carey, Studies on the antiviral activity of 5'-amino-2',5'-dideoxy-5-iodouridine (AIU) against herpes viruses in vivo and in vitro, Antiviral Research 1:393 (1981).

83. K.A. Watanabe, U. Reichman, K. Hirota, C. Lopez and J.J. Fox, Nucleosides. 110. Synthesis and antiherpes virus activity of some 2'-fluoro-2'-deoxyarabinofuranosylpyrimidine nucleosides, J. Med. Chem. 22:21 (1979).

84. C. Lopez, K.A. Watanabe and J.J. Fox, 2'-Fluoro-5-iodo-aracyto-sine, a potent and selective anti-herpesvirus agent, Antimi-crob. Agents Chemother. 17:803 (1980).

85. K.A. Watanabe, T.-L. Su, R.S. Klein, C.K. Chu, A. Matsuda, M.W. Chun, C. Lopez and J.J. Fox, Nucleosides. 123. Synthesis of antiviral nucleosides : 5-Substituted 1-(2-deoxy-2-halogeno-β-D-arabinofuranosyl)cytosines and -uracils. Some structure-activity relationships, J. Med. Chem. 26:152 (1983).

86. J.H. Burchenal, T-C. Chou, L. Lokys, R.S. Smith, K.A. Watanabe, T-L. Su and J.J. Fox, Activity of 2'-fluoro-5-methylarabino-furanosyluracil against mouse leukemias sensitive to and re-sistant to 1-β-D-arabinofuranosylcytosine, Cancer Res. 42: 2598 (1982).

87. T.-C. Chou, J.H. Burchenal, F.A. Schmid, T.J. Braun, T.-L. Su, K.A. Watanabe, J.J. Fox and F.S. Philips, Biochemical effects of 2'-fluoro-5-methyl-1-β-D-arabinofuranosyluracil and 2'-fluoro-5-iodo-1-β-D-arabinofuranosylcytosine in mouse leukemic cells sensitive and resistant to 1-β-D-arabinofuranosylcyto-sine. Cancer Res. 42:3957 (1982).

88. W. Kreis, L. Damin, J. Colacino and C. Lopez, In vitro metabo-lism of 1-β-D-arabinofuranosylcytosine and 1-β-2'-fluoroara-bino-5-iodocytosine in normal and herpes simplex type 1 virus-infected cells, Biochem. Pharmacol. 31:767 (1982).

89. J.L. Ruth and Y.-C. Cheng, Nucleoside analogues with clinical potential in antivirus chemotherapy. The effect of several thy-midine and 2'-deoxycytidine analogue 5'-triphosphates on puri-

fied human (α,β) and herpes simplex virus (types 1, 2) DNA polymerases, Mol. Pharmacol. 20:415 (1981).

90. T.-C. Chou, C. Lopez, J.M. Colacino, A. Grant, A. Feinberg, T.-L. Su, K.A. Watanabe, J.J. Fox and F.S. Philips, Mechanism of antiviral action of 2'-fluoro-arabinosyl pyrimidine nucleosides, Proc. of the American Association for Cancer Research 24:no 1205 (1983).

91. T.-C. Chou, A. Feinberg, A.J. Grant, P. Vidal, U. Reichman, K. A. Watanabe, J.J. Fox and F.S. Philips, Pharmacological disposition and metabolic fate of 2'-fluoro-5-iodo-1-β-D-arabinofuranosylcytosine in mice and rats, Cancer Res. 41:3336 (1981).

92. W.M. Shannon, G. Arnett, L. Westbrook, Y.F. Shealy, C.A. O'Dell and R.W. Brockman, Evaluation of carbodine, the carbocyclic analog of cytidine, and related carbocyclic analogs of pyrimidine nucleosides for antiviral activity against human influenza type A viruses, Antimicrob. Agents Chemother. 20:769 (1981).

93. Y.F. Shealy, C.A. O'Dell, W.M. Shannon and G. Arnett, Carbocyclic analogues of 5-substituted uracil nucleosides: synthesis and antiviral activity, J. Med. Chem. 26:156 (1983).

94. E. Krajewska, E. De Clercq and D. Shugar, Virus-induced nucleoside kinase activities in primary rabbit kidney cells, in: "Translation of Natural and Synthetic Polynucleotides", A.B. Legocki, ed., University of Agriculture in Poznan, p. 105 (1977).

95. B. Rada and M. Dragun, Antiviral action and selectivity of 6-azauridine, Ann. N.Y. Acad. Sci. 284:410 (1977).

96. H.E. Renis, 5,6-Dihydro-5-azathymidine: in vitro antiviral properties against human herpesviruses, Antimicrob. Agents Chemother. 13:613 (1978).

97. G.E. Underwood and S.D. Weed, Efficacy of 5,6-dihydro-5-azathymidine against cutaneous herpes simplex virus in hairless mice, Antimicrob. Agents Chemother. 11:765 (1977).

98. H.E. Renis and E.E. Eidson, Activities of 5,6-dihydro-5-azathymidine against herpes simplex virus infections in mice, Antimicrob. Agents Chemother. 15:213 (1979).

99. G.P. Khare, R.W. Sidwell, J.H. Huffman, R.L. Tolman and R.K. Robins, Inhibition of RNA virus replication in vitro by 3-deazacytidine and 3-deazauridine, Proc. Soc. Exp. Biol. Med. 140:880 (1972).

100. W.M. Shannon, G. Arnett and F.M. Schabel, Jr., 3-Deazauridine : inhibition of ribonucleic acid virus-induced cytopathogenic effects in vitro. Antimicrob. Agents Chemother. 2:159 (1972).

101. W.M. Shannon, R.W. Brockman, L. Westbrook, S. Shaddix and F.M. Schabel, Jr., Inhibition of gross leukemia virus-induced plaque formation in XC cells by 3-deazauridine, J. Natl. Cancer Inst. 52:199 (1974).

102. W.M. Shannon, Selective inhibition of RNA tumor virus replication in vitro and evaluation of candidate antiviral agents in vivo, Ann. N.Y. Acad. Sci. 284:472 (1977).

103. R.P. McPartland,M.C. Wang, A. Bloch and H. Weinfeld, Cytidine 5'-triphosphate synthetase as a target for inhibition by the antitumor agent 3-deazauridine, Cancer Res. 34:3107 (1974).

104. E. De Clercq, J. Descamps, P. De Somer and A. Holy, (S)-9-(2, 3-dihydroxypropyl)adenine : an aliphatic nucleoside analog with broad-spectrum antiviral activity, Science 200:563 (1978).

105. G.B. Elion, P.A. Furman, J.A. Fyfe, P. de Miranda, L. Beauchamp and H.J. Schaeffer, Selectivity of action of an antiherpetic agent, 9-(2-hydroxyethoxymethyl)guanine, Proc. Natl. Acad. Sci. USA 74:5716 (1977).

106. H.J. Schaeffer, L. Beauchamp, P. de Miranda, G.B. Elion, D.J. Bauer and P. Collins, 9-(2-Hydroxyethoxymethyl)guanine activity against viruses of the herpes group, Nature 272:583 (1978).

107. E. De Clercq and A. Holy, Antiviral activity of aliphatic nucleoside analogues: structure-function relationship, J. Med. Chem. 22:510 (1979).

108. J.L. Kelley, J.E. Kelsey, W.R. Hall, M.P. Krochmal and H.J. Schaeffer, Pyrimidine acyclic nucleosides. 1-[(2-Hydroxyethoxy)methyl]pyrimidines as candidate antivirals, J. Med. Chem. 24:753 (1981).

109. P.M. Keller, J.A. Fyfe, L. Beauchamp, C.M. Lubbers, P.A. Furman, H.J. Schaeffer and G.B. Elion, Enzymatic phosphorylation of acyclic nucleoside analogs and correlations with antiherpetic activities, Biochem. Pharmacol. 30:3071 (1981).

110. J.L. Kelley, M.P. Krochmal and H.J. Schaeffer, Pyrimidine acyclic nucleosides. 5-Substituted 1-[(2-aminoethoxy)methyl]uracils as candidate antivirals, J. Med. Chem. 24:472 (1981).

PURINE NUCLEOSIDE ANALOGS AS ANTIVIRAL AGENTS

John C. Drach

School of Dentistry
University of Michigan
Ann Arbor, Michigan 48109 U.S.A.

INTRODUCTION

In a book and conference dealing with "targets" for antiviral agents, a discussion on the "magic bullets"* used to treat virus diseases might seem out of place. The study of existing drugs and active compounds is, however, most appropriate in the context of designing new and better antivirals. As Prusoff[2] has pointed out, the knowledge of a specific chemotherapeutic target has not yet led to the synthesis of a specific drug. Until highly detailed physical-chemical data on sites of inhibition and modes of drug-receptor interaction are obtained and interpreted, this approach will remain a longer term goal. A more achievable goal is the application of mode of action data on existing agents to the design of new and better drugs. In addition, both new and existing drugs can be used to probe known targets and to search for new ones.

This review, therefore, will focus upon the biochemical basis for the action of known purine nucleoside analogs. Emphasis will be upon the few compounds which have proven clinical usefulness (vidarabine, acyclovir, ribavirin) and about which there is a substantial body of knowledge. More abbreviated information will be provided on newer compounds which may or may not be approved for human use but which have exhibited interesting antiviral and biochemical properties. For more extensive coverage of many different antiviral substances, the reader is referred to other chapters in this book and to a number of recent reviews.[3-17]

---

*The term apparently originated by Paul Ehrlich[1] to denote highly selective chemotherapeutic agents.

BIOCHEMICAL BASIS FOR SELECTIVE ACTIVITY

Although hundreds of compounds with antiviral activity have
been identified,[4] few possess sufficient selectivity toward the
virus to give good therapeutic responses *in vivo*. Because many
antiviral agents act by inhibition of macromolecular biosynthesis,
both cellular and viral functions can be affected. Inhibition of
viral or cellular functions in virus-infected cells leads to an inhibition of virus replication. In contrast, inhibition of cellular
functions in uninfected cells produces cytotoxicity *in vitro* that
is manifested as bone marrow suppression, gut ulceration, immune
suppression, etc. *in vivo*. Suppression of the immune system can be
especially devastating in attempted antiviral therapy.[18] The need,
therefore, is for antiviral drugs that inhibit viral replication
without interfering with normal cellular biosynthetic mechanisms.

The key to such inhibition is the existence of virus-associated
and virus-coded or induced enzymes and nucleic acid binding proteins.[19,20] Most of these enzymes are different from corresponding
enzymes found in uninfected cells and are critically involved in
virus replication. Consequently, they are potentially important
chemotherapeutic targets. A partial listing of these enzymes in
viruses important to human health is provided in Table 1.

Virus-associated enzymes differ from virus-induced enzymes in
that the former are contained as such within the virus particle
whereas the latter do not exist *per se*. Rather, virus nucleic acid
contains the genetic information to specify the amino acid sequence
of the enzyme when it is transcribed and/or translated in the virus-
infected host cell. Neuraminidase associated with influenza virus
is the classical example of a virion-associated enzyme and has been
the target of several searches for anti-influenza drugs.

The best known and probably the most thoroughly characterized
virus-induced enzymes are those produced by herpesviruses. There
are five human herpesviruses, namely herpes simplex virus (HSV) type
1, HSV type 2, varicella zoster virus, cytomegalovirus and Epstein
Barr virus. The bulk of the coding potential of herpes DNA can be
described in terms of structural proteins and a few key enzymes involved in nucleic acid biosynthesis.[10,19,20] Several enzymes clearly
have been identified as virus-coded, including DNA polymerase, deoxypyrimidine ("thymidine") kinase, ribonucleoside diphosphate reductase
and deoxyribonuclease. The first two enzymes now are known to be
targets in the selective antiviral activity of several compounds.
These antiviral agents fall into two general classes: (*i*) compounds
that preferentially inhibit HSV-coded DNA polymerase, and (*ii*) fraud-
ulent nucleosides that are converted to the active nucleotide only
by herpes-induced thymidine kinase. The effectiveness of compounds
which successfully utilize these targets is illustrated by the fact

that the only drugs licensed for use in the U.S. to treat systemic
herpesvirus infections (vidarabine, acyclovir) act through these
enzymes.

Table 1. Examples of Virus-Associated and
Virus-Induced Enzymes

Enzyme Activity	Viruses
Virion-associated enzymes	
neuraminidase	orthomyxo- and paramyxoviruses
protein kinase	such as influenza and parain-
RNA polymerase	fluenza
reverse transcriptase	oncoviruses such as mammalian
(RNA-dependent DNA polymerase)	types B and C oncogenic viruses
RNA methylase	
nucleotide phosphohydrolase	
RNA replicase	picornaviruses such as human
	poliovirus
DNA polymerase	hepatitis B virus
RNA polymerase	pox viruses
GTP:mRNA guanylyltransferase	such as vaccinia virus
Virus-induced enzymes	
deoxypyrimidine kinase	herpesviruses such
DNA polymerase	as herpes simplex virus types 1
deoxyribonuclease	and 2, varicella zoster virus
ribonucleoside diphosphate reductase	

ANTIVIRAL ACTIVITY OF PURINE NUCLEOSIDE ANALOGS

    The remainder of this chapter will be devoted to detailing bio-
chemical bases for the antiviral action of selected purine nucleoside
analogs. Although much information is available on screening, pharmacology, toxicology and clinical testing, the primary purpose of this
chapter is to provide information on how these compounds act at the
cellular and molecular level.

Arabinosyl Purine Nucleosides

Vidarabine (9-β-D-arabinofuranosyladenine, ara-A, Vira-A[TM]).
Ara-A (1) has a firmly established role in the management of certain
human herpesvirus infections.[21,22]   It is
effective in the therapy of herpes keratitis,
herpes encephalitis, and varicella zoster
infections in immunosuppressed patients.
The clinical efficacy of ara-A is an extension of its activity against a large number
of DNA viruses in animal models and in cell
culture.[23,24]   The antiviral activity arises
from the capacity of the drug to inhibit
virus replication at concentrations which
do not produce overt cytotoxicity in uninfected cells.[14,25] We have found that this
selectivity toward the virus is related to
a preferential inhibition of viral DNA synthesis compared to cellular
DNA synthesis.[25-27] Although other factors such as inhibition of RNA
polyadenylation[28] may be involved, the best evidence to date indicates that inhibition of DNA synthesis is the primary action of the
drug.[9,10,14,29]

With one exception (discussed below), inhibition of DNA synthesis
requires conversion of ara-A to its 5'-triphosphate (aATP) (see
Fig. 1). LePage and associates[30] were the first to report that mammalian cells were capable of phosphorylating ara-A. Since then
several groups have firmly established that ara-A is converted to
aATP in a number of cell lines[31-34] and in erythrocytes.[35] In addition, we[36] have found that ara-A is converted to its 5'-mono-,
di- and triphosphates in HSV-1-infected cells. Phosphorylation by
HSV-infected cells is of interest because the preferential metabolism
of ara-A by HSV-induced enzymes would provide a mechanism for achieving selective antiviral activity similar to that observed with a
number of pyrimidine antimetabolites[10,17] and acyclovir (see below).
No differences were observed, however, in either the deamination or
phosphorylation of ara-A between HSV-1-infected cells and uninfected
cells. In the presence of an adenosine deaminase inhibitor to block
the degradation of ara-A, aATP accumulated to 130 pmoles/10^6 cells
(approximately 40 μM) when both types of cells were incubated with
50 μM ara-A[36] or to 200-250 pmoles/10^6 cells in the presence of
100 μM ara-A.[37] In both studies,[36,37] there was a direct correlation between intercellular levels of aATP and inhibition of DNA
synthesis and/or HSV replication; the IC50 for inhibition of viral
DNA synthesis was approximately 0.2 μM.[36] In addition to concentration, Shewach and Plunkett[38] proved that time of exposure to aATP
also was critical. That is, cytotoxicity produced by ara-A correlated
with the product of aATP concentration and duration of exposure. The
half-life of aATP in HSV-1-infected cells was 8 to 9 hours[36] compared
to 1.7 or 13 to 15 hours, respectively, in uninfected CHO[39] or human

lymphoblastoid cells.[37] Loss of aATP correlated with a loss of
antiviral activity[36] and a loss of inhibition of DNA synthesis.[36-38]
Indirect evidence suggests acid phosphatase is responsible for degradation of aATP.[40] Although nothing is known about the enzymes
responsible for the phosphorylation of ara-A in HSV-infected cells,
limited work with cell extracts and partially purified enzymes has
established that adenosine kinase,[41,42] deoxycytidine kinase[43,44]
and another deoxyribonucleoside kinase[45] -- possibly deoxyguanosine
kinase[46] -- phosphorylate ara-A to aAMP. The relative importance
of each kinase is not clear; e.g., purified adenosine kinase has
only weak affinity for ara-A[41,42] but in intact cells it appears to
be primarily responsible for phosphorylating ara-A.[47]

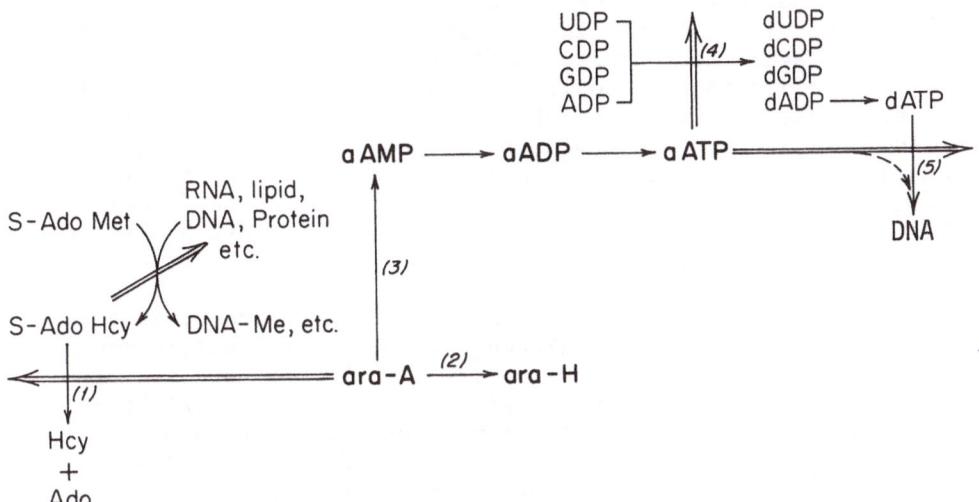

Fig. 1.   Metabolism of ara-A in mammalian cells and principal loci
           of action. Enzymes involved include (1) S-adenosyl-L-
           homocysteine hydrolase, (2) adenosine deaminase, (3) ade-
           nosine and deoxycytidine kinases, (4) ribonucleoside di-
           phosphate reductase and (5) DNA polymerase.

    In comparison to the metabolic activation of ara-A by phosphor-
ylation, its degradation by adenosine deaminase to 9-β-D-arabinosyl-
hypoxanthine (ara-H)[30,48,49] yields a less active compound. In contrast to the very low activity against tumors and neoplastic cells,[30]
ara-H clearly is active against DNA viruses. Miller et al.[50] first
reported that ara-H was about one-tenth as active as ara-A against
HSV in cell culture, but of comparable activity when assayed in
vivo.[24] Other investigators have reported that ara-H is as effective
as ara-A against HSV in some cell culture systems but less potent in
others depending upon different deamination rates of ara-A in

different systems.[23],[27]   Therefore, the metabolic degradation of
ara-A to ara-H reduces antiviral activity but does not eliminate it.
Like ara-A, the antiviral activity of ara-H probably is a consequence
of inhibition of DNA synthesis. We have found that it inhibits both
viral and cellular DNA synthesis *in vitro* [25-27] and that its antiviral activity is antagonized by 2'-deoxyadenosine.[51]   Unlike ara-A,[52]
combination with ribavirin resulted in additive anti-herpes activity.
It is likely that ara-H does not act as such but first must be converted to a nucleotide. It is known, in fact, that the 5'-monophos-
phate of ara-H (aHMP) is active as an antiviral drug[53] and that this
nucleotide is formed metabolically in erythrocytes.[35]   The mono-,
di- and triphosphates also are formed in L5178Y cells.[54]   The triphosphate (aHTP) is an inhibitor of mammalian DNA polymerases α and
β[54] and HSV-1 DNA polymerase.[55]   Inhibition of the former polymerases
by aHTP was competitive with dATP[54] whereas inhibition of the latter
was competitive with dGTP.[55]   In addition to phosphorylation to aHTP,
it is possible ara-H is converted to nucleotides of ara-A[56] or ara-
G.[54],[57]   Metabolism to aATP could explain how an adenine nucleoside
(dAdo) antagonized the activity of a hypoxanthine nucleoside (ara-
H).[51]   In fact, it has been demonstrated that aHMP is a substrate
for adenylosuccinate synthetase[58] and that the resulting arabino-
syladenylosuccinate is cleaved to ara-A-5'-monophosphate (aAMP) by
adenylosuccinate lyase.[59]   Thus, it is possible that ara-H also acts
by conversion to ara-A nucleotides.

Regardless of whether aATP forms from ara-A or ara-H, it does
act to inhibit DNA synthesis. Although a number of possible mechanisms have been explored and reviewed,[9],[10],[29],[49],[60] there are probably only four which have physiological significance: (*i*) Inhibition of viral and cellular DNA polymerases. Inhibition of HSV-1
and HSV-2 DNA polymerases by aATP is potent and competitive with
dATP.[61-63] K_i values for aATP range from 0.1 to 0.4 μM -- very near
the IC50 of 0.2 μM noted above[36] for inhibition of HSV DNA synthesis
in intact cells. The antiviral activity of vidarabine itself is
antagonized in a competitive manner by deoxyadenosine,[51] thereby
providing an additional link between DNA polymerase inhibition and
antiviral activity. DNA polymerases α, β and γ from uninfected
mammalian cells also are inhibited by aATP.[49],[61-68]   On the basis
of K_i and IC50 values, inhibition of DNA polymerase α is more potent
than inhibition of polymerases β and γ, but not as potent as inhibition of HSV DNA polymerases. Therefore, the selective action of
ara-A against HSV may be a consequence of the preferential action of
aATP against HSV-1 and HSV-2 DNA polymerases. On the other hand, if
K_i/K_m ratios derived from highly purified polymerases are considered
more complete measures of inhibition,[69],[70] then there appears to be
little difference in inhibition of HSV DNA polymerases and DNA polymerase α.[63],[67],[68]   Because aATP also inhibits cellular DNA polymerases, the antiviral selectivity of ara-A is not absolute. Inhibition of cellular DNA polymerases undoubtedly is related to the *in*

vitro cytotoxicity[9,71] and *in vivo* teratogenicity[72] and toxicity[73] noted at higher drug concentrations.

(*ii*) Inhibition of ribonucleoside diphosphate reductase -- a mechanism which would potentiate the action of aATP on DNA polymerase.[74] Although it has been known for some time that aATP inhibits this enzyme,[49,57] the recent evidence that it also is virus-coded[75] has added new significance for chemotherapy. Potent inhibition of reductase by aATP (K_i = 4 to 15 μM)[74] has been observed in uninfected cells[57,74,76] but the only reports of studies with the HSV-induced enzyme are in opposition.[77,78] Shannon and associates[78] found aATP was a much more potent inhibitor of the HSV-1 enzyme than of the cellular enzyme but Langelier and Buttin[77] found no inhibition of the viral enzyme. All reports of inhibition of the isolated enzyme are somewhat difficult to interpret because in whole cells ara-A either does not affect[79,80] or initially depresses but then increases dNTP pools.[81] One possible way in which these disparate results might be reconciled is if dNDPs produced by reductase remained in a reductase-kinase-DNA polymerase "replitase"[82], then they might not be detected in soluble dNTP pools.

(*iii*) Incorporation of aAMP residues into DNA. Several possibilities exist as reasons for inhibition of DNA synthesis as a consequence of aAMP incorporation. If the free 3'-hydroxyl group of the incorporated aAMP is ineffective as a primer for continued chain extension, severe inhibition of chain elongation, or, in the extreme case, termination would occur. If, in contrast, the 3'-hydroxyl terminus provided by the added aAMP allows continued chain extension, albeit at a reduced rate, then internal incorporation of aAMP could be extensive and subsequent rounds of replicative DNA synthesis might be blocked by helical aberrations. The available evidence indicates that aAMP residues are incorporated internally into cellular DNA[31,83,84] and such incorporation is related to cytotoxicity.[85] Compared to internal incorporation into cellular DNA, Müller and colleagues[70,86] observed that ara-A acted as a chain-terminating nucleoside in HSV but not cellular DNA and proposed that this was related to the selective antiviral activity of the drug. In contrast, we[84] observed that aAMP residues were incorporated uniformly throughout the HSV-1 genome. In addition, we[87] found that aATP was a substrate (K_m = 0.014 μM) for HSV-2 DNA synthesis in isolated nuclei. The V_{max}, however, was only 2% that of dATP. The reason that aAMP is not an absolute chain terminator may be due to the exonuclease activity of HSV DNA polymerase which will remove some 3'-aAMP termini.[88] aAMP molecules which are incorporated into primer-termini do, in fact, retard subsequent primer elongation. Other studies with mammalian DNA polymerase δ strengthen the latter observation.[89]

(*iv*) Inhibition of S-adenosyl-L-homocysteine hydrolyse (S-Adohcy-ase) by ara-A. Inhibition of this enzyme does not require bioconversion of ara-A to aATP because the nucleoside itself acts as an

apparent suicide-substrate inhibitor.[90]   Inhibition of the hydrolase
leads to accumulation of S-AdoHcy[91,92] which acts as a feedback inhibitor of S-adenosylmethionine-dependent transmethylases[93] (Fig. 1).
This, in turn, leads to decreased methylation of macromolecules[94] and
possibly biogenic amines. Resulting toxicity has been noted *in
vitro*,[94,95] and has been suggested in animals[92] and humans.[96] Neurological toxicity observed in humans[97] and monkeys[98] may be related to
effects on biogenic amines. Additional support for a role of S-
AdoHcyase inhibition in ara-A toxicity comes from the observations
that although B lymphoblasts don't accumulate high amounts of aATP,[99]
they still are somewhat susceptible to ara-A.[100] Thus a nucleoside-
mediated mechanism could be involved. In contrast, I know of no firm
evidence that inhibition of S-AdoHcyase is involved in the antiviral
activity of ara-A.

   (*v*) Whether other sites of action are critically involved in
the antiviral activity or toxicity of ara-A is difficult to determine.
For example, aATP is a potent inhibitor of terminal deoxynucleotidyl
transferase[101] as well as RNA polyadenylation.[28] Since ara-A appears
to have some activity against RNA-containing rhabdoviruses such as
vesicular stomatitis and rabies viruses, the latter mechanism could
be involved.

   In addition to herpesviruses, ara-A has been used clinically
against another DNA virus -- hepatitis B virus.[97] Apparently antiviral activity results from inhibition of Dane particle-associated
DNA polymerase by aATP.[102]

   Conclusions from the foregoing studies have been amplified and
clarified by work on ara-A resistant cells and herpesviruses. The
importance of nucleoside kinases, ribonucleoside diphosphate reductase
and DNA polymerase in the action of ara-A has been proven by the
derivation of cell lines resistant to the drug which have alterations in these enzymes. Studies with cellular kinase mutants have
established that adenosine and deoxycytidine kinases[44,47,103] are
important in the cytotoxicity of ara-A. They also raise the question of the contribution of S-AdoHcyase inhibition by ara-A itself
to overall cytotoxicity. Perhaps the latter is important only in
certain cell types or organs. Resistant cell lines with mutant reductase desensitized to the negative allosteric effector dATP[103,104]
or with increased levels of the enzyme[105] illustrate the importance
of reductase in the action of ara-A. Unlike kinase mutants, these
cells show cross-resistance to 1-β-D-arabinofuranosylcytosine (ara-
C).[104,105] Surprisingly, there has been only one report of an ara-A-
resistant cell line with an apparent alteration in a DNA polymerase.[106]
In contrast, resistance has been documented as a consequence of mutations in bacterial DNA polymerase III[60] and in herpes DNA polymerase.[67,107] The latter studies were crucial in proving that HSV
DNA polymerase is a critical target in the antiviral action of the
drug.

The mechanisms discussed above probably explain most of the biological activity of ara–A. However, whether or not there are sufficient quantitative differences in effects between uninfected and HSV–infected cells to account for the selective antiviral activity observed *in vivo* is open to question. Biochemical and cell culture differences are only on the order of 3 to 10–fold compared to such differences of approximately a thousand–fold for acyclovir and certain pyrimidine nucleoside analogs. Thus, other factors –– such as metabolic disposition[108] –– may be of primary importance in animals and humans.

   Esters and analogs of ara–A. Despite the efficacy and safety of ara–A in its antiviral dose range, the drug has several disadvantages. First, it is metabolized to the less active ara–H by adenosine deaminase; second, it does not readily penetrate intact skin and mucous membranes; and third, its low aqueous solubility necessitates the administration of large fluid volumes for systemic therapy. In attempts to circumvent these problems, a large number of analogs have been synthesized. Adenosine deaminase–resistant nucleosides have been prepared and tested for antiviral and antineoplastic activity. The more interesting deaminase–resistant analogs include 2–fluoro–ara–A,[109] 2'–azido–ara–A,[110] and carbocyclic ara–A (cyclaridine, 2).[111]

These compounds probably are converted to their 5'–triphosphates[109] at least one of which inhibits DNA polymerase.[66] Like ara–A, cyclaridine inhibits the replication of Gross murine leukemia virus and herpesviruses *in vitro*.[111] It and its 5'–methoxyacetate also have shown promising activity *in vivo* against genital herpes and herpes encephalitis.[112]

In attempts to overcome poor topical penetrability of ara–A, a number of O–acyl ester prodrugs have been prepared. One of these, ara–A–5'–valerate, showed good penetrability through skin in a model system.[113] In addition, this prodrug

2

is an inhibitor but not a substrate of adenosine deaminase, thereby providing a powerful combination of activities.[114] Although the valerate ester was not active in HSV–2–infected mice,[115] it and the 2',3'–di–O–acetate[116] were active in the female guinea pig HSV–2 genital infection model.[115,117] In comparison to the possible topical efficacy of the acylesters, the 5'–monophosphate ester (aAMP) has been used as a more soluble dose formulation for systemic administration. Its effectiveness *in vitro* and in animal models[4,5,115–117] has now been demonstrated clinically. Results from initial studies indicate that the water–soluble prodrug is as safe and effective as ara–A in patients with severe herpesvirus infections[118,119] and has been evaluated in type B hepatitis patients.[120] The compound also has been tested as a topical preparation. It is not effective against herpes labialis when used as a 10% cream.[121] However, when aAMP was applied topically and delivered by cathodal iontophoresis, it was highly efficacious in two animal models.[122] The 5'-monophosphate of ara-H (aHMP) also is active against experimental herpesvirus infections.[14,53] Presumably, all the esters act by hydrolysis to ara-A and by bioconversion to aATP.

Additional arabinosyl purine nucleosides. A number of these nucleosides have been prepared and tested as antivirals.[4] Among those showing potent and selective activity, 9-β-D-arabinofuranosyl-2,6-diaminopurine (ara-DAP) and 9-β-D-arabinofuranosylguanine (ara-G) are of particular interest because -- like ara-A and ara-H -- the former is deaminated by adenosine deaminase to the latter.[123] Ara-G and ara-A were prepared at about the same time[9] but the anti-herpes potential of ara-G was recognized much later.[123] Ara-G is less potent but more selective *in vitro* than ara-A[123,124] and is active in several animal models.[123,125] It is phosphorylated to aGMP at least in part by deoxycytidine kinase[43] and presumably is converted to aGTP[123] -- a known inhibitor of mammalian DNA polymerases.[126] Ara-DAP appears to be slightly less potent *in vitro* than ara-A[125] but is considerably less cytotoxic[123] and therefore is more selective. However, it also appears to be more active *in vivo* than either ara-A or ara-G,[123] thus the ara-DAP:ara-G metabolic pair could be therapeutically superior to the ara-A:ara-H pair. The development of ara-DAP apparently was dropped when the Burroughs Wellcome research group concentrated on acyclovir which is more active than ara-DAP.[127] A recent report indicates renewed interest in ara-G for its activity against T-leukemia cells.[128]

Aliphatic Purine Nucleosides

Acyclovir [9-(2-hydroxyethoxymethyl)guanine, acycloguanosine, ACV, Zovirax[TM]]. Acyclovir (3) is a relatively new compound, first described in 1977 by Elion, Schaeffer and coworkers at Burroughs Wellcome.[124,129] The drug has unique biochemical specificity for herpesviruses and recently has been licensed in the U.S. for topical or intravenous treatment of initial (but not recurrent) episodes of genital herpes[130] and initial or recurrent mucosal and cutaneous HSV-1 and HSV-2 infections in immunocompromised patients.[131] Careful clinical trials also have shown promise for the treatment of other herpesvirus infections such as keratitis and varicella zoster.[131] Like ara-A, the drug does not produce significant clinical benefit in herpes labialis[132] but unlike ara-A it is active by the oral route of administration.[133]

In experimental animal models the drug is effective in herpes keratitis and cutaneous herpes and has a protective effect in herpes encephalitis. Efficacy was obtained by parenteral, oral or topical

routes of administration.[129,131,134,135] In several of these studies,
ACV prevented establishment of latent ganglionic infections. As in
the human trials,[131,132]   topical application was not uniformly effective in all cutaneous herpes models. One problem with topical application is the difficulty with which all polar nucleosides cross
the stratum corneum unless a penetrant such as dimethylsulfoxide is
used.[136] Another problem may arise from thymidine in the skin competitively antagonizing acyclovir phosphorylation to its active 5'-
triphosphate (acyclo-GTP).[137] In all the animal studies as in
humans,[130-133] the drug is remarkably non-toxic (acute LD_{50} in mice
>1000 mg/kg).[129,134]

In cell culture, acyclovir is a very potent inhibitor of the
replication of HSV-1 (IC_{50} <0.2 µM)[124,129,138,139] with less potency
against HSV-2, varicella zoster virus and Epstein-Barr virus (IC_{50}
= 2 to 7 µM)[131,138] and least activity against human cytomegalovirus.[131,138,139] In contrast, 100 to >300 µM is required to inhibit
the growth of uninfected host cells[124,131] and progenitor cells or
to adversely affect lymphocyte responses.[139,140] This highly selective antiviral effect also has been observed at the level of DNA
synthesis. In uninfected cells, there was little or no effect on
DNA synthesis at acyclovir concentrations of up to 200 µM.[138,141]In
marked contrast, synthesis of HSV-1 DNA was inhibited by <0.2 µM
drug. Cell DNA synthesis in the same infected cells was less affected but was inhibited more than in uninfected cells.[142] Inhibition of HSV-1 DNA synthesis leads to a reduction in the amount of
late herpes proteins which are synthesized resulting in a loss of
infectivity.[143] Similar conclusions have been drawn from detailed
studies with Epstein-Barr virus DNA replication.[144]

The biochemical basis for these separate effects on viral DNA
synthesis and DNA synthesis in uninfected and HSV-infected cells is
a consequence of the action of two herpes-induced enzymes -- deoxypyrimidine kinase and DNA polymerase (Fig. 2). The large difference
noted between herpes-infected and uninfected cells results from
marked differences in substrate specificity between HSV-induced
deoxypyrimidine kinase and cellular thymidine and purine kinases.
The former recognize this analog whereas the latter do so only
poorly.[124,145] Although the HSV-induced kinase does not have a high
affinity for the drug (K_m = 100 to 200 µM),[145,146] the V_{max} and high
amount of enzyme[19] are responsible for the high concentrations of
acyclovir nucleotides found in HSV-1 and HSV-2-infected cells.[124,147]
Once the 5'-monophosphate is formed, it is converted to the diphosphate by cellular GMP kinase[148] and apparently not by the HSV thymidylate kinase activity of HSV deoxypyrimidine kinase.[149] This is an
unusual instance of substrate specificity; a *purine* nucleoside analog
is phosphorylated by a viral *pyrimidine* kinase but the resulting
monophosphate is a substrate for a cellular but not viral nucleotide kinase. Conversion of the diphosphate to the active metabolite,
acyclo-GTP, is carried out by a number of cellular kinases.[150]

Intracellular pools of acyclo-GTP persist in HSV-1 infected cells with a half-life of 2 hours.[147] Other than nucleotides, no other metabolites have been identified in cell culture studies. Two additional metabolites were identified in the urine of humans receiving [^{14}C]acyclovir. Approximately 10% of the label was 9-carboxymethoxy-methylguanine and 0.5% was 8-hydroxy-acyclovir; the remainder was acyclovir.[151] Additional metabolic disposition and pharmacokinetic studies have been performed -- see reference 131 for details.

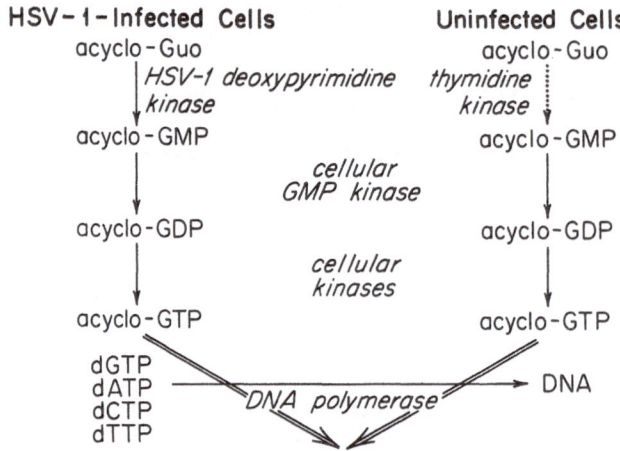

Fig. 2.   Metabolism and primary site of action of acyclovir in un-
           infected and herpes simplex virus-infected mammalian cells.

The sensitivity of uninfected, HSV-1 transformed, TK⁻ cells to ACV provides additional evidence for the importance of herpes deoxypyrimidine kinase. These cells lack cellular thymidine kinase but are very sensitive to the drug because they contain genetic information for the viral kinase.[152] Non-transformed cells are not drug-sensitive. In contrast to HSV-1 and 2, cytomegalovirus (CMV) does not induce a viral kinase[20] and human CMV is relatively insensitive to the drug. Murine CMV, however, is highly drug sensitive (IC$_{50}$ = 0.2 μM)[153] suggesting that either the CMV DNA polymerase is exquisitely sensitive to the small amount of acyclo-GTP formed by cellular enzymes or -- like ara-A -- a nucleoside-mediated action is involved. The former possibility would not be unique since Epstein-Barr virus (EBV) is susceptible to the drug based on this possibility.[154]

In addition to a herpes kinase producing inhibitory levels of acyclo-GTP only in virus-infected cells, HSV-1 DNA polymerase is more sensitive to acyclo-GTP than is mammalian DNA polymerase α.[124,141,155,156] K$_1$ concentrations for the two polymerases differ by 10 to 60 fold depending upon enzyme source and purity. Little or no inhibition of mammalian DNA polymerase β was noted. This significant difference explains how cell and viral DNA both synthesized in the presence of acyclo-GTP can be inhibited to different extents.[141,142] The potent effect on HSV-1 DNA polymerase (K_i = 0.003 to 0.4 μM) is competitive with dGTP and also explains how the drug inhibits DNA replication.[141] In contrast to the different sensitivities of the five human herpesviruses to ACV (see above), all their induced DNA polymerases appear to be sensitive to acyclo-GTP.[155-158] For example, human CMV DNA polymerase is sensitive to the nucleotide (K_i = 0.25 μM) but the virus is insensitive to acyclovir because little acyclo-GTP forms in the infected cell.[155] In contrast, there are conflicting reports on the sensitivity of EBV DNA polymerase to acyclo-GTP and why ACV inhibits EBV replication.[157,158]

Acyclo-GTP also is a better substrate for herpes DNA polymerases than mammalian DNA polymerases and therefore incorporation of acyclo-GMP residues into DNA could be involved in antiviral activity.[141] Such incorporation would lead to chain termination as the molecule has no 3'-hydroxy group. Furman and coworkers[152,159] have provided direct evidence for such termination. In addition to preventing chain elongation, Cheng and associates[156] found that acyclo-GMP-terminated DNA noncompetitively inhibits HSV-1 DNA polymerase (K_i = 0.001 μM) and unlike aAMP, acyclo-GMP is not excised by HSV-1 DNA polymerase-associated 3',5'-exonuclease. Also in contrast to ara-A, ribonucleoside diphosphate reductase was not inhibited by acyclo-GTP (IC_{50} >200 μM)[160] but there were acyclovir-induced increases in dNTP pools.[161]

ACV-resistant mutants of herpesviruses have been investigated both for their potential impact on chemotherapy and as a means of establishing drug mode of action.[131] ACV-resistant mutants of HSV-1, HSV-2 and varicella zoster virus (VZV) have been derived *in vitro*[161-164] and *in vivo*[165] by passage in the presence of the drug. Mutants could not be obtained, however, by passage with ACV plus ara-A.[166] In addition, ACV-resistant HSV are clonally distributed in virus populations that have not been exposed to the drug.[167] Emergence of resistant virus during therapy of immuno-deficient patients also has been reported.[131,168] Depending on the nature of the mutational event, many of these ACV-resistant viruses remain sensitive to other antivirals[131] such as ara-A.[164,168,169] In addition, those which are devoid of herpes deoxypyrimidine kinase (TK⁻ mutants) also are less virulent[170,171] and unable or less likely to establish latency.[131,172] On the other hand, Field and colleagues[173] have described a mutant with altered TK and herpes DNA polymerase that no longer was lethal but was able to replicate in retinal tissue.

Regardless of the clinical consequences, ACV-resistant viruses have provided the means of proving which viral functions are affected by the drug. Two mutational loci have been described, HSV TK and DNA polymerase.[163,174] Mutation that decreases[145] or alters[175,176] TK activity prevents the phosphorylation of ACV thereby rendering the virus resistant to the drug. As discussed above, the former mutants (TK$^-$) have shown decreased virulence but the appearance of a drug (bromovinyl deoxyuridine)-resistant mutant which retained affinity for thymidine (TK$^+$) raises the possibility that such mutants might also retain pathogenicity.[175,176] This latter mutant was sensitive to ACV, which would not occur if it were TK$^-$. A second class of mutants also have been isolated which are resistant to ACV by virtue of a reduced sensitivity of DNA polymerase to acyclo-GTP. Genetic studies have shown that ACV resistance maps with DNA polymerase[163,174] and biochemical experiments established that K$_i$ values for inhibition of mutant polymerases by acyclo-GTP were 9 to 28 fold higher than for the parental strain.[55] Like TK$^+$ mutants, DNA polymerase mutants also have different phenotypes which have different sensitivities to antivirals. For example, DNA polymerase from one mutant (PAA^r 5) is resistant to aATP, acyclo-GTP and phosphonoacetate whereas another (BW^r) is resistant to acyclo-GTP but is sensitive to phosphonoacetate and aATP.[67,177] Although all ACV-resistant mutants appear to result from mutation in the *tk* and/or *pol* genes, the mutants display a variety of phenotypes including different degrees of resistance to ACV and cross-resistance to other antivirals.

    Analogs of acyclovir. The success of acyclovir has stimulated the synthesis of a number of acyclic guanine nucleosides. One with interesting antiviral activity is 9-(1,3-dihydroxy-2-propoxymethyl)guanine (4). This compound was independently invented by three different research groups. It has been termed BIOLF-62,[178] 2'-nor-2'-deoxyguanosine (2'NDG)[179] and DHPG[180] by the different developers; DHPG will be used herein. The *in vitro* antiviral activity of the compound is equal to or greater than that of ACV for all human herpesviruses, including CMV against which ACV has little activity.[180,181] It also produced equal or less cytotoxicity in uninfected cells.[178-181] *In vivo*, DHPG is superior to ACV in HSV-1 and HSV-2 infected animal models.[179-182]

    DHPG is phosphorylated more efficiently than ACV by both HSV TK and cellular GMP kinase but like ACV it is not a substrate for cellular thymidien kinases.[179,180] It is highly synergistic with phosphonoacetate against HSV-1 and HSV-2.[183] TK$^-$ HSV was resistant to DHPG but phosphonoacetate-ACV-resistant HSV was sensitive to the compound.[178] DHPG also was active against TK$^+$, ACG-resistant HSV strains and against HSV variants with altered DNA polymerase resistant to ACV.[181]

    Another compound closely related to both ACV and DHPG recently has been described.[184] (RS)-9-(3,4-dihydroxybutyl)guanine (DHBG, 5)

also is active against HSV-1 and HSV-2 *in vitro*
and *in vivo*. The (R)-enantiomer is the more inhibitory of the two isomers. Both HSV TK and
DNA polymerase appear to be involved in the
action of the compound based upon resistance of
a TK⁻ mutant to DHBG and selective inhibition
of viral DNA synthesis in HSV-1-infected cells.

   Aliphatic adenine nucleosides. These compounds are similar to ACV, DHPG and DHBG in that
they possess aliphatic side chains in the 9-posi-
tion of the purine ring but they differ in their
antiviral spectrum and are in much earlier stages
of development.   In a series of 58 aliphatic nucleosides (S)-9-(2,
3-dihydroxypropyladenine [(S)-DHPA, 6] was the most active congener.[185] It is active against both DNA and RNA
viruses at concentrations that did not adversely
affect cellular functions.[186] A recent study of
effects on lymphocyte function and marrow progenitor cell growth, however, found the compound
to be more toxic than ara-A or acyclovir.[71] It
also appears to be embryotoxic in chicks and
mice.[187] Although little is known about the
antiviral mode of action, preliminary data suggest inhibition of viral mRNA[188] possibly mediated by blockage of S-AdoHcyase.[17] The potency
of (S)-DHPA against vaccinia virus was increased
markedly by combining it with 6-azauridine but
the combination was less active against two RNA
viruses.[189] In preliminary animal work, (S)-DHPA reduced by one-
half mortality caused by intranasal inoculation of mice with vesicular
stomatitis virus.[186] Metabolism studies in mice have revealed the
presence of the hypoxanthine and xanthine analogs in urine samples.[190]

   Antiviral activity also has been observed for other aliphatic
adenosine analogs. The eritadenines possess activity against certain
DNA and RNA viruses.[191] The adenosine deaminase inhibitor EHNA and
several chiral isomers inhibit HSV replication and DNA synthesis
apparently apart from their capacity to inhibit adenosine
deaminase.[192]

Purine-Like Nucleosides

   Ribavirin (1-β-D-ribofuranosyl-1,2,4-triazole-3-carboxamide,
Virazole™). This synthetic truncated nucleoside (7) inhibits the
replication of both RNA and DNA viruses *in vitro* and *in vivo*.[193-195]
Activity has been confirmed against more than 12 DNA viruses and 27
RNA viruses by Sidwell and associates.[193] In both cell culture and
animal studies, ribavirin produced minimal toxicity in its antiviral

dose range.[193-195]   In controlled clinical trials where the drug was given by the oral route, signs and symptoms of influenza A and B improved in some cases but overall effectiveness was marginal.[193-195]   Oral administration of the drug did appear to be of some benefit in genital herpes[196] and hepatitis A[197] although confirming studies are needed. The administration of ribavirin aerosol by inhalation to treat influenza A, B and respiratory syncytial virus infections has been most encouraging.[198-200] These results indicate that broad-spectrum chemotherapy of these infections which are often clinically indistinguishable may be possible. Combination aerosol treatment with ribavirin plus amantadine has been more effective than ribavirin alone in animal studies and suggests such a combination be studied in humans.[194,195]

7

In addition to antiviral activity, extensive animal and human studies have explored the toxicity, teratogenicity and metabolic disposition of the drug.[193-195]   Inadequate drug distribution to the respiratory tract following oral administration, in fact, may be responsible for the variable clinical efficacy of ribavirin against influenza.[194]

The exact biochemical basis for the selective antiviral effect of ribavirin against either RNA or DNA viruses is unknown. The drug is converted to its 5'-mono-, di- and triphosphates in uninfected cells and whole animals and also is degraded to several inactive metabolites[193,195] (Fig. 3). Conversion to the monophosphate is carried out by adenosine or deoxyadenosine kinase.[193,201] No reports have appeared on its metabolism in virus infected cells.

Although reports on the antiviral mode of action of ribavirin are few, it is clear that the action involves inhibition of nucleic acid biosynthesis. We have found that the drug is selective in blocking HSV DNA synthesis compared to cellular DNA synthesis in either uninfected or HSV-infected cells.[25,202] Inhibition of HSV DNA synthesis might be related to the modest inhibition of HSV DNA polymerase by the 5'-triphosphate of ribavirin reported by Öberg and colleagues.[203] Selective inhibition of influenza virus RNA polymerase also has been reported by this group.[204] There is a question, however, whether I_{50} concentrations >100 μM are sufficient to account for the antiviral activity. Ribavirin 5'-monophosphate is a potent inhibitor of IMP dehydrogenase in uninfected cells (K_i = 3 x 10^{-7}).[205] Blockage of this enzyme lowers guanine nucleotide pools[193] and thus could self-potentiate the effect of ribavirin triphosphate on RNA polymerases. Arguing against this mechanism being involved in the antiviral activity is the observation that a related compound -- ribosylthiazole-4-carboxamide (tiazofurin) -- is a potent inhibitor of IMP dehydrogenase[206] but has little antiviral activity.[207] It is, of course, a promising antitumor agent. More recent data that riba- virin triphosphate inhibits capping of mRNA[208] (Fig. 3) provides new insight into why the drug acts against both DNA and RNA viruses. Because blockage of capping was demonstrated by inhibition of vac- cinia virus mRNA guanylyltransferase (K_1 = 32 μM), it is possible that this inhibition also may contribute to antiviral selectivity. To date, no studies on genetics or resistance have appeared, possibly because it seems difficult to develop ribavirin-resistant viruses.[209]

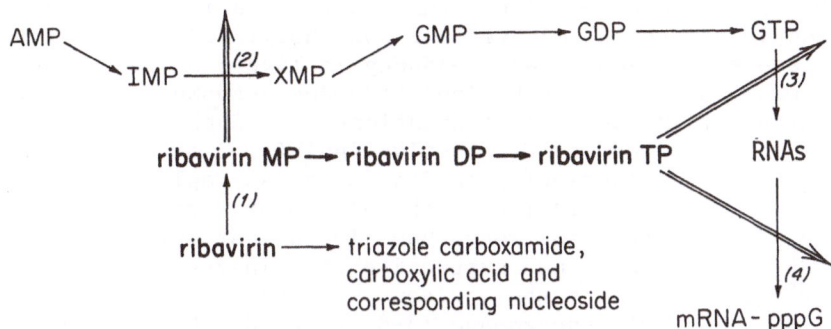

Fig. 3.   Metabolism of ribavirin in mammalian cells and primary
           loci of action. Enzymes involved include (1) adenosine
           kinase, (2) IMP dehydrogenase, (3) viral RNA polymerase
           and (4) GTP:mRNA guanylyltransferase.

Ribavirin Analogs. A large number of these compounds have been prepared in the search for more active compounds.[4,194] Robins and associates[210] recently reported that the selenium analog of tiazo- furin is a highly potent and selective antiviral agent. It appears to be especially active against RNA viruses.[210]

Deazapurine nucleosides. A number of these compounds have been prepared in the search for antineoplastic agents. A limited amount of work has been done in the antiviral area. The antineoplastic agent 3-deazaguanine, its ribosylnucleoside and 5'-nucleotide are active in vitro and in vivo against a spectrum of RNA and DNA viruses.[211] Activity in many cases was more potent than that of ribavirin. Like ribavirin, the monophosphate inhibited IMP dehydro- genase.[212] Another guanosine analog, 5-aza-7-deazaguanosine is active in vitro against HSV.[213]

Certain deazaadenine nucleosides also have antiviral activity. We[214] have examined a series of deaza-ara-A analogs and found that 7-deaza-ara-A and its 5'-nucleotide were about one-tenth as potent as ara-A in blocking HSV-1 and HSV-2 replication and DNA synthesis. Other analogs such as 7-deaza-ara-H and 3-deaza-ara-A[215] were less active. Inhibition of HSV-1, HSV-2 and mammalian DNA polymerases α and β by 7-deaza-aATP (K_i = 17, 5, 72 and >300 μM, respectively) correlated with effects on virus replication and cell growth.

Other deazaadenines also are interesting and probably have more potential of being clinically useful. 3-Deazaadenosine (3-deaza-Ado), a known inhibitor of S-AdoHcyase,[216] inhibits the replication of several RNA viruses at concentrations that do not or only slightly affect cellular DNA, RNA and protein synthesis.[217] The compound also inhibited DNA viruses (HSV-1 and simian virus 40) at concentrations (25 to 50 μM) that did not affect the growth of uninfected cells.[218] 3-Deaza-Ado is neither deaminated nor phosphorylated. It causes elevated levels of S-AdoMet and S-AdoHcy in livers of rats injected with the compound and is metabolized to 3-deaza-AdoHcy. These changes lead to perturbations of transmethylations.[219] Such changes could lead to a selective inhibition of methylation reaction(s) needed for viral replication such as methylation of the 5' cap of viral mRNA.[218] In the dose range used in this study, 3-deaza-Ado was immunosuppresive when injected into mice before the antigen.[220] The carbocyclic analog of 3-deaza-Ado (3-deaza-C-Ado, 8, 3-deazaaristeromycin) also is active against certain RNA and DNA viruses.[221] 3-Deaza-C-Ado is more potent than 3-deaza-Ado, ribavirin or (S)-DHPA against vesicular stomatitis, parainfluenza, measles and other RNA viruses plus vaccinia virus.[222] It is, however, only minimally effective against HSV-1, HSV-2 and polio-1 virus. Like its ribosyl congener, the carbocyclic analog is neither phosphorylated nor deaminated.[222] It is a competitive inhibitor of S-AdoHcyase. In contrast to speculation about other S-AdoHcyase inhibitors, it does not block methylation of vaccinia virus RNA. The metabolism of polyamines is perturbed, however, which may be involved in the antiviral activity.[222]

CONCLUSION

When I began to work with antivirals in 1968, there was only one purine nucleoside of interest, ara-A. It is remarkable that now there are so many purine nucleoside antivirals that it is taxing to write about only the ones of primary interest.

REFERENCES

1. P. Ehrlich, Berichte 42:17 (1909).
2. W.H. Prusoff and P.H. Fisher, in: "Nucleoside Analogs: Chemistry, Biology and Medical Applications," R.T. Walker, E. DeClercq and F. Eckstein, eds., Plenum Press, New York (1979) p. 281.
3. D.L. Swallow, Prog. Drug Res. 22:268 (1978).
4. R.W. Sidwell and J.T. Witkowski, in: "Burger's Medicinal Chemistry, 4th ed., Part II," M. E. Wolff, ed., John Wiley and Sons, New York, (1979) p. 543.
5. G.J. Galasso, T.C. Merigan and R.A. Buchanan, eds., "Antiviral Agents and Viral Diseases of Man," Raven Press, New York, (1979).
6. R.J. Suhadolnik, "Nucleosides as Biological Probes," John Wiley and Sons, New York, (1979).
7. E.C. Herman, Jr. and J.A Hermann, Pharmacol. Therap. 7:35 (1979).
8. E. DeClercq, Arch. Int. Physiol. Biochim. 87:353 (1979).
9. T.W. North and S.S. Cohen, Pharmacol. Ther. 4:81 (1979).
10. J.C. Drach, Ann. Rep. Med. Chem. 15:149 (1980).
11. W.M. Shannon and F.M. Schabel, Jr., Pharmacol. Ther. 11:263 (1980).
12. M.S. Hirsch and M.N. Swartz, N. Engl. J. Med. 302:903 and 949 (1980).
13. F.E. Hahn, ed., "Virus Chemotherapy," Vol. 27 of "Antibiotics and Chemotherapy," S. Karger, Basel (1980).
14. R.A. Smith, R.W. Sidwell and R.K. Robins, Ann. Rev. Pharmacol. Toxicol. 20:259 (1980).
15. L.H. Collier and J. Oxford, eds., "Developments in Antiviral Therapy," Academic Press, London (1980).
16. K.K. Guari, ed., "Antiviral Chemotherapy: Design of Inhibitors of Viral Functions," Academic Press, New York (1981).
17. E. DeClercq, Biochem. J. 205:1 (1982).
18. T.C. Merigan and H.E. Renis, in: "Nucleoside Analogues; Chemistry Biology and Medical Applications," R.T. Walker, E. De Clercq and F. Eckstein, eds., Plenum Press, New York, (1979) p. 395.
19. S. Kit, Pharmacol. Ther. 4:501 (1979).
20. Y.-C. Cheng, K. Nakayama, D. Derse, K. Bastow, J. Ruth, R.-S. Tan, G. Dutschman, S.J. Cardona and S. Grill, in: "Herpesvirus: Clinical, Pharmacological and Basic Aspects," H. Shiota, Y.-C. Cheng and W. Prusoff, eds., Excerpta Medica, Amsterdam (1982) P. 47.
21. R. Whitley, C. Alford, F. Hess and R. Buchanan, Drugs 20:267 (1980).
22. D. Pavan-Langston, R.A. Buchanan, C.A. Alford, Jr., eds., "Adenine Arabinsodie: An Antiviral Agent," Raven Press, New York (1975) p. 197 ff.
23. W.M. Shannon, ibid., p. 1.
24. B.J. Sloan, ibid., p. 45.

25. J.C. Drach and C. Shipman, Jr., Ann. N.Y. Acad. Sci. 284:396 (1977).
26. C. Shipman, Jr., S.H. Smith, R.H. Carlson and J.C. Drach, Antimicrob. Agents Chemother. 9:120 (1976).
27. P.M. Schwartz, C. Shipman, Jr. and J.C. Drach, Antimicrob. Agents Chemother. 10:64 (1976).
28. K.M. Rose, T.B. Leonard and T.H. Carter, Molec. Pharmacol. 22: 517 (1982).
29. W.E.G. Müller, in "Antiviral Agents and Viral Diseases of Man," G.J. Galasso, T.C. Merigan and R.A. Buchanan, eds., Raven Press, New York (1979) p. 77.
30. J.J. Brink and G.A. Le Page, Cancer Res. 24:312 (1964).
31. W. Plunkett and S.S. Cohen, Cancer Res. 35:415 (1975).
32. L.M. Rose and R.W. Brockman, J. Chromatog. 133:335 (1977).
33. W.E.G. Müller, R.K. Zahn, J. Arendes, A. Maidhof and H. Umezawa, Z. Physiol. Chem. 359:1287 (1978).
34. D.E. Bell and A. Fridland, Biochim. Biophys. Acta 606:57 (1980).
35. T. Chang and A.J. Glazko, Res. Commun. Chem. Path. Pharmacol. 14:127 (1976).
36. P.M. Schwartz, J.N. Sandberg, C. Shipman, Jr. and J.C. Drach, 15th Interscience Conf. Antimicrob. Agents Chemother., Washington D.C. (1975) Abst. 357.
37. W. Plunkett, S. Chubb, L. Alexander and J.A. Montgomery, Cancer Res. 40:2349 (1980).
38. D.S. Shewach and W. Plunkett, Cancer Res. 42:3637 (1982).
39. D.S. Shewach and W. Plunkett, Biochem. Pharmacol. 28:2401 (1979).
40. H.W. Spater, J.A. Schnitzer, N. Quintana, S.H. Spater and A.B. Novikoff, J. Histochem. Cytochem. 29:693 (1981).
41. K.J. Pierre, A.P. Kimball and G.A. LePage, Canad. J. Biochem. 45:1619 (1967).
42. H.P. Schnebli, D.L. Hill and L.L. Bennett, J. Biol. Chem. 242: 1997 (1967).
43. T.A. Krenitsky, J.V. Tuttle, G.W. Koszalka, I.S. Chen, L.M. Beacham III, J.L. Rideout and G.B. Elion, J. Biol. Chem. 251: 4055 (1976).
44. V. Verhoef, J. Sarup and A. Fridland, Cancer Res. 41:4478 (1981).
45. C.-H. Chang, R.W. Brockman and L.L. Bennett, Jr., Cancer Res. 42:3033 (1982).
46. W.R. Gower, Jr., M.C. Carr and D.H. Ives, J. Biol. Chem. 254: 2180 (1979).
47. N.B. Katlama and J.C. Drach, unpublished observations.
48. M. Hubert-Habart and S.S. Cohen, Biochim Biophys. Acta 59:468 (1962).
49. S.S. Cohen, Prog. Nucl. Acids Res. Molec. Biol. 5:1 (1966).
50. R.A. Miller, G.J. Dixon, J. Ehrlich, B.J. Sloan and I.W. McLean, Jr., Antimicrob. Agents Chemother. - 1968:136 (1969).
51. S.H. Smith, C. Shipman, Jr. and J.C. Drach, Cancer Res. 38: 1916 (1978).
52. L.B. Allen, L.K. Vander Slice, C.M. Fingal, F.H. McCright, E.F. Harris and P.D. Cook, Antiviral Res. 2:203 (1982).

53. L.B. Allen, C.J. Hintz, S.M. Wolf, J.H. Huffman, L.N. Simon, R.K. Robins and R.W. Sidwell, J. Infect. Dis. 133:A178 (1976).
54. W.E.G. Müller, J. Arendes, A. Maidhof, R.K. Zahn and W. Geurtsen, Chemother. 27:53 (1981).
55. D. Derse, K.F. Bastow and Y.C. Cheng, J. Biol. Chem. 257:10251 (1982).
56. G.A. LePage, Canad. J. Biochem. 48:75 (1970).
57. S.S. Cohen, Med. Biol. 54:299 (1976).
58. T. Spector and R.L. Miller, Biochim. Biophys. Acta 445:509 (1976).
59. T. Spector, Biochim. Biophys. Acta 481:741 (1977).
60. N.R. Cozzarelli, Ann. Rev. Biochem. 46:641 (1977).
61. W.E.G. Müller, R.K. Zahn and D. Falke, Virology 84:320 (1978).
62. C.M. Reinke, J.C. Drach, C. Shipman, Jr. and A. Weissbach in: "Oncogenesis and Herpesviruses III, Part 2," G. de The, W. Henle and F. Rapp, eds., IARC Press, Lyon (1978) p. 999.
63. M. Ostrander and Y.-C. Cheng, Biochim. Biophys. Acta 609:232 (1980).
64. A. Okura and S. Yoshida, J. Biochem. 84:727 (1978).
65. K. Ono, A. Ohashi, A. Yamamoto, A. Matsukago, T. Takahasi, M. Saneyoshi and T. Ueda, Cancer Res. 39:4673 (1979).
66. H.S. Allaudeen, J.W. Kozavich and A.C. Sortorelli, Nucleic Acids Res. 10:1379 (1982).
67. D.M. Coen, P.A. Furman, P.T. Gelop and P.A. Schaffer, J. Virol. 41:909 (1982).
68. C.M. Reinke, J.C. Drach, C. Shipman, Jr. and H.S. Allaudeen, submitted to Antiviral Res.
69. J.L. Webb, in: "Enzyme and Metabolic Inhibitors, Vol I, General Principles of Inhibition," Academic Press, New York (1963) p. 104.
70. W.E.G. Müller, R.K. Zahn, K. Bittlingmaier and D. Falke, Ann. N.Y. Acad. Sci. 284:34 (1977).
71. J.R. Wingard, A.D. Hess, R.K. Stuart, R. Saral and W.H. Burns, Antimicrob. Agents Chemother. 23:593 (1983).
72. J.L. Schardein, D.L. Hentz, J.A. Petrere, J.E. Fitzgerald and S.M. Kurtz, Teratology 15:231 (1977).
73. B. Hafkin, R.B. Pollard, M.L. Tiku, W.S. Robinson and T.C. Merigan, Antimicrob. Agents Chemother. 16:781 (1979).
74. C.-H. Chang and Y.-C. Cheng, Cancer Res. 40:3555 (1980).
75. B.M. Dutia, J. Gen. Virol. 64:513 (1983).
76. E.L. White, S.C. Shaddix, R.W. Brockman and L.L. Bennett, Jr., Cancer Res. 42:2260 (1982).
77. Y. Langelier and G. Buttin, J. Gen. Virol. 57:21 (1981).
78. W.M. Shannon, R.W. Brockman, G. Arnett and S. Shaddix, 1st Amer. Soc. Virol. Meeting, Ithaca, NY (1982).
79. L.W. Dow, D.E. Bell, L. Poulakos and A. Fridland, Cancer Res. 40:1405 (1980).
80. D.S. Shewach and W. Plunkett, Biochem. Pharmacol. 31:2103 (1982).
81. T.W. North, Biochem. Pharmacol. in press (1983).
82. G.P veer Reddy and A.B. Pardee, J. Biol. Chem. 257:12526 (1982).

83. W.E.G. Müller, H.J. Rohde, R. Beyer, A. Maidhof, M. Lachmann, H. Taschner and R.K. Zahn, Cancer Res. 35:2160 (1975).

84. J.C. Pelling, J.C. Drach and C. Shipman, Jr., Virology 104:323 (1981).

85. D.W. Kufe, P.P. Major, D. Munroe, M. Egan and D. Herrick, Cancer Res. 43:2000 (1983).

86. W.E.G. Müller, R.K. Zahn, R. Beyer and D. Falke, Virology 76:787 (1977).

87. J.W. Barnett, C.M. Reinke, S.R. Turk and J.C. Drach, submitted to Biochim. Biophys. Acta.

88. D. Derse and Y.-C. Cheng, J. Biol. Chem. 256:8525 (1981).

89. M.Y.W. Tsang Lee, J.J. Byrnes, K.M. Downey and A.G. So, Biochemistry 19:215 (1980).

90. M.S. Hershfield, J. Biol. Chem. 254:22 (1979).

91. S. Helland and P.M. Ueland, Cancer Res. 42:1130 (1982).

92. T.O. Eloranta, E.O. Kajander and A.M. Raina, Med. Biol. 60: 272 (1982).

93. G.L. Cantoni and P.K. Chiang, in:"Natural Sulfur Compounds Novel Biochemical and Structural Aspects," D. Carallini, G.E. Gaull and V. Zappia, eds., Plenum Press, New York (1980) p. 67.

94. N.M. Kredich and M.S. Hershfield, Proc. Natl. Acad. Sci. USA 76:2450 (1979).

95. C.E. Cass, M. Selner, P.J. Ferguson and J.R. Phillips, Cancer Res. 42:4991 (1982).

96. S.L. Sacks, T.C. Merigan, J. Kaminska and I.H. Fox, J. Clin. Invest. 69:226 (1982).

97. S.L. Sacks, G.H. Scullard, R.B. Pollard, P.B. Gregory, W.S. Robinson and T.C. Merigan, Antimicrob. Agents Chemother. 21: 93 (1982).

98. S.M. Kurtz, in: "Adenine Arabinoside: An Antiviral Agent," D. Pavan-Langstron, R.A. Buchanan and C.A. Alford, Jr., eds., Raven Press, New York (1975) p. 145.

99. N.M. Kredich, J. Biol. Chem. 255:7380 (1980).

100. D.A. Carson, J. Kaye and J.E. Seegmiller, Adv. Exp. Biol. Med. 122B:299 (1980).

101. W.E.G. Müller, R.K. Zahn and J. Arendes, FEBS Lett. 94:47 (1978).

102. G. Hess, W. Arnold and K.-H. Meyer zum Büschenfelde, Antimicrob. Agents Chemother. 19:44 (1981).

103. V.L. Chan and P. Juranka, Somat. Cell Genet. 7:147 (1981).

104. D. Ayusawa, K. Iwata and T. Seno, Cancer Res. 43:814 (1983).

105. K. Iwata, D. Ayusawa and T. Seno, Gann 73:167 (1982).

106. G.A. LePage, Cancer Res. 38:2314 (1978).

107. C.S. Crumpacker, L.E. Schnipper, P.N. Kowalsky and D.M. Sherman, J. Infect. Dis. 146:167 (1982).

108. A.J. Glazko, T. Chang, J.C. Drach, D.R. Mourer, P.E. Borondy, H. Schneider, L. Croskey and E. Maschewske, in: "Adenine Arabinoside: An Antiviral Agent," D. Pavan-Langston, R.A. Buchanan and C.A. Alford, Jr., eds., Raven Press, New York (1975) p. 111.

109. R.W. Brockman, Y.-C. Cheng, F.M. Schabel and J.A. Montogomery, Cancer Res. 40:3610 (1980).

110. C.M. Cermak–Mörth, R. Christian and F.M. Unger, Biochem. Pharmacol. 28:2105 (1979).
111. R. Vince and S. Daluge, J. Med. Chem. 20:612 (1977).
112. W.M. Shannon, L. Westbrook, G. Arnett, S. Daluge, H. Lee and R. Vince, Antimicrob. Agents Chemother. in press (1983).
113. C.D. Yu, N.A. Gordon, J.L. Fox, W.I. Higuchi and N.F. Ho, J. Pharm. Sci. 69:775 (1980).
114. R.A. Lipper, S.M. Machkovech, J.C. Drach and W.I. Higuchi, Molec. Pharmacol. 14:366 (1978).
115. J.T. Richards, E.R. Kern, J.C. Overall, Jr. and L.A. Glasgow, Antiviral Res. 2:27 (1982).
116. D.C. Baker, T.H. Haskell, S.R. Putt and B.J. Sloan, J. Med. Chem. 22:273 (1979).
117. W.M. Shannon, G. Arnett, D.C. Baker and W.I. Higuchi, 22nd Interscience Conf. Antimicrob. Agents Chemother., Miami (1982) Abst. 437.
118. O. Sauer, G.T. Werner, H. Schneider, H.G. Lenard, L.V. Haselberg and H.-J. Nettesheim, Klin. Pädiat. 191:566 (1979).
119. R.J. Whitley, B.C. Tucker, A.W. Kinkel, N.H. Barton, R.F. Pass, J.D. Whelchel, C.G. Cobbs, A.G. Diethelm and R.A. Buchanan, Antimicrob. Agents Chemother. 18:709 (1980).
120. J.H. Hoofnagle, G.Y. Minuk, G.M. Dusheiko, D.F. Schafer, R. Johnson, S. Straus, E.A. Jones, J.L. Gerin and K. Ishak, Hepatol. 2:784 (1982).
121. S.L. Spruance, C.J. Crumpacker, H. Haines, C. Bader, K. Mehr, J. MacCalman, L.E. Schnipper, M.R. Klauber, J.C. Overall and the Collaborative Study Group, N. Engl. J. Med. 300:1180 (1979).
122. J.M. Hill, B.S. Kwon, K.D. Burch, J. Deback, I. Whang, G.T. Jones, B. Luke, R. Harp, Y. Shimomura, D.S. Hull and L.P. Gangarosa, Amer. J. Med. 73:300 (1982).
123. G.B. Elion, J.L. Rideout, P. deMiranda, P. Collins and D.J. Bauer, Ann. N.Y. Acad. Sci. 255:468 (1975).
124. G.B. Elion, P.A. Furman, J.A. Fyfe, P. deMiranda, L. Beauchamp and H.J. Schaeffer, Proc. Natl. Acad. Sci. USA 74:5716 (1977).
125. P. Collins and D.J. Bauer, J. Antimicrob. Chemother. 3A:73 (1977).
126. K. Ono, A. Ohashi, A. Yamamoto, A. Matsukage, T. Takahasi, M. Saneyoshi and T. Ueda, Cancer Res. 39:4673 (1979).
127. H.J. Field, S.E. Bell, G.B. Elion, A.A. Nash and P. Wildy, Antimicrob. Agents Chemother. 15:554 (1979).
128. A. Cohen, J.W.W. Lee and E.W. Gelfand, Blood 61:660 (1983).
129. H.J. Schaeffer, L. Beauchamp, P. deMiranda, G.B. Elion, D.J. Bauer and P. Collins, Nature 272:583 (1978).
130. L. Corey, A.J. Nahmias, M.E. Guinan, J.K. Benedetti, C.W. Critchlow and K.K. Holmes, New Engl. J. Med. 306:1313 (1982).
131. D.H. King and G. Gallaso, eds., Amer. J. Med. 73:1 ff (1982).
132. S.L. Spruance, L.E. Schnipper, J.C. Overall, Jr., E.R. Kern, B. Wester, J. Modlin, G. Wenerstrom, C. Burton, K.A. Arndt, G.L. Chiu and C.S. Crumpacker, J. Infect. Dis. 146:85 (1982).
133. Y.J. Bryson, M. Dillon, M. Lovett, G. Acuna, S. Taylor, J.D. Cherry, B.L. Johnson, E. Wiesmeier, W. Growdon, T. Creagh-Kirk

and R. Keeney, New Engl. J. Med. 308:916 (1983).

134. D.J. Bauer, in:"Developments in Antiviral Therapy," L.H. Collier and J. Oxford, eds., Academic Press, London (1980) p. 43.

135. D. Brigden, P. Fiddian, A.E. Rosling and T. Ravenscroft, Antiviral Res. 1:203 (1981).

136. S. Alenius, M. Berg, F. Broberg, K. Eklind, B. Lindborg and B. Öberg, J. Infect. Dis. 145:569 (1982).

137. J. Harmenberg, Ph.D. Dissertation, Department of Virology, National Bacteriological Laboratory, Stockholm (1982).

138. C.S. Crumpacker, L.E. Schnipper, J.A. Zaia and M.J. Levin, Antimicrob. Agents Chemother. 15:642 (1979).

139. J.R. Wingard, A.D. Hess, R.K. Stuart, R. Saral and W.H. Burns, Antimicrob. Agents Chemother. 23:593 (1983).

140. L.M. Parker, J.M. Lipton, N. Binder, E.L. Crawford, M. Kudisch and M. Levine, Antimicrob. Agents Chemother. 21:146 (1982).

141. P.A. Furman, M.H. St. Clair, J.A. Fyfe, J.L. Rideout, P.M. Keller and G.B. Elion, J. Virol. 32:72 (1979).

142. A. Larsson and B. Öberg, Antimicrob. Agents Chemother. 19:927 (1981).

143. P. Furman and P.V. McGuirt, Antimicrob. Agents Chemother. 23:332 (1983).

144. B.M. Colby, J.E. Shaw, G.B. Elion and J.S. Pagano, J. Virol. 34:560 (1980).

145. J.A. Fyfe, P.M. Keller, P.A. Furman, R.L. Miller and G.B. Elion, J. Biol. Chem. 253:8721 (1978).

146. B.A. Larder and G. Darby, J. Virol. 42:649 (1982).

147. P.A. Furman, P. deMiranda, M.H. St. Clair and G.B. Elion, Antimicrob. Agents Chemother. 20:518 (1981).

148. W.H. Miller and R.L. Miller, J. Biol. Chem. 255:7204 (1980).

149. M.S. Chen, J. Walker and W.H. Prusoff, J. Biol. Chem. 254: 10747 (1979).

150. W.H. Miller and R.L. Miller, Biochem. Pharmacol. 31:3879 (1982).

151. P. deMiranda, S.S. Good, O.L. Laskin, H.C. Krasny, J.D. Connor and P.S. Lietman, Clin. Pharmacol. Ther. 30:662 (1981).

152. P.A. Furman, P.V. McGuirt, P.M. Keller, J.A. Fyfe and G.B. Elion, Virology 102:420 (1980).

153. W.H. Burns, J.R. Wingard, W.J. Bender and R. Saral, J. Virol. 39:889 (1981).

154. B.M. Colby, P.A. Furman, J.E. Shaw, G.B. Elion and J.S. Pagano, J. Virol. 38:606 (1981).

155. M.H. St. Clair, P.A. Furman, C.M. Lubbers and G.B. Elion, Antimicrob. Agents Chemother. 18:741 (1980).

156. D. Derse, Y.-C. Cheng, P.A. Furman, M.H. St. Clair and G.B. Elion, J. Biol. Chem. 256:11447 (1981).

157. A.K. Datta, B.M. Colby, J.E. Shaw and J.S. Pagano, Proc. Natl. Acad. Sci. USA 77:4163 (1980).

158. H.S. Allaudeen, J. Descamps and R.K. Sehgal, Antiviral Res. 2:123 (1982).

159. P.V. McGuirt and P.A. Furman, Amer. J. Med. 73:67 (1982).

160. K. Nakayama, J.L. Ruth and Y.-C. Cheng, J. Virol. 43:325 (1982).

161. P.A. Furman, C.V. Lambe and D.J. Nelson, Amer. J. Med. 73:14 (1982).
162. H.J. Field and G. Darby, Antimicrob. Agents Chemother. 17:209 (1980).
163. C.S. Crumpacker, L.E. Schnipper, P. Chartrand and K.W. Knopf, Amer. J. Med. 73:361 (1982).
164. K.K. Biron, J.A. Fyfe, J.E. Noblin and G.B. Elion, Amer. J. Med. 73:383 (1982).
165. H.J. Field, Antimicrob. Agents Chemother. 21:744 (1982).
166. R.F. Schniazi, J. Peters, C.C. Williams, D. Chance and A.J. Nahmias, Antimicrob. Agents Chemother. 22:499 (1982).
167. D.S. Paris and J.E. Harrington, Antimicrob. Agents Chemother. 22:71 (1982).
168. C.S. Crumpacker, L.E. Schnipper, S.I. Marlowe, P.N. Kowalsky, B.J. Hershey and M.J. Levin, New Engl. J. Med. 306:343 (1982).
169. H. Field, A. McMillan and G. Darby, J. Infect. Dis. 143:281 (1981).
170. C.D. Sibrack, L.T. Gutman, C.M. Wilfert, C. McLaren, M.H. St. Clair, P.M. Keller and D.W. Barry, J. Infect. Dis. 146:673 (1982).
171. T. Hovi, Med. Biol. 60:165 (1982).
172. R.B. Tenser, R.L. Miller and F. Rapp, Science 205:915 (1979).
173. H.J. Field, J.R. Anderson and P. Wildy, J. Gen. Virol. 59:91 (1982).
174. D.M. Coen and P.A. Schaffer, Proc. Natl. Acad. Sci. USA 77:2265 (1980).
175. B.A. Larder, D. Derse, Y.-C. Cheng and G. Darby, J. Biol. Chem. 258:2027 (1983).
176. B.A. Larder, Y.-C. Cheng and G. Darby, J. Gen. Virol. 64:523 (1983).
177. P.A. Furman, D.M. Coen, M.H. St. Clair and P.A. Schaffer, J. Virol. 40:936 (1981).
178. K.O. Smith, K.S. Galloway, W.L. Kennell, K.K. Ogilvie and B.K. Radatus, Antimicrob. Agents Chemother. 22:55 (1982).
179. W.T. Ashton, J.D. Karkas, A.K. Field and R.L. Tolman, Biochem. Biophys. Res. Commun. 108:1716 (1982).
180. D.F. Smee, J.C. Martin, J.P.H. Verheyden and T.R. Mathews, Antimicrob. Agents Chemother. 23:676 (1983).
181. Y.-C. Cheng, E.-S. Huang, J.-C. Lin, E.-C. Mar, J.S. Pagano and G. Dutschman, Proc. Natl. Acad. Sci. USA in press.
182. J.C. Martin, C.A. Dvorak, D.F. Smee, T.R. Mathews and J.P.H. Verheyden, J. Med. Chem. in press (1983).
183. K.O. Smith, K.S. Galloway, K.K. Ogilvie and U.O. Cheriyan, Antimicrob. Agents Chemother. 22:1026 (1982).
184. A. Larsson, B. Öberg, S. Alenius, G.-E. Hagberg, N.-G. Johansson, B. Lindborg and G. Stening, Antimicrob. Agents Chemother. 23:664 (1983).
185. E. DeClercq and A. Holy, J. Med. Chem. 22:510 (1979).
186. E. De Clercq, J. Descamps, P. De Somer and A. Holy, Science 200:563 (1978).

187. R. Jelinek, A. Holy and I. Votruba, Teratol. 24:267 (1981).

188. B. Rada, M. Dragun, I. Votruba and A. Holy, Acta. Virol. 24:433 (1980).

189. B. Rada and A. Holy, Chemother. 26:184 (1980).

190. A. Holy and A. Cihak, Biochem. Pharmacol. 30:2359 (1981).

191. A. Holy, I. Votruba and E. De Clercq, Collect. Czech. Chem. Commun. 47:1392 (1982).

192. T.W. North and L.O'Connor, E. Abushanab and R.P. Panzica, Biochem. Pharmacol. in press (1983).

193. R.W. Sidwell, R.K. Robins and I.W. Hillyard, Pharmacol. Therap. 6:123 (1979).

194. R.A. Smith and W. Kirkpatrick, eds. "Ribavirin, A Broad Spectrum Antiviral Agent," Academic Press, New York (1980) 237 pp.

195. T.-W. Chang and R.C. Heel, Drugs 22:111 (1981).

196. S.M. Bierman, W. Kirkpatrick and H. Fernandez, Chemother. 27:139 (1981).

197. S.A. Patki and P. Gupta, Chemother. 28:298 (1982).

198. V. Knight, S.Z. Wilson, J.M. Quarles, S.E. Greggs, H.W. McClung, B.K. Waters, R.W. Cameron, J.M. Zerwas and R.B. Couch, Lancet 945 (1981).

199. H.W. McClung, V. Knight, B.E. Gilbert, S.Z. Wilson, J.M. Quarles and G.W. Divine, J. Amer. Med. Assn. 249:2671 (1983).

200. C.B. Hall, E.E. Walsh, J.F. Hrusak, R.F. Betts and W.J. Hall, J. Amer. Med Assn. 249:2666 (1983).

201. R.C. Willis, D.A. Carson and J.E. Seegmiller Proc. Natl. Acad. Sci. USA 75:3042 (1978).

202. J.C. Drach, M.A. Thomas, J.W. Barnett, S.H. Smith and C. Shipman, Jr., Science 212:549 (1981).

203. B. Öberg and E. Helgstrand in: "Current Chemotherapy (Proc. 10th Intl. Congr. Chemother., Zurich)," W. Siegenthaler and R. Lüthy eds., ASM, Washington, D.C. (1977) p. 332.

204. B. Eriksson and B. Öberg, Antimicrob. Agents Chemother. 11:946 (1977).

205. D.G. Streeter, J.T. Witkowski, G.P. Khare, R.W. Sidwell, R.J. Bauer, R.K. Robins and L.N. Simon, Proc. Natl. Acad. Sci. USA 70:1174 (1973).

206. D.A. Cooney, H.N. Jayaram, G. Gebeyehu, C.R. Betts, J.A. Kelly, V.E. Marquez and D.G. Johns, Biochem. Pharmacol. 31:2133 (1982).

207. R.K. Robins, P.C Srivastava, V.L. Narayanan, J. Plowman and D.K. Paull, J. Med. Chem. 25:107 (1982).

208. B.B. Goswami, E. Borek, O.K. Sharma, J. Fujitaki and R.A. Smith, Biochem. Biophys. Res. Commun. 89:830 (1979).

209. L.B. Allen and C.M. Fingal, Antimicrob. Agents Chemother. 12:120 (1977).

210. R.K. Robins, G.R. Revankar, P.C. Srivastava, J. Kirsi, J.A. North, B. Murray and P.A. McKernan, Amer. Chem. Soc. Div. Carb. Chem., 185th Natl. Mtg., Seattle (1983) Abst. 45.

211. L.B. Allen, J.H. Huffman, P.D. Cook, R.B. Meyer, Jr., R.K. Robins and R.W. Sidwell, Antimicrob. Agents Chemother. 12:114 (1977).

212. D.G. Streeter and H.H.P. Koyama, Biochem. Pharmacol. 25:2413
     (1976).
213. B. Kojic-Prodic, Z. Ruzic-Toros, L. Golic, B. Brdar and J. Kobe,
     Biochim. Biophys. Acta 698:105 (1982).
214. J.C. Drach, C.M. Reinke, S.H. Smith and P.D. Cook, 21st Inter-
     science Conf. Antimicrob. Agents Chemother., Chicago (1981)
     Abst. 872.
215. J.A. Montgomery, S.J. Clayton and P.K. Chiang, J. Med. Chem.
     25:96 (1982).
216. P.K. Chiang, H.H. Richards and G.L. Cantoni, Molec. Pharmacol.
     13:939 (1977).
217. J.P. Bader, N.R. Brown, P.K. Chiang and G.L. Cantoni, Virology
     89:494 (1978).
218. A.J. Bodner, G.L. Cantoni and P.K. Chiang, Biochem. Biophys.
     Res. Commun. 98:476 (1981).
219. P.K. Chiang and G.L. Cantoni, Biochem. Pharmacol. 28:1897 (1979).
220. J.L. Medzihradsky, T.P. Zimmerman, G. Wolberg and G.B. Elion,
     J. Immunopharmacol. 4:29 (1982).
221. J.A. Montgomery, S.J. Clayton, H.J. Thomas, W.M. Shannon, G.
     Arnett, A.J. Bodner, I.-K. Kion, G.L. Cantoni and P.K. Chiang,
     J. Med. Chem. 25:626 (1982).
222. P.K. Chiang, R.K. Gordon, N.D. Brown, E. DeClercq and J.A.
     Montgomery, Amer. Chem. Soc. Div. Med. Chem., 185th Natl. Mtg.,
     Seattle (1983) Abst. 34.

STRATEGIES IN THE DESIGN OF OLIGONUCLEOTIDES AS POTENTIAL

ANTIVIRAL AGENTS

Paul F. Torrence[1], Jiro Imai[1], Krystyna Lesiak[1],
Jean-Claude Jamoulle[1], Hiroaki Sawai,[2] Johan Warinnier[3]
Jan Balzarini[3] and Erik De Clercq[3]

[1]Laboratory of Chemistry, National Institute of
Arthritis, Diabetes and Digestive and Kidney Diseases,
U.S. National Institutes of Health, Bethesda, Maryland
20205; [2]Faculty of Pharmaceutical Sciences, University
of Tokyo, Bunkyo-ku, Tokyo, Japan; [3]Rega Institute,
University of Leuven, B3000 Leuven, Belgium

Polynucleotides have enjoyed considerable success as antiviral
or antitumor agents due to their interferon-inducing ability[1,2],
their immunoadjuvant activity[3], their reverse-transcriptase
inhibitor properties[4] or their capacity to inhibit virion-
associated transcriptases[5]. On the other hand, the small sequence-
defined oligonucleotides, until recently, have attracted little
interest in this regard. Nonetheless, the potential of oligonucleotides as chemotherapeutic agents has been long appreciated.
Levene and Stollar[6] obtained partial inhibition of systemic lupus
erythematosus sera by tetra- and pentanucleotides, and Shen[7]
suggested the design of high affinity oligonucleotide inhibitors of
antigen-antibody complex formation. Oligonucleotides also have
been employed as prodrug forms[8], but we will not deal with this
particular application.

There are several different strategies that may be envisioned
to employ oligonucleotides as antiviral agents. The first may
be termed the hybridon[9] strategy. In this scenario, a sequence-
specific oligonucleotide might anneal to a complementary region
in the viral genome and either block transcription of the viral
DNA or RNA (of dsDNA viruses or positive- or negative-strand
RNA viruses) or block translation of a positive-strand RNA.
Indeed other possible targets for such a hybridon strategy
exist (vide infra). The literature records several approaches

of this type. A second possibility would be the synthesis of a
<u>fradulent</u> oligonucleotide that would possess the specific sequence
required for recognition by an RNA or DNA binding site of a transcriptase or other protein involved in viral replication. No
reports have appeared that utilize this approach. A third
possibility is to attempt to mimic one of the mechanisms by which
interferon can block virus replication; specifically, to employ
2-5A or its analogs as antiviral (or antitumor) agents.

Probably the earliest successful attempt to capitalize on the
complementarity of nucleic acid interactions came from the group at
Johns Hopkins. In a series of papers[10,11], Miller,, Ts'o and
associates first synthesized the oligodeoxyribonucleotide ethyl
phosphotriesters dTp(Et)dGp(Et)dG and dTp(Et)dTp(Et)dCp(Et)dA
which are complementary to the CpCpA 3'-terminus and -UpGpApA-
anticodon regions, respectively of <u>E. coli</u> or yeast tRNA[Phe].
These triester oligomers formed base-paired complexes with
complementary nucleic acids including the above tRNA's, and
resulting complexes had a greater stability towards dissociation
that the corresponding phosphodiesters, probably due to elimination
of charge repulsion between the phosphates of the polynucleotide and
the phosphates of the oligonucleotide. Such charge neutralized
nucleic acids would presumably have a better opportunity to penetrate the eukaryotic cell membrane; moreover, such oligodeoxy-
nucleotides were found to be chemically stable at pH 7 and ambient
temperatures in aqueous solution and to be completely resistant to
degradation by exonucleases. As an apparent result of the ability
of these oligonucleotide triesters to complex with the corresponding
complementary region of the tRNA, both dTp(Et)dGp(Et)dG and
dTp(Et)dTp(Et)dCp(Et)dA competitively inhibited the aminoacyl-
tRNA synthetase-catalyzed aminoacylation of tRNA at low temperatures;
however, as the incubation temperature was increased from 0° to
37°, the inhibition was decreased as might be expected from the
relatively low Tm of the oligonucleotide-RNA complex. When
different tRNAs were examined for their ability to be amino-
acylated in the presence of the two synthetic oligodeoxynucleotide
triesters, it was found that dTp(Et)dGp(Et)dG was an inhibitor
of aminoacylation of all the tRNA's tested since all these tRNAs
possessed a CCA 3'-terminus. On the other hand, the oligomer
complementary to the anticodon of yeast or <u>E. coli</u> tRNA[Phe]
inhibited tRNA[Phe] aminoacylation most effectively; however, its
specificity was not absolute suggesting that some additional
mode of action. One possibility was related to the observation
that dTp(Et)dTp(Et)dCp(Et)dA non-specifically inhibited pyrophosphate exchange and thereby aminoacyl adenylate generation.

In a second approach,[12] Miller et al. synthesized a
2'-O-methylated oligoribonucleotide triester, Gmp(Et)Gmp(Et)U,
complementary to the -ApCpC-sequence found in the amino acid
accepting stem of most tRNA's. 2'-O-methylated nucleosides
were employed since a) it was discovered that oligoribonucleotides
have tRNA association constants that are eight to twenty times
greater than those of oligodeoxyribonucleotide; b) O-methylation
of the 2'-hydroxyls of the oligonucleotide triesters overcomes
the extreme lability connected with oligoribonucleotide triesters;
c) O-methylation allows retention of predominant ribooligonucleotide
conformation. Like the oligonucleotide triesters described above,
these O-methylated oligomers also formed hydrogen-bonded complexes
with the amino acid accepting stem of yeast tRNA[Phe] as well as
unfractionated tRNA. Similarly Gmp(Et)Gmp(Et)U inhibited aminoacylation of E. coli tRNA[Phe] by phenylalanine even at 37° and
was resistant to degradation by serum or cell lysates. The tritium-
labeled Gmp(Et)Gmp(Et)U was taken up by monolayers of transformed
Syrian hamster fibroblasts. Whereas only intact Gmp(Et)Gmp(Et)U
was found in the medium, the cell lysate contained RNA that was
extensively labeled with uridine and cytidine. Remarkably, nearly
a third of the radioactivity in the lysate was recovered in
sequences of GmGmU containing two, one or no ethyl phosphotriester
residues. Cellular enzymes apparently were able to de-ethylate
the phosphotriesters. The remainder of the radioactivity was
found in various monomeric or dimer species.

The oligonucleotide phosphotriester Gmp(Et)Gmp(Et)U
inhibited protein synthesis (40% at 25 μM) as judged by incorporation of [^{3}H]-leucine into acid-insoluble material. On the other
hand, RNA synthesis (judged by [^{14}C]-adenosine incorporation) was
slightly enhanced during the first few hours after exposure to
the oligomer. Gmp(Et)Gmp(Et)U also caused an inhibition of cell
growth for about 24 hours after its addition to monolayers.
After this time, the treated cells resumed growth at the normal
rate. The oligonucleotide triester had a more dramatic effect
on colony formation by the transformed Syrian hamster fibroblasts:
25 μM oligomer caused a 50% reduction in the number of colonies that
formed as compared to untreated controls. The oligonucleotide
diester, GmpGmpU had no effect on any of the above systems.

Miller et al. allowed that these experiments demonstrated
that the oligonucleotide triester inhibited translation, a
process that involved the oligonucleotide triester targets of
tRNA and mRNA (GpGpU is also complementary to the threonine
codon of mRNA); that the inhibition of translation was reversible

due to the eventual degradation of the triester; that the increased
rate of RNA synthesis represented an attempt by the cell to com-
pensate for the inhibition of protein synthesis.

What might be termed the third generation[13,14] of nonionic
oligonucleotides is represented by the oligonucleotide 3',5'-
methylphosphonates (abbreviated NpN). With complementary
polynucleotides, the analogs form complexes that have Tm's
significantly higher than their natural counterparts, again
likely due to the decreased electrostatic repulsion between
phosphates. The methylphosphonates dApdA, dApdApdA, and
dApdApdApdA, with sequences complementary to the anticodon of
E. coli tRNA[lys], and dGpGpdT, complementary to the ApCpCpApA_{OH}
amino acid accepting stem of tRNA's, were found to block aminoacylation of E. coli tRNA[lys] in an E. coli cell-free system,
but had no effect on aminoacylation of rabbit tRNA[lys] in a
rabbit reticulocyte cell-free system. In an E. coli cell-free
system, the above oligomers inhibited poly(U)-directed phenylalanine
incorporation and poly(A)-directed lysine incorporation. In the
reticulocyte system, poly(U)-directed synthesis also was inhibited,
but globin translation was unaffected. Tritium-labeled dGpdGpdT
was taken up by monolayers of Syrian hamster cells. Of the
material taken up about 30% was associated with the DNA, the
remainder was in the form of dTTP or undegraded dGpdGpdT. Thus,
in distinct contrast to Gmp(Et)Gmp(Et)U, the methylphosphonate
analog was much more stable in the cell. The various synthetic
oligonucleotide methylphosphonates were relatively poor inhibitors
of E. coli colony formation. For instance, it required 160μM
dApdApdApdA to effect a 78% inhibition, however, under these
conditions, no effect on [3H]-leucine incorporation into protein
could be witnessed nor could any effect on the incorporation of
[3H]-thymidine into DNA be seen. Colony formation also was studied
using Syrian hamster cells or human fibroblasts; some marginal
effects on colony formation could be seen but no corresponding
inhibition of translation could be measured.

The first approach that targeted a specific virus for
inhibition was reported by Zamecnik and Stephenson.[9] Their
line of attack was based on the observation that Rous sarcoma
virus 35S RNA possesses a 21 nucleotide sequence near the 5'-cap
and an identical 21 nucleotide sequence next to the poly(A)
terminus some 10,000 nucleotides distant.[15,16] These sequences
are believed to be involved in the circularization of the
proviral DNA before it is incorporated into the cellular genome.
The tridecamer, d(AATGGTAAAATGG) and the corresponding 3',5'-phenyl-
carbamate derivative, originating from reaction of the fully

base-protected triester synthetic intermediate with phenylisocyanate, both inhibited RSV production in chick embryo fibroblasts, although the carbamate derivative was more potent, presumably due to its resistance to degradation by exonuclease. In accord with the idea that it could anneal to regions of the viral RNA, the tridecamer was found to be an efficient primer of the reaction catalyzed by AMV RNA-dependent DNA polymerase when RSV heated 70S or 35S RNA was used as a template.[17]

Although originally prepared to interfere with the circularization step of the proviral DNA intermediate, the oligomer d(AATGGTAAAATGG) also might, by annealing with complementary sequences in the RNA or DNA, block DNA integration, inhibit DNA transcription, interfere with a splicing step involved in mRNA production, block translation initiation or shut-down protein synthesis by association with the ribosome.[17] Stephenson and Zanecnik[17] found that the above oligomer could inhibit translation of denatured RSV 70S RNA in a wheat germ cell-free system. The translation of avian myeloblastosis virus (AMV) RNA as well as globin mRNA and brome mosaic virus RNA also was inhibited by the tridecamer, albeit to a lesser extent than translation of RSV RNA. Other oligonucleotides (an octamer and a decamer) with sequences unrelated to RSV RNA did not cause significant inhibition of translation. The inhibition of AMV RNA translation was held to be due to the partial (10 out of 13 bases) complementarity of the reiterated termini of AMV RNA. The less potent inhibition of globin synthesis observed was held to be due to the partial sequences homology of the 3'terminus of globin mRNA and the 3'-terminus of RSV 35S RNA.

Stridh et al.[18] pursued the possibility that an inhibitor of influenza virus may be obtained through synthesis of an oligomer complementary to the twelve terminal nucleotides of the 3'-end of influenza virus RNA. Influenza virus presents a particularly appealing target for the "hybridon" approach since the virus genome is made up of eight separate segments[19], and each segment has a common sequence at the 5'-terminus and at the 3'-terminus.[20,21] Moreover, these sequences are conserved in several different influenza strains.[20,21] Thus, Stridh et al.[18] prepared d(AGCAAAAGCAGG), complementary to the 3'end of influenza virion RNA, but found that it was not any more active as an inhibitor of influenza RNA polymerase activity than several non-complementary DNA sequences of similar length. Stridh et al.[18] concluded that the 3'end of influenza RNA may be shielded from base-pairing, perhaps by proteins not displaced by the synthesized complementary oligomer.

Two other independent approaches to inhibition of influenza RNA polymerase activity by oligonucleotides have been explored by Stridh et al.[18] The first was based upon the fact that during infection the host cell mRNA serves as a donor of cap structure and several adjoining nucleotides which are incorporated as a primer at the 5'-end of the influenza RNA.[22] This latter phenomenon may be one of the viral targets of ribavirin since as the 5'-triphosphate, ribavirin blocks mRNA guanylyltransferase,[23] the earliest step in mRNA capping. When various cap analogs were evaluated as potential inhibitors of influenza RNA polymerase, it was found that both the 7-methyl and 2'-O-methyl groups were necessary for optimal inhibitory activity in the presence of globin mRNA.[18] Influenza B polymerase seemed to have a preference for caps with two guanosines. These analogs most probably function by competition with the primer, globin mRNA. The second approach to inhibition of influenza RNA polymerase was based upon the observation[24] that dinucleotides could act as initiators of RNA synthesis catalyzed by the influenza virion polymerase. In this case, while the ribonucleotides GpG or ApG were effective stimulators of polymerase activity, the deoxyribonuleotides, dGpdG or dApdG were effective inhibitors. However, even at 100 μM, none of these agents could effect a reduction in plaque formation of influenza A victoria.[18]

Present interest in the 2-5A system[25] can be traced back to some early observations made in Ian Kerr's laboratory. Double-stranded RNA (dsRNA) was found to be a potent inhibitor of translation in extracts of interferon-treated cells but to have no significant effect on protein synthesis in extracts of cells which had not been interferon-treated.[26] This finding could be related to earlier observations that while interferon treatment alone had no effect on translation in resultant cell extracts, the combination of interferon treatment plus virus infection could evoke a translational blockade in subsequent cell extracts.[27] Since dsRNA is generated as a byproduct or intermediate of the replication of at least some viruses, the theory was advanced that interferon treatment generated a latent antiviral state which could be activated by dsRNA.

Presently, two distinct mechanisms have been advanced to explain the inhibition of protein synthesis caused by dsRNA.[25,28,29] The first of these mechanisms is referred to as the 2-5A pathway. According to this scheme, interferon induces the synthesis of an enzyme, 2-5A synthetase, which upon activation by dsRNA, generates 2-5A[30] [ppp5'A2'(p5'A)$_n$] from ATP. This 2-5A activates a latent endonuclease which then degrades mRNA, thereby inhibiting translation. A second separate mechanism that may also be involved in the mechanism of inhibition of translation by dsRNA is referred to as the protein kinase pathway.[25,28,29] Again, a dsRNA-dependent enzyme, protein P_1 kinase, phosphorylates protein P_1 (which may represent an autophosphorylation), and the small α-subunit of eukaryotic initiation factor 2 (eIF-2). This phosphorylation itself may be regulated by a phosphoprotein phosphatase[31] and also may be under dsRNA control.[32] Phosphorylation of the α-subunit of eIF-2 results in a decrease in its ability to engage in initiation complex formation with a 40S ribosomal subunit, GTP and met-$tRNA_f$.[33]

In the mouse L cell-free system, at least two independent lines of evidence suggest that the 2-5A system is the primary mediator of the translational inhibitory effects of dsRNA. First, an established antagonist of the action of 2-5A, p5'A2'p5'A2'p5'A, could prevent most of the protein synthesis inhibitory effects of poly(I)·poly(C) in extracts of interferon-treated L cells.[34] Secondly, an analog of dsRNA, polyadenylic acid·poly(2'-fluoro-2'-deoxyuridylic acid), which was not an activator of the 2-5A synthetase, led to phosphorylation of protein P_1 and eIF-2α, at the same sites as does poly(I)·poly(C), but did not give any inhibition of translation in the mouse L cell-free system. In the reticulocyte lysate system, however, where the protein kinase system is operative, polyadenylic acid·poly(2'-fluoro-2'-deoxyuridylic acid) was as potent an inhibitor of protein synthesis as poly(I)·poly(C).[35]

Several lines of evidence have implicated the 2-5A system in the mechanism of antiviral action of interferon in the intact cell:

i)   when 2-5A is introduced into the cell by various artificial means, such as calcium phosphate co-precipitation,[36] hypertonic salt treatment[37] or microinjection,[38] it gives an inhibition of protein synthesis as well as a blockade of virus replication.

ii) 2-5A, in amounts sufficient to inhibit translation, can be detected in interferon-treated cells that have been treated with poly(I)·poly(C)[39] or infected with reovirus,[40] encephalomyocarditis virus,[41] mengovirus[42] or SV40 virus.[43]

iii) Ribosomal RNA cleavage, peculiar to 2-5A-dependent nuclease action, occurs in interferon-treated cells that have been virus infected.[44]

iv) A cell line, NIH-3T3, deficient in the 2-5A-dependent
     endonuclease, is not protected against EMC virus or
     VSV virus even though interferon induces the 2-5A
     synthetase.[45]

v)   In cells in which the yield of reovirus is inhibited
     by interferon treatment, reovirus mRNA has a much
     shorter half-life than it does in the interferon-
     untreated infected cell.[28] In addition, reovirus
     subviral particles from such interferon-treated cells
     have an ATP-dependent endonuclease activity whereas
     particles from infected cells not treated with inter-
     feron do not have such nuclease activity.[46]

Interferon is certainly a specific antiviral agent.[47] How then
does the 2-5A system attain any specificity of antiviral action?
This question is obviously relevant to any envisioned chemotherapeutic application of the 2-5A system. It seems unlikely that
any specificity may be achieved through specific cleavages by the
2-5A-dependent endonuclease since this enzyme can cleave after
most UpN sequences to give 3'-phosphorylated oligomers terminating
in UpNp.[48,49] In accord with this, the 2-5A system appears to
lead to cleavage of host cell and viral mRNA's without discrimi-
nation. However, when poly(U) was annealled to the poly(A) segment
of VSV RNA or poly(I) was annealled to the poly(C) segment of
EMC RNA, the resulting viral RNA with an attached dsRNA segment
was degraded considerably faster than the same viral RNA without
the added dsRNA segment when these RNA's were added to extracts
of interferon-treated cells supplemented with ATP.[50] Therefore,
the hypothesis[50] has been advanced that 2-5A-dependent endonuclease
activity might be greatest in the area adjacent to the viral
replicative intermediate. A localization of 2-5A action would be
the result of the binding of the 2-5A synthetase to the viral RI
and subsequent 2-5A synthesis combined with the ability of the
2-5A phosphodiesterase to degrade 2-5A. This would presumably
lead to a negative gradient of 2-5A concentration as the distance
from the viral RI increased. Another interesting possibility is
suggested by the studies of Thach[51,52] and colleagues who have
shown that elongation inhibitors such as cycloheximide, emetine,
streptovitacin A and anisomycin may show varying degrees pf
specific antiviral activity and that low concentrations of such
drugs had a specificity similar to interferon itself. This
specificity may reside in the different rates at which different
mRNA's are initiated. Since virulent viruses such as EMC virus
or VSV have mRNA's that, in contrast to most host cell mRNAs,
initiate rapidly, the translation of these viral RNA's is likely
to be limited by elongation. Conversely, the translation of host
cell mRNAs is believed to be limited by initiation so that

initiation inhibitors were observed to block selectively host cell
protein synthesis as opposed to viral protein synthesis.[52]

While dealing with the proposed role of 2-5A in the antiviral
actions of interferon, some caveats need to be added. Not all
evidence is perfectly constant with such a role for 2-5A. For
instance, HEC-1 cells respond with high levels of 2-5A synthetase
(and protein P_1 kinase) upon interferon treatment, but replication
of VSV and EMCV is not blocked in these interferon-treated cells.[53]
Such anomalies are difficult to reconcile with the purported
role of the 2-5A system; however, a better test would include
determination of the _in vivo_ functioning of the 2-5A system, i.e.,
whether 2-5A is synthesized and leads to nuclease activation.
Hela cells present yet another situation. Normally these cells
contain rather high levels of 2-5A synthetase and functional
endonuclease, but they are not protected against virus multiplication since EMCV infection inactivates the endonuclease.[54]
Interferon treatment appears to prevent this inactivation.[54]
This latter phenomenon could represent a limitation in the chemotherapeutic exploitation of the 2-5A system. Finally, it is
likely that interferon may inhibit virus replication by more than
one mechanism,[25,28] and the mechanism which finally obtains may be
a function of both the host cell and the invading virus. Thus,
interferon treatment of rhabdomyosarcoma RD 114 cells suppressed
cell fusion and retrovirus multiplication but did not affect
replication of VSV or RMCV. Neither 2-5A synthetase (nor protein
P_1 kinase) was induced in these cells; heretofore, these enzymes
probably are not connected to the anti-retrovirus effects of
interferon.[55]

Additional interest in 2-5A stems from the possibility that
2-5A may play some role in cell regulation and/or differentiation.
Relatively high levels of the enzyme, 2-5A synthetase, have been
found in cells such as lymphocytes,[56] lymphoblastoid cells,[57]
reticulocytes,[58] cells from dog[59] or mouse liver[60] or estrogen-
stimulated and withdrawn chick oviduct.[61] Conditions other than
interferon treatment may result in an increase in 2-5A synthetase
levels. Thus 2-5A synthetase _and_ 2-5A-dependent endonuclease
underwent substantial increases when JLS-V9R cells were treated with
interferon[62] or when cells proceeded from a subconfluent, actively
growing state to an over-confluent, stationary state.[63] Epidermal
growth factor-urogastrone induced significant elevations of 2-5A
synthetase in quiescent human fibroblasts,[64] dexamethasone induced
the synthetase in lymphoblastoid cells,[65] cortisol induced
synthetase in chick embryo tendon fibroblasts[66] DMSO or sodium
butyrate gave rise to elevated 2-5A synthetase in Maloney sarcoma
virus-transformed murine Balb/c cells,[67] and DMSO induced it in

cultures of Friend erythroleukemia cells.[68] 2-5A itself and/or its
5'-dephosphorylated core, have been detected at various concentrations in a diverse variety of cells including those of mouse spleen,
liver and kidney as well as L1210 cells, Friend cells and human
lymphocytes.[69] Levels of 2-5A synthetase also are elevated in
humans in various situations.[70-73] Of course, interferon treatment and virus (cytomegalovirus, Epstein-Barr, herpes, rubella
and varicella) infection led to rises in 2-5A synthetase levels,
in lymphocytes but so did metastatic breast cancer, severe combined
immunodeficiency disease, systemic lupus erythematosis, and
pregnancy. Lower than normal levels of the enzyme were found in
the PM cells of patients with acute T cell leukemia.

Probably the best evidence that 2-5A has a non-antiviral
function has been provided by experiments of Smekens-Etienne,
et al.[74,75] who have shown that after partial hepatectomy in
rats, intracellular 2-5A synthetase and 2-5A levels dropped
before new DNA synthesis initiated and the liver began to regenerate.
As the DNA synthesis was restored to normal (pre-hepatectomy)
levels and liver regeneration ceased, 2-5A synthetase and 2-5A
levels returned to their original higher levels.

Considerations discussed in the past few paragraphs suggest
that it may be valuable to be able to employ 2-5A or some derivative
thereof in intact cells. This would provide increased opportunities
to pursue the biologic roles of the 2-5A system as well as permit
investigation of a potentially novel approach to virus or cancer
chemotherapy. Moreover such studies need not be confined to 2-5A
itself since antagonists of 2-5A,[34] related to p5'A2'p5'A2'p5'A,
also may be of use in establishing the biological importance of
the 2-5A system and may also find a role in the treatment of
what has been termed interferon-induced disease.[76]

At least two barriers prevent realization of the above goals.
First, the 2-5A molecule has a relatively short half-life (20-30
minutes) in biological systems since it is degraded by a 2',5'-
phosphodiesterase that degrades 2-5A from its 5'end, resulting
in 5'-AMP and 5'-ATP as products.[25,28] Secondly, the 2-5A
molecule, due to its highly ionic character is unable to effectively
penetrate the intact eukaryotic cells. Various drastic procedures
such as calcium phosphate coprecipitation,[36] hypertonic salt[37]
treatment, lysolecithin treatment,[77] or microinjection,[38]
have been employed to allow 2-5A to enter the cell and give an
inhibition of translation or viral growth, but clearly these
procedures are impractical from a chemotherapeutic viewpoint.
One approach to these problems is to chemically modify the 2-5A
molecule so as to obtain the desired properties of phosphodiester-

ase resistance* and enhanced cellular uptake. To pursue this
direction, however, requires a knowledge of what sites on the
2-5A molecule can be altered without adversely affected binding
to or activation of the 2-5A-dependent endonuclease.

1. Relationship of Endonuclease Binding and Activation Ability to Oligoadenylate Chain Length

Several studies have established that for the endonuclease
of several sources (mouse, human), three AMP residues in a
5'-phosphorylated oligonucleotide are needed for maximum interaction with the endonuclease and that high oligomers (> 4 AMP
residues) do not show significantly higher binding.[30,78-83])
This is reflected in the ability of the corresponding oligonucleotide 5'-di- or 5'-triphosphate to activate the 2-5A-
dependent endonuclease. Thus ppp5'A2'p5'A was a poor activator
of the endonuclease since it was bound poorly,[30,78-83] Conversely,
ppp5'A2'(p5'A)$_n$ (n=2-6) were potent endonuclease activators,
albeit the hexamer and heptamer may have been slightly less
effective.[80] Enzyme of rabbit reticulocyte lysates is anomalous
in this regard since it requires a tetramer or higher oligomeric
2-5A for activation even though the 2-5A trimer is bound to the
enzyme.

2. Role of the Oligoadenylate 5'-Substituent

For optimal endonuclease binding ability, a 5'-terminal
mono- or polyphosphate moiety is required. The trimer core,
A2'p5'A2'p5'A, was bound to the endonuclease at least 1000
times less effectively than its corresponding 5'-monophosphate,
p5'A2'p5'A2'p5'A.[78,79,82,83] On the other hand, higher oligomeric
cores, A2'(pA)$_n$ (n=3-5) were bound to the endonuclease only 30-40
times less effectively than their corresponding monophosphates.[79]
One possibility is that the relative increased activity of trimer
core as compared to higher oligomer cores is due to a "slippage"
of the first internucleotide phosphate of the higher oligomer
onto the binding site normally reserved for the 5'-phosphate
moiety.[78,79] That both negative charges of the 5'-phosphate
may not be necessary for endonuclease binding was indicated by
the observation that the 5'-methylester of p5'A2'p5'A2'p5'A,
Mep5'A2'p5'A2'p5'A,[79] and the 5'-S-methylphosphorothioate,

---

* Of course, prolonged stability of a 2-5A derivative also may
be a liability in terms of host cell toxicity. If, however,
in vivo activity could be positively demonstrated, 2-5A analog
stability could be regulated by the proper chemical modification.

MeSp(A2'p)$_2$dA,[80] showed nearly undiminished affinities for the
endonuclease. Endonuclease activation is, however, a different
matter. In this case, a (probably free) 5'-di- or 5'-tri-
phosphate residue is needed for endonuclease activation even
though it has not been established if each are active per se.
A β,γ-methylene analog, pCH$_2$pp5'A2'p5'A2'p5'A, was not active
as a nuclease activator.[84] The 5'-monophosphates, p5'A2'p5'A2'p5'A
and p5'A2'p5'A2'p5'A2'p5'A, were antagonists of endonuclease action
in a cell-free system from mouse L cells, presumably since those
oligomers bound to the 2-5A-dependent endonuclease and, at high
enough concentrations, could prevent 2-5A from binding to and
subsequently activating the enzyme.[34] The 5'-capped 2-5A analogs,
A5'pp5'A2'p5'A2'p5'A, A5'ppp5'A2'p5'A2'p5'A and A5'pppp5'A2'p5'A2'-
p5'A, behave in a manner dependent on the nature of the 5'-5'
polyphosphate linkage.[85] Thus, A5'pp5'A2'p5'A2'p5'A was bound
to the 2-5A-dependent endonuclease as well as 2-5A itself, but did
not activate the endonuclease. The unsymmetrical triphosphate,
A5'ppp5'A2'p5'A2'p5'A was also bound to the endonuclease as
well as 2-5A, but was 100 times less active than 2-5A as an
inhibitor of translation. Finally, the tetraphosphate,
A5'pppp5'A2'p5'A2'p5'A, was equal to 2-5A in binding to and
activation of the 2-5A-dependent endonuclease. The activity of
these capped analogs of 2-5A appeared to be related to their
degradation under protein synthesis conditions.[85] The tetraphosphate, A5'pppp5'A2'p5'A2'p5'A, was degraded ($t_{1/2}$ = 3 min)

to ATP, ADP, AMP, ppp5'A2'p5'A2'p5'A and p5'A2'p5'A2'p5'A
under protein synthesis conditions. The unsymmetrical di- and
triphosphates were, on the other hand, degraded much more slowly
than the tetraphosphate and did not give 2-5A as a degradation
product. Of considerable interest was the observation that when
the degradation of A5'pppp5'A2'p5'A2'p5'A was examined in
incubation mixtures containing human serum or Nalmalwa cell
extract, the tetraphosphate was stable to the action of human
serum, but was rapidly degraded to 2-5A and p5'A2'p5'A2'p5'A
by extracts of the human lymphoplastoid cells.[85] Therefore,
capping (with adenosine) may provide a means of obtaining a
2-5A derivative which would be stable toward cleavage in the
external milieu of the cell, but would be rapidly cleaved by
enzymes of the cytosol to give 2-5A itself.[85]

3. Role of the Oligoadenylate Ribose-Phosphate Backbone

    a. The Nature of the Phosphodiester Linkages

    The existence of the unique 2',5'-phosphodiester bonds in
2-5A, a molecule with such powerful regulatory effects, would
seem to infer their necessity. Nonetheless, the question can

be posed: how critical are the 2',5'-phosphodiester bonds of
2-5A for its interaction with the 2-5A-dependent endonuclease?
To phrase the question another way: Would 3-5A, the 2-5A analog
in which the phosphodiester bonds are 3',5'-linked, possess any
biological activities similar to 2-5A?

     3-5A, ppp5'A3'p5'A3'p5'A, prepared by T4 polynucleotide
kinase phosphorylation of A3'p5'A3'p5'A followed by conversion
to the 5'-triphosphate by means of the 5'-phosphoroimidazolidate
and subsequently characterized by enzymatic digestion and
nucleic magnetic resonance spectrometry, was found to be 10,000
times less effective than 2-5A itself as an inhibitor of translation in extracts of L-cells programmed with EMC virus RNA.[86]
Linkage isomers of 2-5A, ppp5'A2'p5'A3'p5'A and ppp5'A3'p5'A2'p5'A,
were 25-28 times less effective than 2-5A itself as inhibitors of
translation.[86]   Similar results were obtained when these analogs
were examined in a radiobinding assay: ppp5'A3'p5'A2'p5'A,
ppp5'A2'p5'A3'p5'A and ppp5'A3'p5'A3'p5'A were 48, 48 and 13,000
times, respectively less effective than 2-5A in binding to the
2-5A-dependent endonuclease. A nearly identical pattern of
activity was obtained when the corresponding 5'-monophosphates
were evaluated as antagonists of 2-5A action. Finally, the
endonuclease interaction of the above three 2-5A linkage isomers
was evaluated with 2-5A-dependent endonuclease purified ~1,000 x
from Ehrlich ascites tumor cells.[86]   In this case, also, the
quantitative behavior of the three isomers was virtually the same
as other systems which employed crude enzyme, effectively ruling
out the possibility that the failure to observe activity of 3-5A
was due to its premature degradation by phosphodiesterases. Thus,
substitution of either of the phosphodiester linkages of
ppp5'A2'p5'A2'p5'A with 3',5'-bonds results in an order of
magnitude or more loss of binding or activation ability. If
binding at one phosphodiester bond site is independent from
binding at the other, then these data could be used to predict
that replacement of both phosphodiester bonds of (pp)p5'A2'p5'-
A2'p5'A should lead to a greater than two orders of magnitude
loss in binding or activation ability, in accordance with the
experimental observations on 3-5A itself.[86]

     The above conclusion is, however, challenged by a report by
Rappaport et al.[87]   The diphosphate pp5'A3'p5'A3'p5'A was prepared
by chemical phosphorylation of the product of digestion of poly(A)
with nuclease P$_1$. No chemical characterization of this product
was reported. This material was compared with the corresponding
diphosphate of 2-5A, pp5'A2'p5'A2'p5'A, for its ability to inhibit
translation after calcium phosphate coprecipitation in an intact
L cell system. Under these conditions, the pp5'A3'p5'A3'p5'A
was 5-10X less active than pp5'A2'p5'A2'p5'A giving 50% inhibition

of protein synthesis at $4 \cdot 10^{-6}$M. When the ribosomal RNA was extracted from such treated cells, the characteristic 2-5A-induced
pattern of cleavage was seen with both the isomers, pp5'A2'p5'A2'-
p5'A and pp5'A3'p5'A3'p5'A. Consistent with the previously
discussed experiments[86] the 3',5'-phosphodiester bond isomer of
2-5A (or pp5'A2'p5'A2'p5'A) could activate the 2-5A-dependent
endonuclease. The major difference between the conflicting studies
is the magnitude of the difference. Studies of inhibition of translation or radiobinding assays in L cell-free systems indicated
that 3-5A is $\approx$ 10,000 times less active than 2-5A. Studies in a
calcium phosphate coprecipitation system in intact L cells
suggested a maximum tenfold difference. It would seem that in
the latter system, the exact concentrations of oligoadenylate
present in the cell may be rather difficult to control and this
could account for some of the observed differences.

b. Role of the 3'-Hydroxyl Groups of 2-5A

Recently Doetsch et al.[88] reported the preparation of a
putative cordycepin analogue of 2-5A; namely, ppp5'(3'dA)2'p5'-
3'dA)2'p5'(3'dA). They claimed that in rabbit reticulocyte
lysate this analogue was a more potent inhibitor of translation
than was 2-5A trimer triphosphate and that the new analogue
was equipotent with 2-5A tetramer triphosphate. Furthermore,
the "core" of this analogue, i.e., (3'dA)2'p5'(3'dA)2'p5'(3'dA),
was reported[89] to block transformation of human lymphocytes
infected by Epstein-Barr virus. Based on such results, it was
suggested that cordycepin "core" or 2-5A core itself might
replace or augment interferon treatment.

Doetsch et al.[88] prepared the above cordycepin analogue of
2-5A by use of the enzyme 2-5A synthetase. Although Charubala
and Pfleiderer[90] reported an unambiguous chemical synthesis of the
cordycepin core, they did not convert it to the 5'-triphosphate,
a step necessary to determine its interaction with the 2-5A-
dependent nuclease. Sawai et al.[91] have reported an unambiguous
synthesis and structure proof of the cordycepin analogues of 2-5A
trimer and tetramer, specifically, ppp5'(3'dA)2'p5'(3'dA)2'p5'(3'dA)
and ppp5'(3'dA)2'p5'(3'dA)2'p5'(3'dA)2'p5'(3'dA). These materials
were thoroughly characterized by enzymic digests and proton and
phosphorus NMR.[91] Moreover, the bacterial alkaline phosphatase
digest product of the trimer analogue was identical to the trimer
core prepared by independent means by Charubala and Pfleiderer.[90]
Surprisingly, none of the cordycepin 2-5A analogues could inhibit
protein synthesis in L cell extracts in concentrations up to 10^{-5}M
under conditions where 2-5A itself gave a 50% reduction in trans-

lation at 10^{-9} M. In reticulocyte lysates, the trimer analogue
was devoid of activity; the cordycepin tetramer, ppp5'(3'dA)-
2'p5'(3'dA)2'p5'(3'dA)2'p5'(3'dA) was at least[91] 20-100 times
less active than 2-5A tetramer triphosphate. Nonetheless,
although these cordycepin 2-5A analogues could not inhibit
protein synthesis effectively, they were able to interact with
the 2-5A-activated endonuclease only 10X less effectively as
2-5A itself as determined by radiobinding assays and ability
to antagonize 2-5A action. These data demonstrated the i) the
3'-hydroxyl groups of 2-5A are not essential in determing binding
to the endonuclease, but are critical determinants of enzyme
activation; ii) it is unlikely that the reported ability of
cordycepin trimer core to inhibit transformation of lymphocytes
by EBV is due to activation of the 2-5A-dependent endonulcease;
iii) since cordycepin analogues of 2-5A are poor activators of
the 2-5A-dependent endonuclease, there is no theoretical basis
to expect that they could replace interferon in chemotherapeutic
situations.

A collaborative effort[80] between groups at Kings College
and the Imperial Cancer Research Fund Laboratories in London
has verified the findings of Sawai et al.[91] The cordycepin
analog 5'-diphosphate, pp5'(3'dA)2'p5'(3'dA)2'p5'(3'dA) was
synthesized by an independent unambiguous stepwise method and
was found to be unable to activate the endonuclease of either
mouse L cells or human (Daudi) cells although it did bind to
the endonuclease of mouse L-cells, Daudi cells and rabbit
reticulocytes, albeit less effectively than 2-5A itself.[80]

4. Role of the Adenine Basis of 2-5A

Thus far, this laboratory has examined the biological activity
of 2-5A analogs with the following nucleosides substituted for
adenosine: uridine,[78,79] cytidine,[78,79] inosine,[78,79] 1,N[6]-
ethenoadenosine[92] and tubercidin (7-deazaadenosine).[93] All
of these analogs were prepared by chemical synthesis and all of
the proposed structures have been corroborated by enzymatic
digestion patterns, ^{1}H-NMR and ^{31}P-NMR. The base-modified
analogs, p5'C2'p5'C2'p5'C, p5'U2'p5'U2'p5'U, p5'I2'p5'I2'p5'I
and p5'(εA)2'p5'(εA)2'p5'(εA), showed dramatically reduced
antagonistic and endonuclease binding capacities.[78,79,92]
Apparently, just the simple elimination of the adenosine amino
group was sufficient to lead to a >2000-fold loss of endonuclease
binding affinity. Moreover, since the inosinate analog showed
comparable activities to the cytidylate and uridylate analogs, it
would appear that the adenine 6-amino group and possibly its
N-1 nitrogen may be the greatest contributors to the binding of the

adenine rings to the endonuclease. The triphosphate, ppp5'-
(εA)2'p5'(εA)2'p5'(εA),[92] has also been prepared and was devoid
of endonuclease activation ability, in distinct contrast to a
previous report which provided no proof of structure.

Three reports have appeared concerning oligoinosinate analogs
of 2-5A.[94-96] In these studies, the preparation of the inosinate
analog was claimed by either nitrous acid deamination[94] of 2-5A
or by enzymatic synthesis using the 2-5A synthetase from mouse
L cells,[95] although the former group of investigators[94] found that
ITP was not a substrate for the reticulocyte enzyme.[94] No
structure proof of the putative oligoinosinate products has been
forwarded. When the 5'-triphosphate, ppp5'I2'p5'I2'p5'I, was
prepared in our laboratory by standard chemical procedures, it
was found to be bound to the 2-5A-dependent endonuclease of
mouse L cells at least 500 times less effectively than 2-5A
and to be 10,000 times less active than 2-5A as an inhibitor of
protein synthesis.[79]

The tubercidin analog of 2-5A, however, presents a somewhat
different story.[93] When evaluated for its ability to bind to the
2-5A-dependent endonuclease of mouse L cells, the tubercidin
analog of 2-5A, namely ppp5'(c^{7}A)2'p5'(c^{7}A)2'p5'(c^{7}A), was bound
as effectively as 2-5A itself; nonetheless, it and the corresponding tetramer, ppp5'(c^{7}A)2'p5'(c^{7}A)2'p5'(c^{7}A)2'p5'(c^{7}A), failed
to activate the 2-5A-dependent endonuclease as judged by its
inability to inhibit translation in extracts of mouse L cell
programmed with EMCV RNA or to give rise to ribosomal RNA
cleavage in the same cell system under conditions where 2-5A
showed activity at 10^{-9} M.[93] As would be expected from such
behavior, ppp5'(c^{7}A)2'p5'(c^{7}A)2'p5'(c^{7}A) was an antagonist of
2-5A action in the L cell extract.[93] In the lysed rabbit reticulocyte system, both the trimer, ppp5'(c^{7}A)2'p5'(c^{7}A)2'p5'(c^{7}A), and
tetramer, ppp5'(c^{7}A)2'p5'(c^{7}A)2'p5'(c^{7}A)2'p5'(c^{7}A), were bound to
the endonuclease as well as 2-5A, but while the trimer species
was inactive as it was in the L cell system, the tetramer was
just as potent an inhibitor of translation as 2-5A tetramer
itself.[93] Moreover, the inhibition of protein synthesis caused
by ppp5'(c^{7}A)2'p5'(c^{7}A)2'p5'(c^{7}A)2'p5'(c^{7}A) was antagonized by
p5'A2'p5'A2'p5'A, demonstrating that the inhibition was mediated
via the 2-5A-dependent endonuclease. Therefore, for the endonuclease of mouse L cells, human (Daudi) cells and rabbit
reticulocytes, it appears that the purine N7's of the adenine rings
of 2-5A are not needed for effective endonuclease binding.
Activation of the mouse cell L cell endonuclease, but not rabbit
reticulocyte endonuclease, is dependent on one or more of the purine
N7 moieties for activation of its nuclease function.[93]

5. Modifications at the 2'-Terminus of 2-5A and the Development
of Analogs Resistant to Degradation by Phosphodiesterase Action

        The relative importance of the third or 2'-terminal residue
of 2',5'-(pA)$_3$ or 2-5A itself can be ascertained by partly
dissecting the terminal residue.[79] When adenosine was removed from
the tetramer, p5'A2'p5'A2'p5'A2'p5'A, the resulting diphosphorylated
oligonucleotide, p5'A2'p5'A2'p5'A2'p, became a more effective
antagonist of 2-5A action and possessed the same apparent
affinity for the endonuclease as did the tetramer p5'A2'p5'A2'p5'-
A2'p5'A itself.[79] On the contrary, a similar shortening of the
trimer, p5'A2'p5'A2'p5'A, to give the diphosphorylated dimer,
p5'A2'p5'A2'p, produced a compound with dramtically reduced
binding activity. Thus, the third or 2'-terminal adenosine
moiety must make a significant contribution to endonuclease
binding.[79] Most of this contribution may reside in the terminal
adenine ring since conversion of the 2'-terminal ribose of the
trimer p5'A2'p5'A2'p5'A to an N-hexylmorpholine derivative
increased antagonistic capacity relative to the parent compound,
thereby implying the lack of importance of the ribose ring of
the 2'-terminal adenosine residue in binding to the endonuclease.[97]
To the contrary, Drocourt et al.[98] found that while compounds of
the general formula ppp5'A2'p5'A2'p5'N (where N is any of the
common nucleosides) could not activate the endonuclease, but
could antagonize 2-5A action as effectively as p5'A2'p5'A2'p5'A
itself, implying an equally efficient binding to the endonuclease.
This question deserves reexamination since the latter data were
obtained in a calcium phosphate coprecipitation system wherein
the actual concentrations of oligonulceotide entering the cell
are impossible to ascertain.

        From the foregoing, it has been reasoned that certain regions
of the 2-5A molecule are strategically involved in binding to
and/or activation of the 2-5A-dependent endonuclease.[99] These
regions include the 5'-phosphate or 5-triphosphate residue, the
heterocyclic bases and the ribose phosphate backbone. It is
likely, however, that the 2-terminus will tolerate considerable
changes without loss of ability to bind to the endonuclease.
Thus, this area of the molecule may be expected to be subject to
chemical modifications which would not affect adversely the binding
to or activation of the 2-5A-dependent endonuclease. In line with
this, it has been found that conversion of the 2'-terminal ribose
of 2-5A tetramer to an N-hexylmorpholine gave a new 2-5A analog
with 5-10 times the translational inhibitory activity of natural
2-5A.[97] The increased biological activity of this analog is

most probably related to its increased resistance to degradation[97]
since it had a similar binding affinity to the endonuclease as
2-5A itself,[100] but, under conditions of protein synthesis, was
undergraded after 5 hours incubation whereas 2-5A was degraded
with a half-life of 20 minutes.[97]

Other derivatives of 2-5A with modifications at the 2'-terminus
also have been shown to possess increased resistance to degradation.
While the permethylated 2-5A analog, ppp5'(Am)2'p5'(Am)2'p5'(Am),
did not activate the 2-5A-dependent nuclease of He La cells, the
terminally methylated analog, ppp5'A2'p5'A2'p5'(Am) was approximately three times more effective than 2-5A itself in this regard,[84]
apparently due to its increased resistance to degradation.
Silverman et al. found that 3'-terminally phosphorylated derivatives
of 2-5A were more resistant to degradation; however, while
pppA2'p5'A2'p5'A2'p5'A3'p5'Cp was as active as 2-5A tetramer
itself in extracts of rabbit reticulocytes, in extracts of mouse
L cells or Ehrlich ascites tumor cells, it was $\geq$ 30-fold less
active than 2-5A.[101] Haugh et al.[80] also have reported that various
oligomers modified at the 2'-terminal residue [pp5'A2'p5'A2'p5'(3'dA),
pp5'A2'p5'A2'p5'A and pp5'A2'p5'A2'p5'A2,3'CH$_2$ (the last residue
is 2',3'-isopropylidene adenosine) had half-lifes 6-10 times longer
than 2-5A in mouse or human cell extracts. The analogs terminating
in 3'dA or A$_m$ were able to bind to and activate the 2-5A-dependent
nuclease about as well as 2-5A; however, the analog terminating in[80]
isopropylidene adenosine was significantly less active.

One other aspect of 2-5A analogs may be mentioned at this point
even though the 2-5A-dependent endonuclease is not involved.
2-5A has been reported to inhibit both purified viral and cellular
(guanine-7)-methyl-transferases:[102] the 3'-O-methylated analogs
of 2-5A, with a methyl group only at the 3'-terminus or completely
methylated at all ribose 3'-hydroxyl groups, was reported to
preferentially inhibit the virus enzyme as compared to the cellular
enzyme, probably as a competitor of mRNA substrate.[102] Significantly, the concentrations of 2-5A or analog need to bring about
such inhibitions were 1,000-10,000X greater than the concentrations
of 2-5A needed to block protein synthesis.

Parenthetically, it may be noted that an inhibition of
(guanine-7)methyltransferase activity has been observed in extracts
of interferon-treated cells when such extracts were incubated with
capped but unmethylated reovirus mRNA.[103] Others have found that
in interferon-treated He La cells, a fraction of VSV mRNA did not
associate with the polysomes and did not bind to reticulocyte
ribosomes probably due to the lack of a 7-methyl group in the

terminal guanosine in the cap.[104] This inhibition may be explained
by the increased concentrations of S-adenosylhomocysteine found
in interferon-treated HeLa cells.[105] On the other hand, in the case
of SV40 virus infection in interferon-treated cells, the SV40 mRNA
and the cellular RNA were overmethylated at internal m^6A residues.

In view of the established antiproliferative and antitumor
effects of interferon,[107] and because of the inability of 2-5A
itself to penetrate the intact cell, considerable interest has
greeted the reports that 2-5A core (A'2p5'A2'p5'A) exhibited antimitogenic activity in mouse splenic leukocytes,[108,109] Balb/c
3T3 cells,[110] lymphoblastoid cells[111] and Swiss 3T3 cells,[112] just
as did interferon itself. Although never rigorously established,
the suggestion was advanced that such activity was due to conversion of the 2-5A core to 2-5A itself by phosphorylation in
the intact cell.[17] Recently, several investigations have suggested
that this latter hypothesis is not a viable one. Eppstein et al.[113]
could find no evidence of 2-5A-characteristic ribosomal RNA
cleavages in rRNA extracted from Swiss 3T3 cells treated with 2-5A
trimer core under conditions that gave rise to an antimitogenic
effect. Torrence et al.[114] (vide infra) noted that the structure-
activity requirements for activation of the 2-5A-dependent
endonuclease and the antimitogenic effect of oligoadenylates were
not the same.

Some reports have appeared regarding the antimitogenic effects
of 2-5A analogs. A xylo analog of 2-5A, in which the 2'-OH
group is of inverted ("up") configuration, was dramatically
more stable to degradation in cell extracts than 2-5A core itself
and had a potency 5X greater than 2-5A core in inhibiting [^{3}H]-
thymidine or [^{3}H]-leucine incorporation into DNA and protein,
respectively, of Swiss 3T3 cells.[112] This xylofuranosyladenosine
analog of 2-5A core had an antiviral activity against herpes
simplex-1 and 2 similar to acycloguanosine.[113] The distinct
possibility exists, however, that this analogue owes its
action to a release of xylo A itself. It is unlikely that the
xylo A core owes its activities to prior phosphorylation at the
5'-terminus and subsequent activation of the 2-5A-dependent
endonuclease since the 5'-monophosphate of the xylo A core is
bound poorly to the endonuclease (Torrence and Lesiak, unpublished).

Other analogs which have been examined for their antimitogenic activity include one in which the internal phosphodiester bonds of 2-5A were completely methylated to give methyl
triesters as the internucleotide linkages.[112] This analog was
devoid of antimitogenic activity. The cordycepin analog of
2-5A core, i.e., (3'dA)2'p5'(3'dA)2'p5'(3'dA), possessed

significantly greater antimitogenic activity than 2-5A itself,
but when the 2'-terminal residue was replaced with 2'-deoxy-[112]
adenosine or adenosine, the resulting analogs lost activity.
In this instance also, the role of released cordycepin or cordycepin mononucleotides in mediating the observed effects needs
to be ascertained. Kimchi et al.[110] have reported on the antimitogenic activity of a series of acetylated or butyrylated
2-5A core derivatives. When all the free hydroxyl groups of 2-5A
trimer core were substituted with acetyl or butyryl groups, the
resulting analogs lost most of their antimitogenic activity. When
only the 5'-hydroxylbutyryl group of the butyrylated derivative was
removed, antimitogenic activity was retained. Finally, when the 5'-
hydroxyl of 2-5A core trimer was unsubstituted, but all other
hydroxyl and adenine amino functions were butyrylated, there was a
complete loss of antimitogenic activity. This suggested to Kimchi
et al.[110] that a free 5'-hydroxyl residue was essential for activity
in support of the hypothesis that it becomes phosphorylated.

A recent study[114] has examined the antimitogenic activity (in
synchronized Balb/c 3T3 cells) of a series of 2-5A core congeners
which differed in oligonucleotide chain length and the nature
of the substituent at the 5'-terminus. Some of the conclusions of
this study[114] included the findings that i) in the class of 2',5'-
linked core oligoadenylates, the most potent antimitogen was 2-5A
trimer core, A2'p5'A2'p5'A. Oligomers of shorter or longer chain
length were less active. The oligomers could be arranged in the
following order of (descending) activity: A2'p5'A2'p5'A >
A2'p5'A2'p5'A2'p5'A > A2'p5'A2'p5'A2'p5'A2'p5'A > A2'p5'A >
A2'p5'A2'p5'A2'p5'A2'p5'A2'p5'A. ii) 5'-monophosphorylated
2',5'-core oligoadenylates also were potent antimitogenic agents;
however, the most potent of this series was the tetramer,
p5'A2'p5'A2'p5'A2'p5'A, and the 5'-phosphorylated oligoadenylates
could be arranged in the following order of (descending) activity:
p5'A2'p5'A2'p5'A2'p5'A > p5'A2'p5'A2'p5'A2'p5'A2'p5'A >
p5'A2'p5'A2'p5'A > p5'A2'p5'A2'p5'A2'p5'A2'p5'A2'p5'A > p5'A2'p5'A.
iii) Surprisingly, 5'-triphosphorylated 2',5'-core oligomers,
including 2-5A trimer, possessed some apparent antimitogenic
properties. In this case, the dimer triphosphate, ppp5'A2'p5'A,
was the most potent agent followed by the trimer triphosphate.
The tetramer triphosphate was nearly devoid of activity. (v)
P^2-Phenyl-substituted 2',5'-linked core 5'-diphosphates showed
antimitogenic behavior which was maximal with the trimer,
Øpp5'A2'p5'A2'p5'A. The dimer, Øpp5'A2'p5'A, expressed some
marginal activity, but the derivative of adenosine itself, ØppA was
without activity. Several compounds emerged from this study as
either equipotent with or even more potent than 2-5A core,

A2'p5'A2'p5'A, as antimitogens. These included p5'A2'p5'A2'p5'A,
p5'A2'p5'A2'p5'A2'p5'A, p5'A2'p5'A2'p5'A2'p5'A2'p5'A,
ppp5'A2'p5'A2'p5'A and Øpp5'A2'p5'A2'p5'A. Outstanding among these
was the tetramer 5'-monophosphate which was definitely more active
than 2-5A trimer or tetramer cores (A2'p5'A2'p5'A or A2'p5'A2'p5'-
A2'p5'A). These results[114] implied that chemical modification
does not mitigate but rather potentiates the antimitogenic
activity of 2',5'-linked oligoadenylates. Whether the mechanism
behind this phenomon shares any common basis with the antiviral
action of either interferon or 2-5A and whether these observations
can be developed into useful agents for use in the whole animal
remain to be established.

Assuming, rather optimistically, that problems associated
with uptake or premature degradation can be solved eventually,
what could be considered some of the most important barriers to
the application of oligonucleotides as antiviral agents or chemotherapeutic agents in particular?

1. In regards to the hybridon strategy, the anticomplementary
oligonucleotide may need either to displace proteins bound to
the specific target sequence or may need to outcompete an anticomplementary sequence in the same molecule if that sequence has
engaged in secondary structure formation with the target sequence.
Approaches to these problems could include the use of modified
nucleosides that would give a thermodynamic edge to the oligonucleotide hybridon; for instance, substitution of bromine for
methyl at the 5-position of the uracil ring leads to a significant
increase in the Tm's of resulting polynucleotide duplexes. In
addition, to insure that once the oligonucleotide hybridon achieved
interaction with its target sequence, and would not be subsequently
displaced, the oligonulceotide might be chemically crosslinked to
the target strand.

2. In regards to the potential antiviral applications of
the 2-5A system, a serious question arises concerning how these
agents might show specific activity. Unless they follow a course
similar to the low concentrations of elongation inhibitors used
in the experiments (vide supra) from Thach's laboratory, then,
on the basis of present knowledge they may not be expected to
exhibit selectivity in blockade of host versus viral functions.
Nonetheless, the basis for the selectivity of interferon action
and the part that 2-5A plays in this is at best dimly understood,
so that predictions are difficult to make. The best course in
this regard is an empirical one; developing 2-5A analogs with in
vivo activity--then determining what the limitations of such agents
may be.

3. Another serious question regarding antiviral applications of the 2-5A system pertains to the possibility that virus infection may inactivate the 2-5A-dependent endonuclease (vide supra). The generality of this phenomenon with respect to virus and host cell needs to be determined.

Oligonucleotides represent a substantially untapped resource so far as the development of chemotherapeutic agents is concerned. Presently, however, the most rudimentary problems foil progress. By concentrating on a particular scheme, the 2-5A system, with potent biological activity, it may be possible to develop strategies and guidelines which could be applied to numerous other situations. In all this, of course, serendipity would be a welcome research companion.

REFERENCES

1. P. F. Torrence and E. De Clercq, Pharmacol. Therapeu. 2, 1-88 (1977).
2. E. De Clercq and P. F. Torrence, Tex. Rep. Biol. Med. 41, 76-83 (1982).
3. A. G. Johnson, Springer Seminar, Immunopathol. 2, 149-168 (1979).
4. E. De Clercq, Biochem. J. 205, 1-13 (1982).
5. E. M. Round and N. Stebbing, Antiviral Res. 1, 237-248 (1981).
6. L. Levine and B. D. Stollar, Progr. Allergy 12, 161 (1968).
7. T. Y. Shen, Angew. Chem. Internat. Edit. 9, 678-688 (1970).
8. C. G. Smith, H. H. Buskirk and W. L. Lummis, J. Med. Chem. 10, 774-776.
9. P. C. Zamecnik and M. L. Stephenson, Proc. Natl. Acad. Sci. U.S.A. 75, 280-284 (1978).
10. P. S. Miller, J. C. Barrett and P.O.P. Ts'o, Biochemistry 13, 4887-4896 (1974).
11. J. C. Barrett, P. S. Miller and P.O.P. Ts'o, Biochemistry 13, 4897-4906 (1974).
12. P. S. Miller, L. T. Braiterman and P.O.P. Ts'o, Biochemistry 16, 1988-1996 (1977).
13. P. S. Miller, J. Yano, E. Yano, C. Carroll, K. Jayaraman, and P.O.P. Ts'o, Biochemistry 18, 5134-5142 (1979).
14. P. S. Miller, K. B. McParand, K. Jayaraman and P.O.P. Ts'o, Biochemistry 20, 1874-1880 (1981).
15. D. E. Schwartz, P. C. Zamecnik and H. L. Weith, Proc. Natl. Acad. Sci. U.S.A. 74, 994-998 (1977).
16. W. A. Haseltine, A. M. Maxam and W. Gilbert, Proc. Natl. Acad. Sci. U.S.A. 74, 989-993 (1977).

41. B. R. G. Williams, R. R. Golgher, R. E. Brown, C. S. Gilbert
    and I. M. Kerr, Nature 282, 582-586 (1979).
42. C. Vaguero, J. Sanceau and R. Falcoff, Biochem. Biophys. Res.
    Comm. 107, 974-980 (1982).
43. C. L. Hersh and G. R. Stark, J. Cell. Biochem. Supp 6, 98
    (1982).
44. D. H. Wreschner, T. C. James, R. H. Silverman and I. M. Kerr,
    Nucleic Acids Res. 9, 1571-1578 (1981).
45. D. A. Epstein, C. W. Czarniecki, H. Jacobsen, R. M. Friedman
    and A. Panet, Eur. J. Biochem. 118, 9015 (1981).
46. R. L. Galster and P. Lengyel, Nucleic Acids Res. 3, 581-598
    (1976).
47. M. I. Johnston and P. F. Torrence, in: "Interferon: Mechanisms
    of Production and Action," R. M. Friedman, ed., Elsevier,
    Amsterdam, in press.
48. D. H. Wreschner, J. N. McCauley, J. J. Skehel and I. M. Kerr,
    Nature 289, 414-417 (1981).
49. G. Floyd-Smith, E. Slattery and P. Lengyel, Science 212,
    1030-1032 (1981).
50. T. W. Nilsen and C. Baglioni, Proc. Natl. Acad. Sci. U.S.A.
    76, 2600-2004 (1979).
51. P. M. P. Yan, T. Godefroy-Colburn, C. H. Birge, T. V.
    Ramakhadran and R. E. Thach, J. Virol. 27, 648-658 (1978).
52. T. V. Ramakhadran and R. E. Thach, J. Virol. 34, 293-296
    (1980).
53. M. Verhagen, M. Divizia, P. Vandenbussche, T. Kuwata, and
    J. Content. Proc. Natl. Acad. Sci. U.S.A. 77, 4479-4483
    (1980).
54. R. H. Silverman, P. J. Cayley, M. Knight, C. S. Gilbert and
    I. M. Kerr, Eur. J. Biochem. 124, 131-138 (1982).
55. Y. Tomita, J. Nishimaki, F. Takahashi and T. Kuwata. Virology
    120, 258-263 (1982).
56. N. Shimizu and Y. Sokawa, J. Biol. Chem. 254, 12034-12037 (1979).
57. M. I. Johnston, K. C. Zoon, R. M. Friedman, E. De Clercq and
    P. F. Torrence, Biochem. Biophys. Res. Comm. 97, 375-383
    (1980).
58. A. G. Hovanessian and I. M. Kerr, Eur. J. Biochem. 84, 149-159
    (1978).
59. M. Etienne-Smekens, G. Vassart, J. Content and J. Dumont,
    FEBS Lett. 125, 146-150 (1981).
60. T. W. Nilsen, D. L. Wood and C. Baglioni, J. Bio. Chem. 256,
    10751-10754 (1981).
61. G. R. Stark, W. J. Dower, R. T. Schimke, R. E. Brown and
    I. M. Kerr, Nature 278, 471-473 (1979).
62. H. Jacobsen, C. W. Czarniecki, D. Krause, R. M. Friedman and
    R. H. Silverman, Virology 125, 496-501 (1983).

17. M. L. Stephenson and P. C. Zamecnik, Proc. Natl. Acad. Sci. U.S.A. 75, 285–288 (1978).
18. S. Stridh, B. Oberg, J. Chattopadhyaya and S. Josephson, Antiviral Res. 1, 97–105 (1981).
19. P. Palese, Cell 10, 1–10 (1977).
20. J. J. Skehel and A. J. Hay, Nucleic Acids Res. 5, 1207–1219 (1978).
21. J. S. Robertson, Nucleic Acids Res. 6, 3745–3757 (1979).
22. S. J. Ploteh, M. Bouloy and R. M. Krug, Proc. Natl. Acad. Sci. U.S.A. 76, 1618–1622 (1979).
23. B. B. Goswami, E. Borek, O. K. Sharma, J. Fujitaki and R. A. Smith, Biochem. Biophys. Res. Commun. 89, 830–836 (1979).
24. S. J. Plotch, J. Tomasz and R. M. Krug, J. Virol. 28, 75–83 (1978).
25. P. F. Torrence, Mol. Aspects Med. 5, 129–171 (1982).
26. I. M. Kerr, R. E. Brown and L. A. Ball, Nature 250, 57–59 (1974).
27. R. M. Friedman, D. H. Metz, R. M. Esteban, D. R. Tovell, L. A. Ball and I. M. Kerr, J. Virol. 10, 1184–1198 (1972).
28. P. Lengyel, Ann. Rev. Biochem. 51, 251–282 (1982).
29. G. C. Sen, Progr. Nucleic Acid Res. Mol. Biol. 27, 105–156 (1982).
30. I. M. Kerr and R. E. Brown, Proc. Natl. Acad. Sci. U.S.A. 75, 256–260 (1978).
31. A. Kimchi, A. Zilberstein, A. Schmidt, L. Shulman and M. Revel, J. Biol. Chem. 254, 9846–9853 (1979).
32. D. A. Epstein, P. F. Torrence and R. M. Friedman, Proc. Natl. Acad. Sci. U.S.A. 77, 107–111 (1980).
33. R. Jagus, W. F. Anderson and B. Safer, Progr. Nucleic Acids Res. Mol. Biol. 25, 127–185 (1981).
34. P. F. Torrence, J. Imai and M. I. Johnston, Proc. Natl. Acad. Sci. U.S.A. 78, 5993–5997 (1981).
35. H. Jacobsen, D. A. Epstein, R. M. Friedman, B. Safer and P. F. Torrence, Proc. Natl. Acad. Sci. U.S.A. 80, 41–45 (1983).
36. A. G. Hovanessan and J. N. Wood, Virology 101, 81–90 (1980).
37. B. R. G. Williams, R. R. Golgher and I. M. Kerr, FEBS Lett. 105, 47–52 (1979).
38. Y. Higashi and Y. Sokawa, J. Biochem. (Tokyo) 91, 2021–2028 (1982).
39. T. W. Nilsen, P. A. Maroney and C. Baglioni, J. Biol. Chem. 256, 7806–7811 (1981).
40. T. W. Nilsen, P. A. Maroney and C. Baglioni, J. Virol. 42, 1039–1045 (1982).

63. H. Jacobsen, D. Krause, R. M. Friedman and R. H. Silverman,
       Proc. Natl. Acad. Sci. U.S.A., in press.
64. S. L. Lin, P.O.P. Ts'o and M. D. Hollenberg, Life Sciences
       32, 1479–1488 (1983).
65. I. Krishnan and C. Baglioni, Proc. Natl. Acad. Sci. U.S.A. 77,
       6506–6510 (1980).
66. J. Oikarinen, Biochem. Biophys. Res. Comm. 105, 876–881 (1982).
67. F. Besancon, M. F. Bourgeade, J. Justesen, D. Ferbus, and
       M. N. Thang. Biochem. Biophys. Res. Comm. 103, 16–24
       (1981).
68. A. Kimchi, J. Interferon Res. 1, 559–569 (1981).
69. H. Cailla, L. Laurence, C. LeBorgne de Kaouel, D. Roux and
       J. Marti, in: "Radioimmunoassay and Related Procedures
       in Medicine 1982," International Atomic Energy Agency,
       Vienna, Austria, pp. 33–44.
70. D. Wallach, A. Schattner, G. Merlin, A. Kimchi, M. Fellous and
       M. Revel, in: "Interferons," T. Merigan, R. M. Friedman
       and C. F. Fox, eds., Academic Press, N. Y., pp. 449–463
       (1982).
71. A Schattner, D. Wallach, G. Merlin, T. Hahn, S. Levin, B.
       Ramot and M. Revel, J. Interferon Res. 2, 355–361 (1982).
72. J. A. Merritt, E. C. Borden and L. A. Ball, J. Cell. Biochem.
       Suppl. 6, 97 (1982).
73. B. R. G. Williams, S. E. Read, M. H. Freedman, D. H. Carver
       and E. W. Gelfand, in: "Interferons," T. Merigan,
       R. M. Friedman and C. F. Fox, eds., Academic Press, N. Y.,
       pp. 253–267 (1982).
74. M. Smekens-Etienne, J. Goldstein, H. A. Ooms, and J. Dumont,
       Eur. J. Biochem. 130, 269–273 (1983).
75. M. Etienne-Smekens, P. Vandenbrussche, J. Content and J.E.
       Dumont, Proc. Natl. Acad. Sci. U.S.A. 80, in press.
76. I. Gresser, L. Morel-Maroger, Y. Riviere, J.–C. Guillon,
       M. G. Tovey, D. Woodrow, J. C. Sloper and J. Moss, Ann.
       N. Y. Acad. Sci. 350, 12–20 (1980).
77. A. Panet, C. W. Czarniecki, H. Falk and R. M. Friedman,
       Virology 114, 567–571 (1981).
78. P. F. Torrence, J. Imai, K. Lesiak, M. I. Johnston, H. Jacobsen,
       R. M. Friedman, H. Sawai and B. Safer, in: "Interferons,"
       T. C. Merigan, R. M. Friedman and C. F. Fox, eds., Academic
       Press, N. Y., pp. 123–142 (1982).
79. P. F. Torrence, J. Imai, K. Lesiak, J.–C. Jamoulle and H. Sawai,
       submitted for publication.
80. M. C. Haugh, P. J. Cayley, H. Serafinowska, D. G. Norman, C.B.
       Reese and I. M. Kerr, Eur. J. Biochem. 132, 77–84 (1983).
81. B. R. G. Williams and I. M. Kerr, Nature 276, 88–90 (1978).

82. M. Knight, P. J. Cayley, R. H. Silverman, D. H. Wreschner, C. S. Gilbert, R. E. Brown and I. M. Kerr, Nature 288, 189-197 (1980).

83. M. Knight, D. H. Wreschner, R. H. Silverman and I. M. Kerr, Methods Enz. 79, 216-227 (1981).

84. C. Baglioni, S. B. D'Alessandro, T. W. Nilsen, J. A. J. den Hartog, R. Crea and J. H. van Boom, J. Biol. Chem. 256, 3253-3257 (1981).

85. J. Imai and P. F. Torrence, submitted for publication.

86. K. Lesiak, J. Imai, G. Floyd-Smith and P. F. Torrence, submitted for publication.

87. S. Rappoport, G. Arad, Y. Lapidot and A. Panet, FEBS Lett. 149, 47-50 (1982).

88. P. Doetsch, J. M. Wu, Y. Sawada and R. J. Suhadolnik, Nature 291, 355-358 (1981).

89. P. W. Doetsch, R. J. Suhadolnik, Y. Sawada, J. D. Mosca, N. L. Reichenbach, A. W. Dang, J. M. Wu, R. Charubala, N. Pfleiderer and E. E. Henderson, Proc. Natl. Acad. Sci. U.S.A. 78, 6699-6703 (1981).

90. R. Charubala and W. Pfleiderer, Tetrahedron Lett. 21, 4077-4080 (1980).

91. H. Sawai, J. Imai, K. Lesiak, M. I. Johnston and P. F. Torrence, J. Biol. Chem. 258, 1671-1677 (1983).

92. K. Lesiak and P. F. Torrence, FEBS Lett. 151, 291-296 (1983).

93. J.-C. Jamoulle and P. F. Torrence, submitted for publication.

94. R. J. Suhadolnik, D. W. Doetsch and N. L. Reichenbach, Fed. Proc. 41, 1455 (1982).

95. B. G. Hughes, P. C. Srivastava, D. D. Muse and R. K. Robins, Biochemistry 22, 2116-2126 (1983).

96. B. G. Hughes and R. K. Robins, Biochemistry 22, 2127-2135 (1983).

97. J. Imai, M. I. Johnston and P. F. Torrence, J. Biol. Chem. 257, 12739-12745 (1982).

98. J.-L. Droucourt, C. W. Dieffenbach, P.O.P. Ts'o, T. Justesen and M. N. Thang, Nucleic Acids Res. 10, 2163-2174 (1982).

99. P. F. Torrence, K. Lesiak, J. Imai, M. I. Johnston and H. Sawai, Proceedings of the Fifth International Round Table on Nucleosides, Nucleotides and Their Biological Application, Academic Press, N. Y., in press.

100. P. F. Torrence and J. Imai, unpublished observations.

101. R. H. Silverman, D. H. Wreschner, C. S. Gilbert and I. M. Kerr, Eur. J. Biochem. 115, 79-85 (1981).

102. B. B. Goswami, R. Crea, J. H. van Boom and O. K. Sharma, J. Biol. Chem. 257, 6867-6870 (1982).

103. G. C. Sen, S. Shaila, B. Lebleu, G. E. Brown, R. C. Desrosiers and P. Lengyel, J. Virol. 21, 69-83 (1977).

104. F. de Ferra and C. Baglioni, Virology 112, 426–435 (1981).

105. F. de Ferra and C. Baglioni, J. Biol. Chem. 258, 2118–
        2121 (1983).

106. C. Kahana, E. Yakobson, M. Revel and Y. Groner, Virology
        112, 109–118 (1981).

107. I. Gresser and M. Tovey, Biochim. Biophys. Acta 516, 231–
        242 (1978).

108. A. Kimchi, H. Shure and M. Revel, Nature 282, 849–851 (1979).

109. A. Kimchi, H. Shure and M. Revel, Eur. J. Biochem. 114, 5–10
        (1981).

110. A. Kimchi, H. Shure, Y. Lapidot, S. Rapoport, A. Panet and
        M. Revel, FEBS Lett. 134, 212–216 (1981).

111. T. Leanderson, R. Nordefelth, and E. Lundgren, Biochem.
        Biophys. Res. Comm. 107, 511–517 (1982).

112. D. A. Eppstein, Y. V. Marsh, B. B. Sehryver, M. A. Larsen,
        J. W. Barnett, J. P. H. Verheyden and E. J. Priske,
        J. Biol. Chem. 257, 13390–13397 (1982).

113. D. A. Eppstein, B. B. Sehryver, Y. V. Marsh, M. A. Larsen and
        C. G. Kurahara, J. Interferon Res., in press.

114. P. F. Torrence, J. Imai, K. Lesiak, J. Warinnier, J. Balzarini
        and E. De Clercq, submitted for publication.

115. D. A. Eppstein, J. W. Barnett, Y. V. Marsh, G. Gosselin
        and J.-L. Imbach, Nature 302, 723–724 (1983).

OLIGOPEPTIDES AS SPECIFIC ANTIVIRAL AGENTS

Purnell W. Choppin, Christopher D. Richardson, and
Andreas Scheid

The Rockefeller University
New York, New York 10021

INTRODUCTION

The occurrence of this course is testimony to the recent
progress toward the development of antiviral chemotherapy. Much
of this progress has been due to the identification and characterization of viral enzymes, particularly those involved in nucleic
acid synthesis, topics that are discussed extensively in this volume.
In addition to the elucidation of the roles of viral enzymes, a
great deal has also been learned about the biological activities
of other viral proteins, and in some cases detailed information has
been obtained on structure-function relationships of proteins with
biological activities that are targets for inhibition. This presentation will deal with the development of specific oligopeptide
inhibitors of paramyxoviruses and myxoviruses, inhibitors that were
designed on the basis of information gained on the structure and
biological functions of viral surface glycoproteins, including
information that indicated that a specific region of the protein
was involved in biological activity. The proteins are the fusion
(F) protein of paramyxoviruses and the hemagglutinin (HA) protein
of myxoviruses, and the step in virus replication that is inhibited
is penetration of the virus into the cell, which occurs by fusion
of the viral membrane with the cell membrane. These findings have
been described in detail in several previous publications
(Richardson et al., 1980, 1981, 1983; Choppin et al., 1981, 1983)
and will be summarized briefly in this communication.

These studies provide an example of a specific antiviral agent
that acts by inhibiting a viral protein other than a replicative
enzyme and that was designed on the basis of basic research on the
structure and biological activity of the protein.

THE F PROTEIN OF PARAMYXOVIRUSES

The paramyxovirus family contains a large number of viruses, including the parainfluenza viruses, mumps, Newcastle disease virus (NDV), measles, and canine distemper. These viruses cause a wide variety of diseases in man and lower animals, ranging from mild respiratory infections to chronic neurological diseases. Our laboratory has been engaged for several years in the isolation and biological and biochemical characterization of the paramyxovirus glycoproteins. Two glycoproteins, designated HN and F, are associated with the paramyxovirus envelope, and these have been isolated and purified in biologically active form using non-ionic detergents and various chromatographic and sedimentation methods (Scheid et al., 1972; Scheid and Choppin, 1973, 1974a,b; Tozawa et al., 1973; Hsu et al., 1979). These proteins form spike-like projections on the surface of the virion $\sim$ 10 nm long, and are anchored in the viral membrane by a hydrophobic portion at the C-terminal end of the protein.

The HN protein has a molecular weight (mol wt) of $\sim$ 65,000-70,000 depending on the virus strain, is present on the surface of the virion as a dimer, which is held together by disulfide bonds in the hydrophilic region and hydrophobic bonds at the base of the protein (Scheid et al., 1978). The isolated HN protein has receptor binding activity, which is manifested by hemagglutination when the erythrocyte is used as the target cell, and neuraminidase activity, which is capable of destroying the receptors for the virus (Scheid et al., 1972; Scheid and Choppin, 1973). Thus, the HN protein is responsible for the first step in infection, the adsorption of the virus to the cell. Under appropriate conditions, the hydrophobic portion of the HN protein of the parainfluenza virus SV5 ($\sim$ 5,000 daltons) can be removed by a proteolytic enzyme, resulting in a water-soluble protein with a mol wt of $\sim$ 59,000. This polypeptide retains receptor-binding and neuraminidase activity, although it cannot function as an agglutinin because it is monovalent (Scheid et al., 1978). The intact HN molecules aggregate by their bases when detergent is removed, forming rosette-like clusters which are thus multivalent and hemagglutinate (Scheid et al., 1972). Because neuraminic acid-containing macromolecules that can act as receptors for paramyxoviruses or myxoviruses are ubiquitous on the surfaces of vertebrate cells, the adsorption of these viruses to receptors is not a step in replication that plays an important role in determining host range or tissue tropism, in contrast to the situation with some other viruses, such as picornaviruses with which virus-receptor interactions play a decisive role in host-cell specificity (Crowell, 1980). Thus, the HN protein of paramyxoviruses, although it mediates the important step of virus adsorption, is not a significant factor in determining host range, tropism, and virulence. In contrast, the other viral surface protein, F, plays a crucial role in these viral properties.

The biological Activity of the F protein

The evidence for the biological activity of the F protein came initially from studies showing that proteolytic cleavage of a precursor protein (F_o) to yield two polypeptides (F_1 and F_2) activated virus-induced cell fusion and hemolysis and the initiation of infection at the step of virus penetration, which is accomplished by fusion of the viral and cell membranes (Homma and Ouchi, 1973; Scheid and Choppin, 1974a, 1976, 1977). Each of these biological activities, virus penetration, cell fusion, and hemolysis reflects the ability of the F protein to cause membrane fusion. The proteolytic cleavage that activates the F protein is normally accomplished by a host protease and some cells lack an appropriate enzyme to cleave the F protein of certain viruses. For example, bovine kidney (MDBK) or mouse fibroblasts (L) cells lack an enzyme capable of cleaving the Sendai virus F_o. These cells produce non-infectious virions that are incapable of causing cell fusion and hemolysis. These biological properties can be activated in vitro by cleavage of the F protein with trypsin. Mutants have been isolated that require different proteases from wild type (wt) viruses for activation, and that exhibit a different host range at levels of both cultured cells and the chick embryo (Scheid and Choppin, 1976).

Because the ability of the virus to initiate infection, and therefore to undergo multiple-cycle replication, depends on the activation of the F protein by a host protease, these results indicated that the host range and tissue tropism of the virus, and its ability to spread in the host and cause disease are dependent on the availability of an appropriate activating protease in the host (Scheid and Choppin, 1975, 1976; Choppin and Scheid, 1977). Based on these observations in cell culture and the chick embryo, we postulated that virus virulence in the natural animal host would also depend on the susceptibility of the viral F protein to cleavage by host enzymes (Scheid and Choppin, 1975). We cited as possible examples the virulent and avirulent strains of NDV that had been isolated from chickens over the years, because retrospective analysis of the gels published previously by Lomniczi et al. (1971) suggested the presence of F_o in cells infected with the less virulent but not virulent strains of NDV. The role of proteolytic cleavage in viral pathogenesis was soon clearly established in chickens in extensive studies by Nagai et al. (1976, 1977), who examined many strains of NDV that were virulent or avirulent, and found that virulence correlated with the ability of chicken fibroblasts to cleave the F protein of the virus.

In addition to the F protein being a major determinant of virus virulence because of its role in tropism and spread in the host, it is also involved in pathogenesis of viral infection in other ways. It has been observed by pathologists for over a century (Virchow, 1858) that paramyxoviruses such as measles virus cause the formation

of giant cells in infected tissues, and soon after the introduction
of cell culture for virus propagation, it was found that paramyxoviruses caused fusion of infected cells in vitro (Enders and
Peebles, 1954; Henle et al., 1954). As described above, cell fusion
is caused by the F protein, and studies have indicated that the
membrane-fusing activity of the F protein is responsible for cell
injury and death in certain virus-host systems. With the parainfluenza virus SV5, it was shown that the death of cells, or the
survival of cells without inhibition of cell metabolism, was
dependent on the susceptibility or resistance of the plasma membrane
of the cell to the fusing activity of the virus (Holmes and Choppin,
1966; Compans et al., 1966). The role of the membrane fusing
activity of the F protein in the killing of paramyxovirus-infected
cells was subsequently demonstrated clearly in studies with an
inhibitor of the action of the F protein (Graves et al., 1978).
Measles virus infected cells that were prevented from fusing by
the inhibitor survived for many days without cytopathology while
continuing to produce virus, whereas infected cells not treated
with the inhibitor fused and died within a day. The direct action
of the F protein on cell membranes is thus an important factor
in cell damage induced by paramyxoviruses. Other factors can also
play a role in cell injury by paramyxoviruses, and the action of
the F protein on cell membranes may lead to lysis and death in the
absence of recognizable cell fusion, as exemplified by the hemolysis reaction.

The F protein has also been found to be involved in viral
pathogenesis in the form of an immunopathological reaction under
certain conditions. Using monospecific antibodies to the HN and
F protein, we found (Merz et al., 1980) that antibody to the HN
protein, although neutralizing virus released into the medium,
could not inhibit the spread of a paramyxovirus infection from one
infected cell to adjacent cells, because the membrane fusing action
of the F protein allowed spread of the virus genome through fusion
of the cell membranes. Anti-F antibody not only neutralized
released virus at the level of virus penetration, but also prevented spread to adjacent cells by inhibiting the action of the
F protein. These findings demonstrated that for a paramyxovirus
vaccine to be effective it must stimulate immunity to F. Furthermore, when coupled with the finding of Norrby and coworkers (Norrby
et al., 1975; Norrby and Gollmar, 1975; Norrby and Penttinen, 1978)
that formalin inactivates antigenicity of the F protein of some
paramyxoviruses, our results provided an explanation (Merz et al.,
1980) for not only the failure of previous formalin inactivated
vaccines, but also for the atypical and severe infections that
occurred in some patients who received formalin-inactivated measles
vaccine. We suggested the following explanation for the atypical
measles syndrome. The inactivated vaccine induces antibodies to
the hemagglutinin protein and other internal viral proteins, but
not to the F protein. When such individuals were subsequently

exposed to the virus, some cells in the respiratory tract could be infected and the infection could spread to adjacent cells by membrane fusion due to the lack of immunity to F. As the infection spreads, viral antigens are produced and released, stimulating a secondary immune response to H and the other viral proteins except F. The situation would then exist in which there was continuing production of viral antigens in the presence of a hyperimmune response to these antigens, a situation that could lead to a pathological immune reaction. A similar explanation may apply to the severe respiratory infections that occurred in children receiving a formalin-inactivated respiratory syncytial (RSV) virus vaccine (Chanock et al., 1968).

The structure of the F protein

The biological importance of the F protein prompted extensive studies of its structure. The biologically inactive precursor protein (F_0) has a mol wt of $\sim 65,000$. The proteolytic cleavage that activates it yields two disulfide-linked polypeptides (F_1 and F_2) with mol wts of $\sim 50,000$ and $\sim 15,000$ respectively (Scheid et al., 1972; Homma and Ohuchi, 1973; Scheid and Choppin, 1974a, 1976, 1977). The C-terminus of F_0, which is the C-terminus of the F_1 polypeptide, is embedded in the viral membrane, and the activating cleavage generates a new N-terminus on F_1 (Scheid and Choppin, 1977; Gething et al., 1978). The original N-terminus of F_0 is the N-terminus of F_2, and F_2 is linked to F_1 by disulfide bonds. The F protein is a glycoprotein, and F_1 and F_2 contain 3 and 1 carbohydrate chains, respectively, linked to asparagine. The structure of the carbohydrate moiety of the F protein of the parainfluenza virus SV5 grown in bovine kidney (MDBK) cells has been described in detail (Prehm et al., 1979).

The new N-terminal region of F_1 has been of particular interest because it is generated by the cleavage that activates the biological functions of the F protein, and as discussed below much evidence has accumulated which suggests that this region of the F_1 polypeptide is directly involved in the membrane fusing activity of the protein. The amino acid sequences of three different paramyxoviruses (Scheid et al., 1978; Gething et al., 1978; Richardson et al., 1980) are shown in Table 1. The primary structure of the N-terminal region of F_1 is highly conserved among different viruses. In the first 20 amino acids there are only 6 positions at which differences occur among the three viruses shown in Table 1, and in each case one hydrophobic amino acid is substituted for another. In addition, as mentioned above, mutants of Sendai virus have been isolated that are cleaved and activated by proteases different from the enzyme that activates wt virus (Scheid and Choppin, 1976), and the N-terminus of such a mutant is the same as that of wt virus, i.e., Phe-Phe-Gly (Scheid et al., 1978). This suggests that the mutation

occurred to the left of the cleavage site, and supported the concept
of a specific amino acid sequence at the N-terminus of F_1 being
required for biological activity. This region is also highly
hydrophobic; there are no charged amino acids in the first 20
residues in each virus, and the first charged amino acid detected
does not appear until residue 26.

Table 1. N-terminal Amino Acid Sequences of the F
          Polypeptides of Three Paramyxoviruses

	1	5	10
Sendai	Phe-Phe-Gly-Ala-Val-Ile-Gly-Thr-Ile-Ala-		
SV5	Phe-Ala-Gly-Val-Val-Ile-Gly-Leu-Ala-Ala-		
NDV	Phe-Ile-Gly-Ala-Ile-Ile-Gly-Gly-Val-Ala-		
	11	15	20
Sendai	-Leu-Gly-Val-Ala-Thr-Ala-Ala-Gln-Ile-Thr-		
SV5	-Leu-Gly-Val-Ala-Thr-Ala-Ala-Gln-Val-Thr-		
NDV	-Leu-Gly-Val-Ala-Thr-Ala-Ala-Gln-Ile-Thr-		

Underlining indicates residues that differ from the
residue in that position in Sendai virus

Structure-Function Relationships of the F Protein

The findings that the biological activities of the F protein,
i.e., virus penetration and virus-induced cell fusion and hemolysis,
appeared when the new N-terminus on the F_1 polypeptide was generated
by cleavage, and that the primary structure of this region is highly
conserved, suggested that this region of the molecule was involved
in the membrane fusing activity of the protein (Scheid and Choppin,
1977, Scheid et al., 1978). The highly hydrophobic nature of this
region raised the possibility that it could interact with the lipid
bilayer of the target membrane (Scheid et al., 1978; Gething et al.,
1978; Richardson et al., 1980; Hsu, et al., 1981). Such a hydrophobic interaction could result in the F_1 polypeptide facilitating
fusion by bringing the bilayers of the cell and viral membranes
together, i.e., the N-terminus of F_1 would be inserted in the cell
membrane while its C-terminus was anchored in the viral membrane.
This hypothesis was supported by the finding of a conformational
change in the F protein upon cleavage (Hsu et al., 1981). The
change in conformation was demonstrated by an alteration in the
circular dichroism spectrum (which showed an increase in α-helical
content), and by an increase in the amount of detergent (Triton
X-100) bound to the cleaved protein as compared to the uncleaved
protein (from 27 molecules of Triton per protein molecule to 67).

These changes indicated an increased hydrophobicity compatible with the exposure of the hydrophobic N-terminus of F_1. Following our demonstration of a conformational change in the F protein of Sendai virus (Hsu et al., 1981), a similar finding was obtained on the hemagglutinin protein of influenza virus, in which cleavage and acid pH expose a hydrophobic region (Skehel et al., 1982).

Paramyxoviruses can cause cell fusion and hemolysis at neutral pH as well as over a wide pH range (pH 5-9). Acid pH is thus not required for fusion of paramyxoviruses with membranes, in contrast to the situation with Semliki forest, vesicular stomatitis, and influenza viruses (Helenius et al., 1980; Lenard and Miller, 1980; White et al., 1981). The evidence suggests that these viruses penetrate by fusion of the viral membrane with the cell membrane in an endocytic vesicle at acid pH, rather than by fusion at the plasma membrane as occurs with paramyxoviruses. It has been found in our laboratory that paramyxovirus fusion activity not only does not require acid pH, it is enhanced at alkaline pH (Hsu et al., 1982). An irreversible enhancement of activity of the F protein was found upon exposure of Sendai virus to alkaline pH (optimum pH, 9.0) followed by returning the virus to neutral pH before assaying cell fusing or hemolysing activities. The increased activity was correlated with a conformational change in the protein as shown by its circular dichroism spectrum (Hsu et al., 1982). The findings with different enveloped viruses suggest that a similar type of hydrophobic interaction could mediate fusion of viral and cell membranes, with the variation among viruses in the site of membrane fusion being dependent on the pH at which the viral protein involved in membrane fusion acquires the optimal conformation for activity.

Specific Inhibition of the Activity of the F Protein

The characteristics of the F protein discussed above suggested that the N-terminus of F_1 was directly involved in membrane fusing activity. On the basis of these findings we thought it might be possible to competitively inhibit the action of the protein with synthetic oligopeptides that mimicked the N-terminal region of the F_1 polypeptide. An additional reason for thinking this might be possible was that we noted the similarity between the sequence of this region and peptides that had been found to inhibit measles virus in the screening of a large number of compounds for antiviral activity (Miller et al., 1968), and which were later shown to inhibit penetration and hemolysis by measles virus (Norrby et al., 1971) and SV5 (Choppin, 1972, unpublished experiments).

To examine this hypothesis we synthesized a large number of oligopeptides using the sequence of the N-terminus of Sendai virus as the primary model, but varying several parameters, including the sequence and steric configuration of the amino acids and various substitutions on both the N- and C-terminal amino acids of the oligopeptides (Richardson et al., 1980, 1981, 1983; Choppin et al., 1983). Some of these oligopeptides were tested against measles, canine distemper, SV5, and Sendai viruses, and oligopeptides with the appropriate structure were found to inhibit the membrane-fusing activity of the F protein, as manifested by virus penetration and virus-induced cell fusion and hemolysis. Measles virus was found to be the most sensitive to inhibition by the oligopeptides, and there-fore it was chosen for extensive structure-activity studies. Although cell fusion and hemolysis were also inhibited, the most sensitive and efficient method of quantitating inhibitory activity was to employ a plaque assay to determine inhibition of infectivity at the level of virus penetration. African green monkey kidney (CV-1) cells were used for most assays, and in each experiment, the ability of varying concentrations of the oligopeptide to inhibit plaque formation was determined. Dose response curves were obtained and 50% effective inhibitory concentrations were derived from these curves. It should be emphasized that these experiments were carried out under conditions in which the F protein is cleaved; the oligopeptides do not act as protease inhibitors but as inhibitors of the membrane-fusing activity of the F protein. Our studies with these oligopeptides are described in detail elsewhere (Richardson et al., 1980, 1981, 1983; Choppin et al., 1983) and will therefore be summarized briefly here.

Structure-activity studies with a large number of different oligopeptides led to the following conclusions. 1. Oligopeptides with the correct amino acid sequence are highly active specific inhibitors. The longer the oligopeptide the more effective it is, up to a length of seven amino acids, the longest tested; this hepta-peptide had a 50% effective plaque inhibiting concentration of 0.02 μM. 2. The addition of certain hydrophobic groups to the N-terminal amino acid significantly increased inhibitory activity. A carbo-benzoxy group (Z) greatly enhanced activity as compared to that of the same oligopeptide without an addition. Other hydrophobic additions such as a dansyl or a t-butyloxycarbonyl group also increased activity. 3. Esterification of the C-terminal amino acid with a methyl group decreased activity. 4. The steric configuration of the first and second amino acids was found to significantly affect activity, e.g., Z-D-Phe-L-Phe-Gly was more active than Z-L-Phe-L-Phe-Gly or Z-D-Phe-D-Phe-Gly. Table 2 shows the 50% effective con-centrations derived from dose-response curves carried out with different oligopeptides which illustrate these points.

These studies with a variety of oligopeptides with different structures have established that the correct amino acid sequence is of crucial importance for maximum inhibitory activity. Further evidence for this was obtained using a measles virus mutant that was selected by repeated passage in the presence of a tripeptide (Z-D-Phe-L-Phe-L(NO$_2$)Arg). As shown in Table 3, the mutant was resistant to inhibition by this peptide but remained relatively sensitive to another peptide that differed only in the third amino acid (Richardson et al., 1980). These tripeptides were also tested for inhibition of canine distemper virus (CDV), which is a member of the same subgroup of paramyxoviruses (the morbillivirus subgroup) as measles virus and is immunologically related to measles virus. CDV exhibited a similar pattern of susceptibility to the peptides as the mutant measles virus, suggesting that the F protein of CDV may be more similar in structure to the mutant than wild type measles virus.

Table 2. Inhibition by Oligopeptides of Plaque Formation by Measles Virus

Peptide	50% Effective Concentration, μM
Z-D-Phe-L-Phe-Gly-D-Ala-D-Val-D-Ile-Gly	0.02
Z-D-Phe-L-Phe-Gly	0.20
Z-D-Phe-L-Phe	28
Z-L-Phe-L-Ser	141
Z-Gly-L-Phe-L-Phe	530
Z-D-Phe-L-Phe-Gly	0.20
DNS-D-Phe-L-Phe-Gly	0.34
t-BOC-D-Phe-L-Phe-Gly	2.0
D-Phe-L-Phe-Gly	180
Z-D-Phe-L-Phe-Gly(methyl ester)	20
Z-D-Phe-L-Phe-Gly	0.20
Z-D-Phe-D-Phe-Gly	10
Z-L-Phe-L-Phe-Gly	23
Z-D-Phe-L-Phe-Gly(chloromethylketone)	0.20
Z-D-Phe-L-Phe-L-(azido)Phe	0.28

Z indicates a carbobenzoxy group; DNS, a dansyl group; t-BOC, a tertiary butoxylcarbonyl group.

The reason for the effects on inhibitory activity of the additions at the termini of the oligopeptides is not yet clear, however the added groups may have an effect on the positioning of the oligopeptide at its site of action. The carbobenzoxy and dansyl groups add hydrophobicity to the N-terminus, and the esterification of the C-terminus decreases the polarity of the peptide. Such changes could affect the orientation of the peptide at its site of action at the membrane. The site of action of the inhibitors will be discussed further below. The explanation for the effect of the steric con-figuration of the amino acids is also not completely understood. The observed greater inhibitory activity of an oligopeptide with D-phenylalanine at the N-terminus, as compared to the naturally-occurring L-amino acid may in part be explained by protection of the peptide from proteolytic activity. However this is probably not the only factor involved, because the same structure-activity relation-ship was found when inhibition of hemolysis was used to assay acti-vity, and proteolytic digestion of the peptide is less likely to occur in a short period of time in a hemolysis assay than in a plaque assay in cultured cells which requires several days. It therefore appears that the steric configuration of the terminal amino acids has an effect on inhibitory activity per se.

Table 3. Inhibition of Wild Type and Mutant Measles Viruses and Canine Distemper Virus (CDV) by Oligopeptides

| | 50% Effective Concentration, μM | | |
| | Measles Virus | | |
Peptide	Wt	Mutant	CDV
Z-D-Phe-L-Phe-L-(NO$_2$)Arg	0.2	>1000	>1000
Z-D-Phe-L-Phe-Gly	0.2	21	1.5

Although the oligopeptides with the appropriate structure are highly active, specific inhibitors of the membrane-fusing activity of the F protein, and therefore of the replication of the virus due to inhibition of the virus penetration step, they do not appear to have significant toxic effects on cells. The cultured cells used in these studies can survive for many days with no detectable cyto-pathic effects, and the cells can divide and grow normally in the presence of the peptides in concentrations many fold higher than those that can inhibit virus replication, hemolysis, and cell fusion. Preliminary studies have suggested that the oligopeptides are also not significantly toxic for mice.

Inhibition by Oligopeptides of Influenza Virus Replication

Influenza viruses (myxoviruses) also possess two glycoproteins that form spike-like projections on their surface; however, unlike paramyxoviruses, the hemagglutinating and neuraminidase functions of influenza viruses reside on separate proteins, designed HA and NA,

respectively. The influenza virus HA protein serves two functions,
receptor binding and penetration. Like the F protein of paramyxoviruses, the HA protein of influenza virus is cleaved by a host cell
protease to yield two disulfide-bonded subunits (Lazarowitz et al.,
1971). This cleavage has no effect on hemagglutinating activity
(Lazarowitz et al., 1973), but it activates the infectivity of the
virus (Klenk et al., 1975; Lazarowitz and Choppin, 1975) at the level
of viral penetration. The similarities between the HA protein of
influenza viruses and the F protein of paramyxoviruses with respect
to function in penetration and activation by proteolytic cleavage
by a host enzyme to yield two disulfide-linked polypeptides was
pointed out previously (Scheid et al., 1978). Recent evidence has
shown that at acid pH, influenza virus can induce hemolysis and cell
fusion (Huang et al., 1981; Lenard and Miller, 1981; White et al.,
1981, 1982; Maeda et al., 1981). These results suggested that both
cleavage and an acid pH are needed to activate the membrane fusing
activity of influenza virus, whereas cleavage of the F protein of
paramyxoviruses alone can activate fusion by these viruses at physiological pH. As indicated above, this was supported by the finding
that cleavage and acid pH cause a conformational change in the HA
protein with exposure of a hydrophobic region (Skehel et al., 1982),
a result analogous to that found by cleavage alone of the paramyxovirus F protein (Hsu et al., 1981). The structure and function of
the HA protein are discussed in the paper by Skehel in this volume,
however for the present discussion it is pertinent to point out that
not only is the membrane fusing activity of the HA protein activated
by cleavage, as is the F protein, but also that the amino acid
sequence of the new N-terminus generated on the HA_2 polypeptide by
cleavage resembles that of the paramyxovirus F_1 polypeptide, except
that glycine is the N-terminal amino acid. The sequences in this
region of the HA_2 polypeptide were first determined by Skehel and
Waterfield (1975), and have since been determined for many strains
(Laver and Air, 1980). The sequence of the influenza B Lee HA_2
polypeptide is Gly-Phe-Phe-Gly-Ala-Ile-Ala-Gly-Phe-Leu; comparison
of this with the sequence of Sendai virus shown in Table 1 will
demonstrate their similarity. With influenza A virus strains, the
sequence is the same, except that leucine is substituted for the
first phenylalanine and isoleucine for the terminal leucine.

The structural and functional similarities between the HA and
F polypeptides, and the success that we had with the specific
inhibition of paramyxoviruses, led us to synthesize oligopeptides
that mimicked the N-terminus of the HA_2 polypeptide, and to test
their ability to inhibit influenza virus replication. With influenza
virus also, inhibition was obtained that was amino acid-sequence
specific. As shown in Table 4, oligopeptides that mimicked the
N-terminal sequence of the HA_2 polypeptide of influenza A and B
viruses (Z-Gly-L-Leu-L-Phe-Gly and Z-Gly-D-Phe-L-Phe-Gly, respectively) were active against influenza virus, but inactive against
measles virus, whereas as shown in Table 2, the reverse was true

with Z-D-Phe-L-Phe-Gly, which resembles the N-terminus of paramyxoviruses.

Table 4.   Comparison of the Inhibition of Influenza A and
           Measles Viruses by Different Oligopeptides

| | 50% Effective Concentration, µM | |
Peptide	Measles	Influenza A
Z-Gly-L-Leu-L-Phe-Gly	20	1000
Z-Gly-L-Phe-L-Phe-Gly	53	870
Z-Gly-D-Phe-L-Phe-Gly	23	-
Z-D-Phe-L-Phe-Gly	290	0.2

Site of Action of the Oligopeptide Inhibitors

    The precise site of action of the oligopeptides is obviously
a question of great interest. They could conceivably act on the
cell membrane, on the virus, or on the virus-cell complex. For
example, they could compete with the N-terminus of the F_1 polypeptide for a site on the target cell membrane or on the viral
membrane. Alternatively, they might compete for a site on the F
protein itself, into which the N-terminus might fold as a result of
the conformational change that is induced by the activating proteolytic cleavage (Hsu et al., 1981). The most straightforward
possibility which has been considered from the beginning of these
studies, is that the oligopeptides act at the cell membrane. However, the finding that influenza virus as well as paramyxoviruses
were inhibited, but by different oligopeptides, and that a mutant of
measles virus was inhibited by different oligopeptides than wt virus,
suggested that if the oligopeptides were competing for sites on the
cell membrane, then the different viruses would appear to react with
somewhat different sites on the membrane. On the other hand, if the
oligopeptides acted on the virus, the differences between viruses
would be readily understood. To investigate the site of action of
the oligopeptides several experimental approaches were taken
(Richardson et al., 1983). These involved the use of oligopeptides
that were radioactively labeled, or that had been modified so that
they bound irreversibly.

    Oligopeptides were synthesized that were labeled with ^{3}H or
[125]I and the location of their binding sites was investigated. The

labeled oligopeptides were reacted at 4° C with purified virus, mock-infected cells, or infected cells, and after washing, samples were assayed for radioactivity to quantitate binding. Such studies suggested that the oligopeptides bound to cells, whether infected or uninfected, but not to virus (Richardson, et al., 1983; Choppin, et al., 1983). The inhibitory action of these oligopeptides is reversible at 37° C; if the medium containing the oligopeptides is removed and the cells washed at 37° C, inhibitory activity is greatly reduced. To further investigate their site of action, oligopeptides were synthesized that could bind irreversibly. A chloromethylketone (CK) derivative of Z-D-Phe-L-Phe-Gly (ZPPGCK) was synthesized. As shown in Table 2, this oligopeptide retained its virus inhibitory activity, as demonstrated by inhibition of plaque formation, and it also inhibited hemolysis by measles virus (Richardson et al., 1983). It has the irreversible binding properties of the protease inhibitors, N-p-tosyl-L-phenylalanine-chloromethylketone (TPCK) and N-p-tosyl-L-lysine-chloromethylketone (TLCK), which can alkylate histidine, serine, methionine, or cysteine residues near the active sites of proteases. The irreversible inhibitor, Z-PPGCK, was used to investigate its site of action. Z-PPGCK was preincubated with monkey erythrocytes (RBC), or RBC plus adsorbed measles virus, or purified measles virus, and after extensive washing, hemolysis assays were performed (Richardson et al., 1983). It was found that ZPPGCK irreversibly inhibited hemolysis when it was preincubated with RBC or RBC plus virus, but not when it was preincubated with virus alone. These results provide further evidence that the oligopeptides act on the cell and not on the virus.

Another approach to determining the site of action of the peptides involved the use of an oligopeptide inhibitor containing an azido group as a photoaffinity probe, i.e., Z-D-Phe-L-Phe-(azido) Phe (Z-APPP!). As shown in Table 1 this peptide retained activity when the azido group was added. RBC or measles virus were incubated with this peptide and exposed to ultraviolet light for various times to cross-link the oligopeptide in situ. The samples were then washed extensively, and hemolysis assays were carried out. Pre-treatment of virus with Z-APPP! had no effect, but pre-treatment of RBC with this oligopeptide inhibited hemolysis (Richardson et al., 1983). These studies on the binding of the oligopeptide inhibitors indicate that their site of action is on the target cell membrane, not on the virus. The results further suggest that the inhibitors compete with the N-terminus of the F_1 polypeptide for a site on the cell membrane. The exact location and nature of this presumed "receptor" for the F_1 protein is under investigation. If such a receptor exists, there may be several related but different classes of receptors, because, as discussed above, different oligopeptides inhibit different viruses to different extents and mutants of measles virus can be isolated that differ markedly from wt virus in sensitivity to specific oligopeptides.

Kinetics of Binding of Oligopeptides

The kinetics binding of a radioactively-labeled tripeptide (Z-D-Phe-L-Phe[^{125}I]-Tyr) to African green monkey kidney cells was investigated. This tripeptide is an active inhibitor of F protein action; its 50% effective plaque inhibitory concentration for measles virus is 0.20 μM. Saturation binding kinetics were observed and a Skatchard plot revealed that the labeled peptide bound to the cells with a dissociation constant of 1.2×10^{-7} M and that there were approximately 3×10^6 binding sites per cell (Richardson et al., 1983). The results are comparable with the other findings suggesting that the inhibitors act on the cell membrane, and that there is some kind of a "receptor" site on the cell for the F protein for which the oligopeptides compete.

Significance of the Oligopeptide Inhibitors of Membrane Fusing Activity

The studies that have been summarized in this paper have resulted in the development of highly active specific inhibitors of virus penetration and virus-induced cell fusion and hemolysis. Thus they have provided a possible new approach to the chemical inhibition of virus replication. It is not yet clear whether these oligopeptides will have any use in the chemotherapy or chemoprophylaxis of virus infections in vivo. An important point that is clear, however, is that they provide an example of a specific inhibitor of virus replication that has been designed on the basis of detailed information on the structure and biological function of viral proteins, indeed on evidence suggesting that a specific region of the protein was involved in biological activity. This knowledge was gained from basic research on the viruses and their replication that was not initially directed toward the development of antiviral agents. The viral proteins whose action is being inhibited by these oligopeptides are not enzymes involved in the synthesis of viral nucleic acids, and they thus provide an example of specific and effective inhibition of another type of viral activity. These studies have also elucidated mechanisms involved in virus penetration and virus-induced hemolysis and cell fusion, and yielded information on membrane fusion in general, a process of wide biological importance. The availability of these oligopeptide inhibitors and the ability to attach specific probes to them will enable further studies on the nature of the sites on the cell membrane with which they and the viral proteins interact, and on the precise mechanism of action of the proteins as well as the inhibitors.

SUMMARY

The F glycoprotein of the paramyxoviruses is responsible for virus penetration and for virus-induced cell fusion and hemolysis, functions that are related to the ability of the protein to induce membrane fusion. The activity of the F protein is activated by a specific proteolytic cleavage of a precursor to yield two disulfide-linked polypeptides (F1 and F2). Because the cleavage is accomplished by a host protease, the susceptibility of the F protein to cleavage by an enzyme available in the host is a major determinant of the host range, tissue tropism, and virulence of the virus. The F protein is also involved in viral pathogenesis in other ways. In some virus-cell systems, cell killing is due to the membrane injury caused by the F protein, and lack of immunity to the F protein has been responsible for failure of inactivated paramyxovirus vaccines and the development of atypical infections with a pathological immune reaction in some individuals who received such vaccines.

The cleavage that activates the F protein generates a new N-terminus on F_1, and the amino acid sequence in this region is extremely hydrophobic and highly conserved among different paramyxoviruses. Several lines of evidence, including a conformational change upon cleavage characterized by the exposure of a new hydrophobic region, suggested that the N-terminus of the F_1 polypeptide was directly involved in membrane fusion, possibly through a hydrophobic interaction with the membrane of the target cell. To explore this possibility, oligopeptides were synthesized to resemble the N-terminus of the F_1 polypeptide, and were found to be highly active, specific inhibitors of the membrane-fusing activity of the virus. These peptides inhibited virus infectivity at the penetration step and virus-induced cell fusion and hemolysis. Structure-activity studies with a large number of peptides that varied in amino acid sequence, length, steric configuration, and N- and C-terminal additions have elucidated the structural requirements for maximum inhibitory activity. To investigate the precise site and mechanism of action of the inhibitory peptides, and of the F protein, inhibitory oligopeptides were synthesized that were radioactively-labeled to enable quantitation of specific binding. Inhibitors that bound irreversibly were synthesized by adding a chloromethylketone group or a photoactivable azido group to the peptide. Studies with these oligopeptides have indicated that their site of action is on the cell membrane. These results support the conclusion that the N-terminus of the F_1 polypeptide interacts with specific sites on the cell membrane during the fusion process.

The hemagglutinin (HA) protein of influenza virus mediates membrane virus penetration by membrane fusion and is also activated by a specific cleavage yielding two disulfide-linked polypeptides (HA1 and HA2). The new N-terminus on the HA2 polypeptide, like that of F_1, is hydrophobic and highly conserved. Oligopeptides synthesized to mimic the N-terminus of HA2 specifically inhibited the replication of influenza virus.

These studies have yielded information on the mechanism of membrane fusion, as well as providing a possible new approach to chemical inhibition of virus replication in which the activity of a

viral protein other than a synthetic enzyme is inhibited. These
studies also provide an example of a specific antiviral agent that
has been designed on the basis of basic knowledge of the structure
and function of a viral protein.

ACKNOWLEDGMENTS

Research by the authors was supported by research grants
AI-05600 from the National Institute of Allergy and Infectious
Diseases, PCM80-13464 from the National Science Foundation, and
CA-18213 from the National Cancer Institute.

REFERENCES

Chanock, R. M., Parrott, R. H., Kapikian, A. Z., Kim, H. W., and
    Brandt, C. D., 1968, Possible role of immunological factors in
    pathogenesis of RS lower respiratory tract disease, in:
    "Perspectives in Virology VI", M. Pollard, ed., Academic Press,
    New York.
Choppin, P. W., and Scheid, A., 1977, The biologic role of host-
    dependent proteolytic cleavage of a paramyxovirus glycoprotein,
    in: "Slow Virus Infectious of the Central Nervous System",
    V. ter Meulen and M. Katz, eds., Springer Verlag, New York.
Choppin, P. W., Richardson, C. D., Merz, D. C., and Scheid, A.,
    1981, Functions of surface glycoproteins of myxoviruses and
    paramyxoviruses and their inhibition, in: "Adhesion and
    Microorganism Pathogenicity", Ciba Foundation Symposium,
    M. O'Connor, ed., Pitman Medical, London.
Choppin, P. W., Richardson, C. D., and Scheid, A., 1983, Analogues
    of viral polypeptides which specifically inhibit viral replica-
    tion, in: "Problems of Antiviral Therapy", C. Stuart-Harris,
    ed., Academic Press, London and New York.
Compans, R. W., Holmes, K. V., Dales, S., and Choppin, P. W., 1966,
    An electron microscopic study of moderate and virulent virus-
    cell interactions of the parainfluenza virus SV5, Virology
    30:411.
Crowell, R. L., 1980, Receptors as determinants of cellular tropism
    in picornavirus infections, in: "Receptors and Human Diseases",
    A. G. Bearn and P. W. Choppin, eds., Josiah Macy, Jr. Founda-
    tion, New York.
Enders, J. F., and Peebles, T. C., 1954, Propagation in tissue
    cultures of cytopathogenic agents from patients with measles
    virus, Proc. Soc. Exp. Biol. Med. 86:277.
Gething, M. J., White, J. M., and Waterfield, M. D., 1978, Purifica-
    tion of the fusion protein of Sendai virus: analysis of the NH_2-
    terminal sequence generated during precursor activation, Proc.
    Nat. Acad. Sci. U.S.A. 75:2737.
Graves, M. C., Silver, S. M., and Choppin, P. W., 1978, Measles virus
    polypeptide synthesis in infected cells, Virology, 86:254.

Helenius, A., Kartenbeck, J., Simono, K., and Fries, E., 1980, On the entry of Semliki Forest virus into BHK-21 cells, J. Cell Biol., 84:404.

Henle, G., Deinhardt, F., and Girardi, A., 1954, Cytolytic effects of mumps virus in tissue cultures of epithelial cells, Proc. Soc. Exp. Biol. Med., 87:386.

Holmes, K. V., and Choppin, P. W., 1966, On the role of the response of the cell membrane in determining virus virulence. Contrasting effects of the parainfluenza virus SV5 in two cell types, J. Exptl. Med. 124:501.

Homma, M., and Ohuchi, M., 1973, Trypsin action on the growth of Sendai virus in tissue culture cells. III. Structural difference of Sendai viruses grown in eggs and in tissue culture cells, J. Virol. 12:1457.

Hsu, M.-C., Scheid, A., and Choppin, P. W., 1979, Reconstitution of membranes with individual paramyxovirus glycoproteins and phospholipid in cholate solution, Virology, 95:476.

Hsu, M.-C., Scheid, A., and Choppin, P. W., 1981, Activation of the Sendai virus fusion protein (F) involves a conformational change with exposure of a new hydrophobic region, J. Biol. Chem. 256:3357.

Hsu, M.-C., Scheid, A., and Choppin, P. W., 1982, Enhancement of membrane fusing activity of Sendai virus by exposure of the virus to basic pH is correlated with a conformational change in the fusion protein, Proc. Natl. Acad. Sci. U.S.A., 79:5862.

Huang, R. T. C., Rott, R., and Klenk, H.-D., 1981, Influenza viruses cause hemolysis and fusion of cells. Virology, 110:243.

Klenk, H.-D., Rott, R., Orlich, M., and Blödorn, J., 1975, Activation of influenza A viruses by trypsin treatment, Virology, 68:425.

Laver, G., and Air, G., eds., 1980, "Structure and Variation in Influenza Virus", Elsevier/North Holland, New York.

Lazarowitz, S. G., and Choppin, P. W., 1975, Enhancement of the infectivity of influenza A and B viruses by proteolytic cleavage of the hemagglutinin polypeptide, Virology, 68:440.

Lazarowitz, S. G., Compans, R. W., and Choppin, P. W., 1971, Influenza virus structural and non-structural proteins in infected cells and their plasma membranes, Virology, 46:830.

Lazarowitz, S. G., Goldberg, A. R., and Choppin, P. W., 1973, Proteolytic cleavage by plasmin of the HA polypeptide of influenza virus: Host cell activation of serum plasminogen, Virology, 56:172.

Lenard, J., and Miller, D. K., 1981, pH-dependent hemolysis by influenza, Semliki Forest virus, and Sendai virus, Virology, 110:479.

Lomniczi, B., Meager, A., and Burke, D. C., 1971, Virus RNA and protein synthesis in cells infected with different strains of Newcastle disease virus, J. Gen. Virol., 13:111.

Maeda, T., Kawasaki, K., and Ohnishi, S.-I., 1981, Interaction of
    influenza virus hemagglutinin with target membrane lipids is a
    key step in virus-induced hemolysis and fusion at pH 5.2,
    Proc. Nat. Acad. Sci. U.S.A., 78:4133.
Merz, D. C., Scheid, A., and Choppin, P. W., 1980, The importance
    of antibodies to the fusion glycoprotein (F) of paramyxoviruses
    in the prevention of spread of infection, J. Exptl. Med.
    151:275.
Miller, F. A., Dixon, G. J., Arnett, G., Dice, J. R., Rightsel, W.
    A., Schabel, F. M., and McLean, J. W., 1968, Antiviral activity
    of carbobenzoxy di-and tripeptides on measles virus, Appl.
    Microbiol. 16:1489.
Nagai, Y., Klenk, H.-D., and Rott, R., 1976, Proteolytic cleavage
    of viral glycoproteins and its significance for the virulence
    of Newcastle disease virus, Virology, 72:494.
Nagai, Y., and Klenk, H.-D., 1977, Activation of precursors to both
    glycoproteins of Newcastle disease virus by proteolytic cleavage,
    Virology, 77:125.
Norrby, E., 1971, The effect of carbobenzoxy tripeptides on the bio-
    logical activity of measles virus, Virology, 44:599.
Norrby, E., and Gollmar, Y., 1975, Identification of measles virus-
    specific hemolysis-inhibiting antibodies separate from
    hemagglutination-inhibiting antibodies, Infect. Immun. 11:231.
Norrby, E., and Penttinen, K., 1978, Differences in antibodies to
    the surface components of mumps virus after immunization with
    formalin-inactivated and live virus vaccines, J. Infect. Dis.
    138:672.
Norrby, E., Enders-Ruckle, G., and ter Meulen, V., 1975, Differences
    in the appearance of antibodies to structural components of
    measles virus after immunization with inactivated and live
    virus, J. Infect. Dis. 132:262.
Prehm, P., Scheid, A., and Choppin, P. W., 1979, The carbohydrate
    structure of the glycoproteins of the paramyxovirus SV5 grown
    in bovine kidney (MDBK) cells, J. Biol. Chem., 254:9669.
Richardson, C. D., Scheid, A., and Choppin, P. W., 1980, Specific
    inhibition of paramyxovirus and myxovirus replication by
    oligopeptides with amino acid sequences similar to those at
    the N-termini of the F_1 or HA_2 viral polypeptides, Virology,
    105:205.
Richardson, C. D., Scheid, A., and Choppin, P. W., 1981, Specific
    inhibition of paramyxovirus and myxovirus replication by
    hydrophobic oligopeptides, in: "The Replication of Negative
    Strand Viruses", D. H. L. Bishop and R. W. Compans, eds.,
    Elsevier-North Holland, New York.
Richardson, C. D., Scheid, A., and Choppin, P. W., 1983, Studies on
    the site of action of oligopeptides that specifically inhibit
    the membrane fusion activity of paramyxoviruses, Virology,
    to be submitted.
Scheid, A., and Choppin, P. W., 1973, Isolation and purification of

the envelope proteins of Newcastle disease virus, <u>J. Virol</u>. 11:263.

Scheid, A., and Choppin, P. W., 1974a, Identification of biological activities of paramyxovirus glycoproteins. Activation of cell fusion, hemolysis and infectivity by proteolytic cleavage of an inactive precursor protein of Sendai virus, <u>Virology</u>, 57:475.

Scheid, A., and Choppin, P. W., 1974, The hemagglutinating and neuraminidase protein of a paramyxovirus: Interaction with neuraminic acid in affinity chromatography. <u>Virology</u>, 62:125.

Scheid, A., and Choppin, P. W., 1975, Isolation of paramyxovirus glycoproteins and identification of their biological properties, <u>in</u>: "Negative Strand Viruses", B.W.J. Mahy and R.D. Barry, eds., Academic Press, London.

Scheid, A., and Choppin, P. W., 1976, Protease activation mutants of Sendai virus: Activation of biological properties by specific proteases, <u>Virology</u>, 69:265.

Scheid, A., and Choppin, P. W., 1977, Two disulfide-linked polypeptide chains constitute the active F protein of paramyxoviruses, <u>Virology</u>, 80:54.

Scheid, A., Caliguiri, L. A., Compans, R. W., and Choppin, P. W., 1972, Isolation of paramyxovirus glycoproteins. Association of both hemagglutinating and neuraminidase activities with the larger SV5 glycoprotein. <u>Virology</u>, 50:640.

Scheid, A., Graves, M. C., Silver, S. M., and Choppin, P. W., 1978, Studies on the structure and function of paramyxovirus glycoproteins, <u>in</u>: "Negative Strand Viruses and the Host Cell", B.W.J. Mahy and R.D. Barry, eds., Academic Press, London.

Skehel, J. J., and Waterfield, M. D., 1975, Studies of the primary structure of the influenza virus hemagglutinin, <u>Proc. Nat. Acad. Sci. U.S.A.</u> 72:93.

Skehel, J. J., Bayley, P. M., Brown, E. B., Martin, S. R., Waterfield, M. D., White, J. M., Wilson, I. A., and Wiley, D. C., 1982, Changes in the conformation of influenza virus hemagglutinin at the pH optimum of virus-mediated membrane fusion, <u>Proc. Natl. Acad. Sci. U.S.A.</u>, 79:968.

Tozawa, H., Watanabe, M., and Ishida, N., 1973, Structural components of Sendai virus. Serological, and physicochemical characterization of hemagglutinin subunit associated with neuraminidase activity, <u>Virology</u>, 55:242.

White, J., Matlin, K., and Helenius, A., 1981, Cell fusion by Semliki Forest influenza, and vesicular stomatitis, viruses, <u>J. Cell Biol</u>., 89:674.

White, J., Helenius, A., and Gething, M. J., 1982, Haemagglutinin of influenza virus expressed from a cloned gene promotes membrane fusion, <u>Nature</u>, 300:658.

VIRUS ASSOCIATED DNA POLYMERIZING ACTIVITIES: THEIR ROLE IN

DESIGNING ANTIVIRAL AND ANTITUMOR AGENTS

P. Chandra, I. Demirhan, U. Ebner and B. Kornhuber

Center of Biological Chemistry (Lab. of Molecular Biology)
and Dept. of Pediatric Oncology, Univ. School of Medicine
Theodor-Stern-Kai 7
D-6 Frankfurt (Main) 70, W. Germany

INTRODUCTION

Inhibitors of nucleic acid biosynthesis have an ho-
norable place in the history of chemotherapy. Nucleic
acids, or nucleic acid polymerizing enzymes, have served
as extremely useful targets in the development of chemotherapeutic agents to combat infectious and malignant
diseases. In some cases, such as antiviral agents, this
approach has provided virtually the only means to design
effective chemotherapeutic compounds.

The strategical role of nucleic acids, or nucleic
acid polymerases as targets, involves two distinct features. The investigational drug must have the capacity
to recognize distinct bases or base-pair sequences, either by direct interaction between functional groupings
on the base-pairs and the drug molecule, or indirectly
via recognition of the conformational peculiarity of the
nucleic acid molecule. The other alternative is the specific affinity of the drug for one or the other nucleic
acid polymerizing enzymes. From the strategical stand
point this is a more specific approach, since a number
of enzymes are involved in the polymerization of nucleic
acids.

Although the first eukaryotic DNA polymerase, now
known as DNA polymerase-α, was discovered in 1958 (1), it
was not until 1970 that the existence of multiple forms
of DNA polymerases was established in these cells. The
search for these multiple species in eukaryotic cells

was, at that time, implicated by the data in prokaryotic
cells which documented several types of DNA polymerases
present in these cells (2,3,4,5). We now know that several
DNA-polymerizing activities are normally present in an
eukaryotic cell, and that additional DNA polymerase activities are brought into the cell or induced by RNA tumor
viruses (Retroviruses) and by some DNA viruses. These
additional or foreign DNA polymerases are coded by the
virus genome, and are biochemically as well as serologi-
cally different to the constitutive DNA polymerases of
eukaryotic cells. Thus, virus-associated DNA polymerases
offer an attractive target for designing antiviral and
antineoplastic agents.

DNA-POLYMERIZING   ACTIVITIES   IN   NORMAL   MAMMALIAN   CELLS

     The normal mammalian cells contain at least three
DNA polymerases, designated as DNA polymerases α, β and
γ (6). These polymerases can be distinguished by the differences in their biochemical properties and serological
behavior. A number of excellent reviews have appeared in
the past covering these details (7,8,9,10,11,12).Based
on these reviews, a spectrum of biochemical features distinct for the constitutive eukaryotic DNA polymerases is
presented in Table 1.

TABLE 1

Some Important Features of Eukaryotic DNA Polymerases

	POLYMERASE-α	$-\beta$	$-\gamma$
LOCATION	Cytoplasma[11]*	nucleus	nucleus mitochondria
Mass,kdal	120-220	30-50	100-160
S value	6-8	3-4	7-9
Isoelectric Point	5.6-6	9-9.4	5.4-6.1
Template-Primer preferred	Act.DNA	$(dA)_n \cdot (dT)_{12}$	$(rA)_n \cdot (dT)_{12}$
pH-optimum	7.2	8.5-9	8.0

* evidence from immunological studies reported by Bollum
  (11). All other data were accumulated from references
  7-12.

While the in vivo role of virally induced DNA polymerases is more or less predictable, the physiological functions of normal cellular DNA polymerases are not yet well understood. Studies with mutants have been specially useful in elucidating the functional activity of DNA polymerases (I,II and III) in prokaryotic cells. For lack of polymerase mutants in eukaryotic systems, great reliance is placed on specific inhibitors of various DNA polymerases to characterize their activity and functional role under in vivo conditions. For example, the role of DNA polymerase α as the principal replicating enzyme was established by the specific inhibitory action of aphidicolin, a tetracyclic diterpenoid obtained from Cephalosporium - aphidocola .Isolation of an aphidicolin-resistent DNA polymerase α mutant of Drosophila(13) has been very useful in studying in vivo role of this polymerase activity.Another experimental approach to study the in vivo functions of various cellular DNA polymerases has been developed by Pardee's group. Miller, Castellot and Pardee (14,15) have examined roles of DNA polymerases using lysolecithin-permeabilized baby hamster kidney(BHK) cells. According to their results, this technique yields a cell preparation which can carry out normal, replicative DNA synthesis at in vivo rates when supplied with deoxyribonucleoside triphosphates (14,15). They have also investigated the effect of known inhibitors of DNA polymerases on the replication and repair DNA synthesis in permeabilized cells (16). The different sensitivities of replication and repair synthesis to various inhibitors (N-ethylmaleimide, high KCl and aphidocolin etc.) indicated that DNA polymerase α plays a rate-limiting role in replication, whereas DNA polymerase ß plays a major role in repair.

Whereas the DNA polymerases α,β and γ are present in all animal cells and tissues, a novel DNA-polymerizing activity located only in the thymus tissue, was discovered by Bollum (17).This enzyme, terminal deoxynucleotidyl-transferase (TdT), catalyzes the polymerization of deoxynucleotides onto the 3´-OH ends of oligo or polydeoxynucleotide initiators in the absence of a template.Thus, TdT, in the strictest sense, is not a DNA polymerase as it does not direct the genetic information of a template DNA. Nevertheless, in a broader sense it does catalyze the incorporation of deoxyribonucleoside monophosphates into DNA and thus we consider it here as one of the DNA synthetic enzymes. Like retroviral reverse transcriptase, terminal transferase merits considerable attention as a useful marker for various types of human leukemia(18,19 20,21,22,23,24). For this reason, terminal transferase, though an enzyme of cellular origin, offers an attractive target site for developing antineoplastic agents.

VIRUS   ASSOCIATED   DNA   POLYMERIZING   ACTIVITIES

    The DNA polymerase patterns in virus-infected cells
are qualitatively and quantitatively different. To understand this, we will describe the virus-associated DNA
polymerase activities in two sections:

    1. <u>DNA Virus-associated DNA Polymerases</u>:The animal
DNA viruses can be divided into three groups with respect
to the nature of DNA polymerase involved in their replication: a) The small transforming viruses(polyoma,SV40)
have a limited coding capacity in their genomes and therefore, these viruses cannot specify their own enzymes to
support all the replicative and other events involved in
their life cycle. However, cells infected by such viruses
produce virus coded antigens, such as the T-antigen of
SV40, which serve as regulatory signals for the replication of viral DNA(25). It has been known for more than
fifteen years that infection of resting cells by polyoma
virus leads to an induction of cellular DNA synthesis
along with the viral DNA replication, and it has been recognized that these induced functions are accompanied by
increaes in the activities of cellular enzymes that are
involved in DNA synthesis (26,27,28,29). Using a replication complex isolated from adenovirus-2-infected KB
cell nuclei, Ito et al. (30) studied the question of the
DNA polymerase involved in the replication of viral DNA.
The only polymerase that could be extracted and characterized from the nuclear membrane complex from these cells
was an activity resembling host DNA polymerase γ.However,
recently an active complex of 80-kdal protein-adenoviral
DNA with a 140-kdal DNA polymerase could be isolated from
cytosol extracts of infected cells. The associated DNA
polymerase in the complex resembled DNA polymerase α in
its template-primer preference and sensitivity to N-ethylmaleimide, araCTP,and salt, but differed in its insensi-
tivity to aphidicolin.The unusual nature of this strange
DNA polymerase needs to be clarified.b) To the second
group belong vaccinia and various types of herpes viruses
which induce specific DNA polymerases, coded by the virus
genome.DNA polymerase activities associated to herpes
viruses has been detected in cells infected by HSV-1
(31,32,33), HSV-2 (34,35), equine herpesvirus (36,37),
human cytomegalovirus (38,39,40), Marek´s disease herpesvirus (41) and Epstein-Barr virus (42).One distinguishable
property of the DNA polymerases induced by all the herpes
viruses is the characteristic stimulation of viral enzyme
by high salt concentration. Therefore, the viral induced

DNA polymerases can be conveniently assayed in crude extracts in the presence of host enzymes by supplementing
the assay systems with 100-150 mM ammonium sulfate. Under
these conditions host DNA polymerases are almost totally
inhibited.The molecular weights of purified herpes-induced
DNA polymerases (31-44) are in the range 150-kdal-200-k-
dal. These enzymes have a pH optimum of 8-8.5 and require
Mg-ions and sulfhydryl compounds for activity (32,44).
Among the synthetic template-primers, $(dG)_{12}.(dC)_n$ was
by far the best primer-template although at very high
concentrations this synthetic oligomer-polymer was slightly inhibitory. Weissbach et al. believe(32) that the preferential ability of HSV-induced enzyme to copy (dG)12.
(dC)n may be due to high G+C content (67%) of the DNA of
this virus. The herpesvirus-induced DNA polymerases are
strongly inhibited by phosphonoacetate & phosphonoformate
(45), and acycloguanosine (46,47,48). Aphidicolin, the
inhibitor of DNA polymerase α, also inhibits herpes-induced DNA polymerase activity (49).The target for phosphonoacetic acid inhibition of the herpes-induced polymerase
was thought to be the inorganic pyrophosphate binding
site of the enzyme (51). However, the concurrence of resistance to phosphonoacetate, acycloguanosine, and araA
bring this simple mechanism into question. The basis for
acycloguanosine action is: i) its conversion to the monophosphate by the virus-coded thymidine kinase only, and
ii) its inhibition as a nucleoside triphosphate of the
viral DNA polymerase which is ten to thirty times more
sensitive than the cellular polymerases (50). c) To the
third group belong some small viruses (hepatitis B Dane
particle) which carry within the virion a DNA polymerase
whose origin is not yet clear.Resistance of Dane particles to solubilization has made the purification and
characterization of this DNA polymerase difficult. Analysis of the banding patterns in sucrose density gradients
of NP-40-treated plasma concentrates rich in Dane particles showed DNA polymerase activities in the density range
covered by the hepatitis B cores (52,53).The only primer-
template that has been known to be utilized by this enzyme is the endogenous Dane particle DNA.

In summary, we can conclude three types of DNA-
polymerase activities associated to the replication of
DNA viruses. The smallest class of DNA viruses do not
specify new DNA polymerases in the infected cell. Instead
they stimulate the synthesis of cellular DNA polymerases
and make use of these to replicate their DNA. On the other
hand, the large DNA viruses induce new DNA polymerases
coded by their genomes, and therefore these viruses do
not depend on the replicative machinery of the host. This

class of viruses offers a selective approach to design
antiviral compounds which act via inhibition of the viral
induced DNA polymerases. The strategy to combat such viral
infections where the replication of viral DNA is solely
dependent on cellular enzymes is difficult, and the inhibitors of DNA polymerases may have only a limited application.

2. Retrovirus Associated DNA Polymerases

Perhapse not as fundamental as the discovery of the
DNA double helix, but most important to our understanding
of the replicative cycle of retroviruses, has been the
discovery of an RNA-dependent DNA polymerase(54,55), the
reverse transcriptase.Reverse transcriptase is located
in the core fraction of the virus as a complex with the
viral RNA genome. These core structures can be obtained
by treatment of the virus particles with nonionic detergents and fractionated by equilibrium density centrifugation on sucrose gradients, taking advantage of their
high density characteristics (approx. 1.25 g/cc). In the
ribonucleoprotein complex the reverse transcriptase (RT)
catalyzes the transcription of the genomic RNA into a
complimentary DNA. The viral genome itself is diploid,
composed of two single stranded, identical subunits of
35S.Each of these subunits of the genomic RNA has features of eukaryotic messenger RNA, being capped at the
5´end and polyadenylated at the 3´end.Most crucial to
the viral genome are five domains (Fig. 1): U5, 76-120
nucleotides at the 5´terminus of the genome; U3,150-200
nucleotides at the 3´end of the genome; TR (terminal
repeats), a direct redundancy of 16-79 nucleotides at
both ends of the viral genome; PBS, a sequence of about
18-21 nucleotides that base pairs with the 3 -stem region
of the tRNA primer; and a short sequence of purines at
the 5´boundary of U3 region. These domains account for
the genesis of long terminal repeats (LTR), the replication of both ends of a linear template without loss of
the genetic content, the initiation and propagation of
both strands of viral DNA, and the role of terminal repeats in viral DNA synthesis.

RNA-directed DNA synthesis is unique to the life
cycle of retroviruses. No similar reaction is known to
occur in normal uninfected cells. Reverse transcriptase
is a multifunctional enzyme, with at least four domains
that mediate RNA-dependent DNA synthesis, RNase-H activity, DNA-dependent DNA polymerase activity, and a site
for nucleic acid binding. All these domain mediated
functional activities offer attractive targets for in-

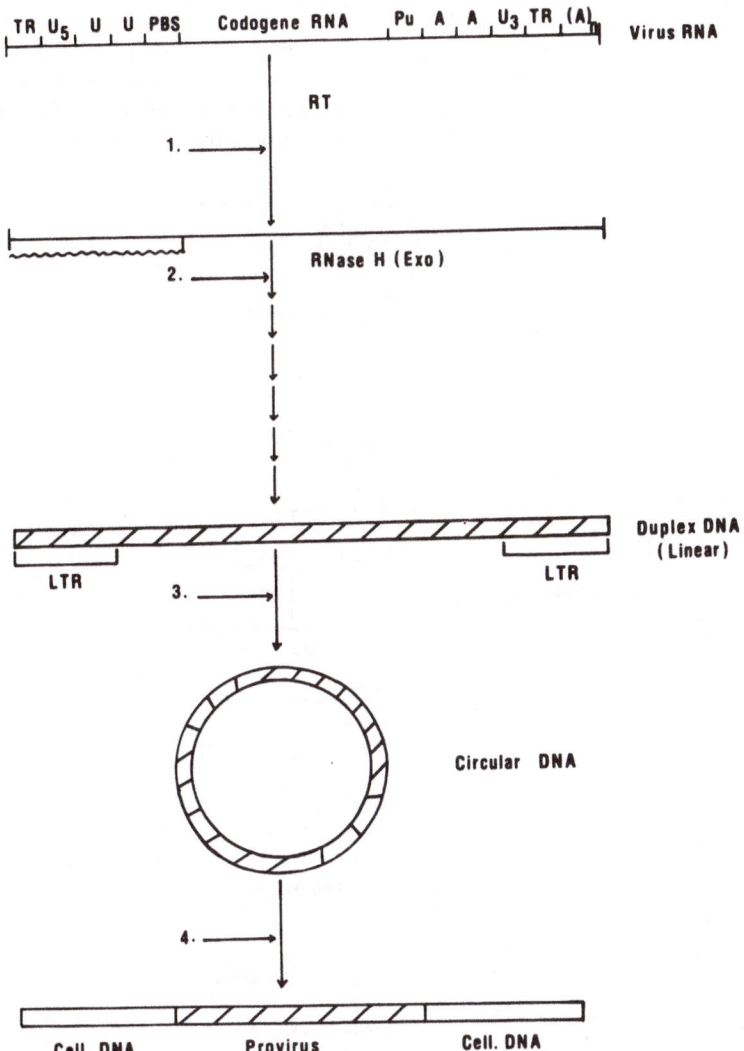

Figure 1:   Schematic presentation of molecular events
            in the provirus integration into host DNA.

hibiting the synthesis of DNA in retroviruses.

The final product of the reactions mediated by

reverse transcriptase is a linear duplex DNA (Fig.1). This
linear DNA is longer than the 35S subunit of genomic RNA
and contains long terminal repeats (whose composition
can be written as 5´-U3 -TR -U5- 3´) present at both ends
(56,57).The linear duplex viral DNA is synthesized in the
cytoplasma, which then migrates to the nucleus where it
is converted a closed circular species (Fig. 2).Apparent-
ly, two mechanisms are involved in the formation of cir-

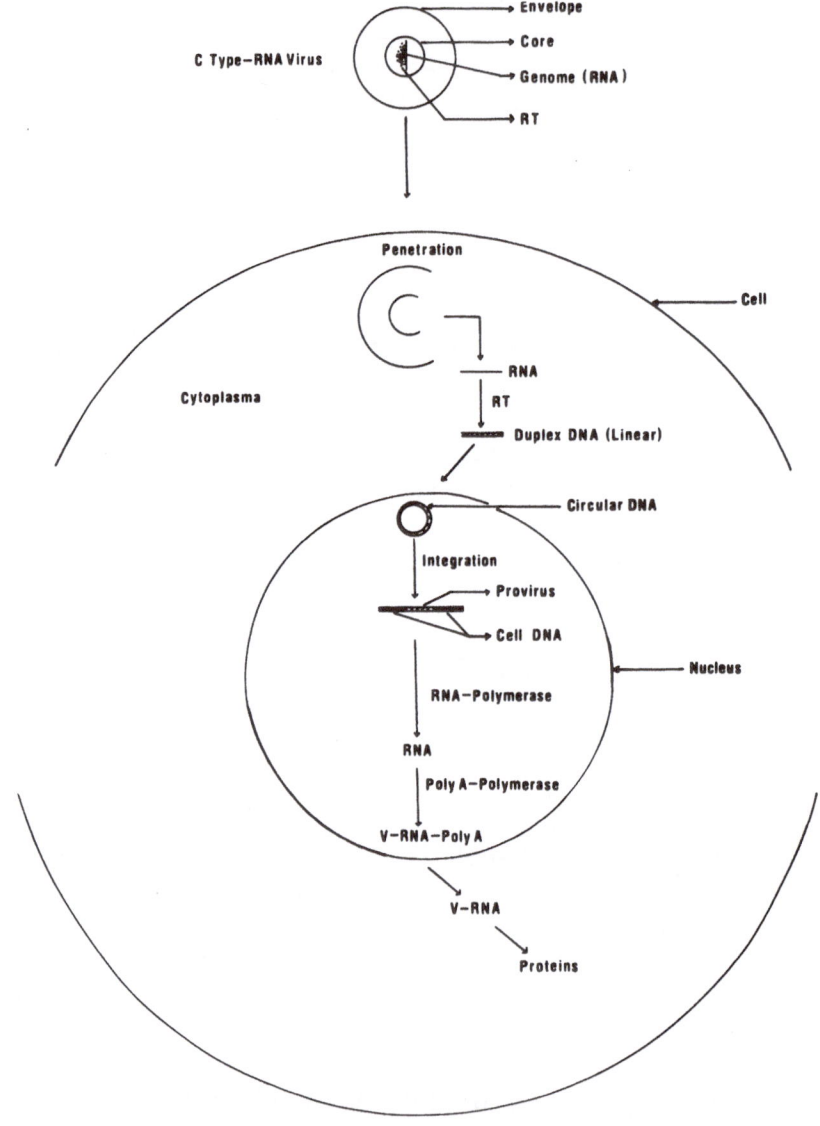

Figure 2: Replicative cycle of retroviruses

cular duplexes of viral DNA:(i) circularization may occur
by homologous recombination between the redundant ends,
leaving a single complete LTR in the circle (58,59), or
(ii) by direct ligation of ends of linear duplex DNA (56,
57,60).

Inhibitor studies by Guntaka et al. (61) indicate
that the circularization of duplex DNA is a prerequisite
for the integration of proviral DNA.The current views on
the mechanism and nature of integrated DNA can be summarized as follows (62,63,64,65): (i) The integrated proviral DNA is almost colinear with the linear form of viral
DNA, synthesized in the cytoplasma. This means that the
integrated proviral DNA has two LTR on terminal ends.(ii)
The integration site on viral DNA is specified by direct
repeats of short cellular sequences (4-6 nucleotide pairs)
(iii) Integration into the host genome can occur in many
different regions of cellular DNA, and apparently, there
is no homology between the interacting sites on viral and
cellular DNA.

The integrated proviral DNA is endowed with specific
squences or domains which mediate signals for the initiation of RNA synthesis, splicing of RNA during the genesis
of viral mRNAs, polyadenylation of the mature mRNA, and
in case of MMTV (mouse mammary tumor virus), a site exists
which signals the induction of viral RNA by steroid hormones (66).

CHEMICAL   INHIBITORS   OF   RETROVIRAL   DNA   POLYMERASE

Soon after the discovery of an RNA-dependent DNA
polymerase (reverse transcriptase) activity in retroviruses (54,55), Gallo and his associates (67) made an
exciting observation, that human leukemic cells possess
an enzymic activity that biochemically resembles the DNA-
polymerase activity associated with retroviruses. This
obervation infected the idea that retroviruses may be involved in human cancer. Immunological probing of reverse
transcriptase has further contributed to our knowledge
on the expression of retroviral information in human tumors (68-81). However, the recent discovery of a retrovirus (HTLV= Human T-cell Leukemia or Lymphoma Virus) in
T cells (fresh and in culture) from a lymphnode biopsy
of a patient with cutaneous T-cell lymphoma is a direct
evidence that retroviruses are associated to human cancer,
at least to this particular type of human leukemias (82,
83).The search for a specific inhibitor of retroviral
reverse transcriptase was, as a matter of fact, induced
by these studies which indicated as association of retro-

viruses to human malignancy.In view of the importance of
reverse transcription in cancer, and as a biological process, attempts have been made in our laboratory to develop inhibitors of DNA synthesis in RNA tumor viruses.

It may be possible to inhibit proviral DNA formation
at several stages (cf. Figures 1 and 2): (a) blocking the
penetration of the membrane by the virus;(b) blocking the
release of viral components inside the host cell (uncoating); inhibiting the formation of linear duplex DNA by
blocking one of the domain-mediated activities of reverse
transcriptase; and finally, inhibiting the integration of
viral DNA by blocking the circularization of linear duplex DNA. Of all these possibilities, the inhibition of
viral DNA polymerase is most attractive since this enzyme
is biochemically and immunologically quite different to .
all the known cellular DNA polymerases, as mentioned in
the earlier section.

The *in vitro* assay systems employed for reverse
transcriptase determination reveal a variety of substrates,
template-primers and interacting compounds which can mo-
dulate the catalytic rate of DNA synthesis(84). Some examples of this type of modulation are: the detergent effect
on the activity of rifamycins (85), influence of divalent
cations (Mg^{++} or Mn^{++}) on the rate of DNA synthesis with
different substrates and the role of chelating agents or
cation binders in buffers (84,86), and the interaction
of thiols with some potential inhibitors or their direct
influence on the measured DNA synthesis (87). Thus slight
variations in assay conditions may lead to wrong interpretations with respect to the specificity of a aprti-
cular inhibitor in the viral DNA-polymerase system.

The second problem is the interpretation of the
enzymatic data with respect to the antiviral activity
of these inhibitors (88). This is particularly the case
with such compounds which exert their inhibitory action
by complexing with one or more synthetic templates. Such
an effect can not be very specific for the viral enzyme.
Drugs exhibiting cytotoxic activities to the extent of
causing a delayed death, or those which intervene in the
replicative cycle of the host cell may give erroneous
information about the specific antiviral activity of the
compound. Thus antiviral studies *in vitro* should be
carried out under conditions and at inhibitor concentration which have little or no effect on the replicative
cycle of the cell. Molecular manipulations of parent
compounds have proved to be very useful in several instances to develop inhibitors of viral DNA synthesis

which exhibit a low cytotoxicity, and at the same time a
high antiviral potential. This is evidenced by our earlier
studies on distamycin derivatives(89,90) and tilorone congeners (91,92,93).

Further efforts to develop compounds which inhibit
viral DNA polymerases by interacting with templates may
lead to the discovery of useful compounds exhibiting a
higher therapeutic index, i.e. low cytotoxicity and high
antiviral activity. However, this approach will not lead
to the development of a specific inhibitor of the viral
enzyme, unless one finds a compound which binds specifically to 70S RNA. Though 70S RNA is a novel feature of
oncornaviruses, from the chemical and physical standpoint
it does not appear to offer any uniqueness to distinguish
it from cellular nucleic acids. Thus, the strategical approach of developing an inhibitor of this type is, at the
present state of our knowledge, unthinkable.

The second approach, which has proved to be more useful and relatively specific in developing such inhibitors,
is to design compounds that bind to the viral enzyme. In
searching for this type of inhibitor, the enzymes chosen
for comparison are important. Many studies have been done
with either avian or mammalian reverse transcriptase, since
these two types of enzymes have some different characteristics. A number of studies claim specificity (or selectivity) of a compound by comparing an inhibitory response
to a bacterial DNA polymerase with the viral reverse transcriptase. Such a comparison has no relevance to the selective nature of the compound. The most imortant approach
to demonstrate selectivity of a compound is to compare inhibitory effects against various cellular DNA polymerases,
such as α, β and γ DNA polymerases.

Our efforts to develop compounds that inhibit viral
DNA polymerases by interacting directly to the enzyme led
to the discovery of a polycytidylic acid analog, containing 5-mercapto-substituted cytosine bases, a partially
thiolated polycytidylic acid (94,95). This compound, abbreviated as MPC (marcapto-polycytidylic acid)was found to
inhibit the retroviral DNA polymerase in a very specific
manner (94,95,96,97,98,99,100,101).In addition to studies
with MPC, we have recently evaluated the effects of several
polyadenylic acid analogs on various DNA polymerases. In
the following pages we would like to describe the mode of
action of polynucleotide analogs on viral reverse transcriptase, and discuss the biological and clinical implica-
tions of these studies. In order to document the specificity of action of these analogs, we have compared their

activities against purified cellular DNA polymerases α ,
β and γ, and terminal deoxynucleotidyl transferase (TdT).

  1. Studies with 5-mercapto polycyctidyl acid:The
inhibition of DNA polymerases from RNA tumor viruses by
5.mercapto polycytidylic acid (MPC) was described earlier
(94-101). Partially thiolated polycytidylic acid preparations (MPC I-III), containing 1.7%, 3.5% and 8.6% 5-mer-
captocytidylate units inhibited the DNA polymerase activity of Friend leukemia virus (FLV) in the endogenous reaction as well as in the presence of exogenous template-
primers, such as (rA)n.(dT)12 and (rC)n.(dG)12. The inhibitory activities were directly related to the percent
of thiolation. A maximum inhibition was observed with preparations containing 15-17% of the thiolated cytosine bases. In these experiments, non-thiolated samples of polycytidylic acid showed no inhibition of the reverse transcriptase reaction.

  Polyuridylic acid thiolated by the same procedure
(94) was also found to inhibit this reaction strongly(102);
however, the non-thiolated sample of polyuridylic acid
also inhibited the reaction significantly. Besides, the
inhibitory effect of thiolated samples of polyuridylic
acid was not strictly related to their degree of thiolation, as was observed for the partially thiolated polycytidylic acid.The inhibitory effect of non-thiolated
polyuridylic acid is, presumably, due to hydrogen bonding
between poly-U and the added template, (rA)n. The inhibition of the endogenous reaction by non-thiolated poly-U
is explanable by the fact, that genomic RNA contains large
stretches of polyadenylic acid at 3 -terminal. For this
reason, it was difficult to designate the role of thiolation on the inhibitory effect of MPU samples.

  The mode of action of MPC on the viral DNA synthesis
was investigated by product analysis of the DNA polymerase
reaction (endogenous) in the absence or in the presence
of MPC, as described by Kotler and Becker (103). The reaction mixtures were dissolved with Na-dodecyl sulfate
(1%, wt/wt, final concentration), loaded on a hydroxylapatit column (1g, Bio-Rad Lab., Munich), eluted with a
Na-Phosphate gradient (0.05-0.4 M), collected into about
40 tubes (total volume approx. 100 ml), and the TCA-in-
soluble activity collected on GF/C filters (Whatman) and
counted in a liquid scintillation counter.

  Analysis of the endogenous products of the detergent disrupted virions exhibits three DNA species: single
stranded DNA (ss DNA), RNA.DNA hybrids (hy-DNA) and the

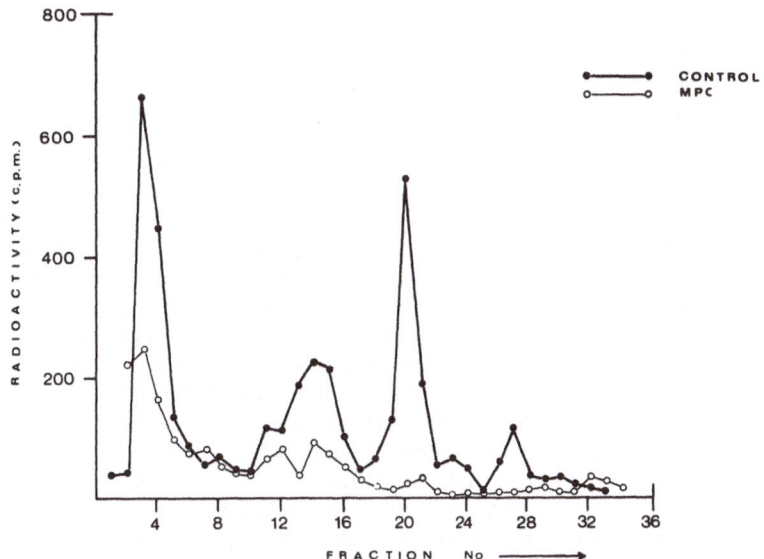

Figure 3: Analysis of the DNA species synthesized by FLV-
DNA polymerase by elution from hydroxylapatit
column. Experimental details are described in
the text. The first species to be eluted from
the column contained ss-DNA, the second contained
hy-DNA and finally, the ds-DNA, eluted in the
last peak. The concentration of MPC in the reaction mixture = 20 µg (95).

double stranded DNA (ds-DNA). As follows from Figure 3,
in the presence of MPC (open circles) there is an overall inhibition of ^{3}H-dTMP incorporation, indicating that
the formation of all the 3 species is blocked. This is to
be expected since the inhibitor binds to the enzyme. This
has been confirmed by the ultracentrifugation studies in
which the binding of ^{35}S-labeled MPC to a purified FLV-
DNA polymerase was investigated (95).

In view of the fact that all of the oncornaviral DNA
polymerases examined so far, do require a primer-template-
like double stranded structure for the initiation of DNA
synthesis, it is not surprising that single stranded synthetic polynucleotides (unprimed templates) can act as inhibitors of the polymerization reaction. This, presumably,
is due to hydrogen bonding of the base sequences between
the added polymer and the functional template. Thus, the

specificity of inhibition by such polymers is not limited to the viral enzyme system only. On the other hand,
minor modification in the chemical structure of synthetic polynucleotides might be useful to develop inhibitors
that interact directly with the enzyme but fail to be
transcribed , i.e. they function as "dead template" for
the enzyme. The data from our laboratory have shown that
the partially thiolated polycytidylic acid is functioning
as a "dead template" in the DNA polymerase system of FLV
(95). The results of these studies (95) can be sammarized
as follows:(a) The incorporation of ^{3}H-dGMP into DNA by
the viral enzyme is stimulated to about 9-fold (compared
to endogenous value) in the presence of (rC)n.(dG)12. However, under similar conditions a hybrid of (MPC).(dG)12
failed to stimulate the incorporation of ^{3}H-dGMP into DNA;
(b) In the presence of MPC.(dG)12, the increasing concentrations of (rC)n.(dG)12 in the reaction mixture have
no effect on the activity of the enzyme; however, at higher enzyme concentrations the stimulatory effect of (rC)n.
(dG)12 gradually reappears. These data indicate that the
viral enzyme has a higher affinity towards MPC than to its
optimal template (rC)n. The presence of zinc in reverse
transcriptase molecule makes it attractive to suggest that
the mercapto group may undergo an interaction with zinc
to form a stable complex with the enzyme.

     In order to determine the selectivity of MPC action
cellular DNA polymerases (α , β and γ) were purified from
normal and FLV-infected mouse spleen. The inhibitory effects of MPC on the cellular polymerases, and on the activity of reverse transcriptase from FLV-infected mouse
spleen were compared (101). None of the cellular DNA polymerases, from normal as well as from FLV-infected mouse
spleen, was inhibited by MPC. On the other hand, (rA)n.
(dT)12 and (rC).(dG)12-catalyzed activities of FLV-rev-
erse transcriptase were highly sensitive to MPC action.
Kinetic studies on the FLV-DNA polymerase reaction revealed that MPC inhibition is of a non-competitive nature
(Fig. 4).

     The studies on the selectivity of MPC action were
substantiated using a purified preparation of terminal
deoxynucleotidyl transferase (TdT) from calf thymus. The
comparative effects of MPC on the (dA)n-catalyzed reaction
of TdT, and on the (rC)n.(dG)12-dependent activity of reverse transcriptase is shown in Figure 5. The non-thiolated samples of polycytidylic acid (PC) were, in both cases
without any inhibitory effect; on the contrary, at lower
concentrations a significant stimulation of the reactions
was observed. The thiolated, sample (SH= 18.2%) of poly-

cytidylic acid(MPC) inhibited both the enzymic reactions;
however the reverse transcriptase reaction was almost 15
times more sensitive to the action of MPC than the TdT-
reaction.

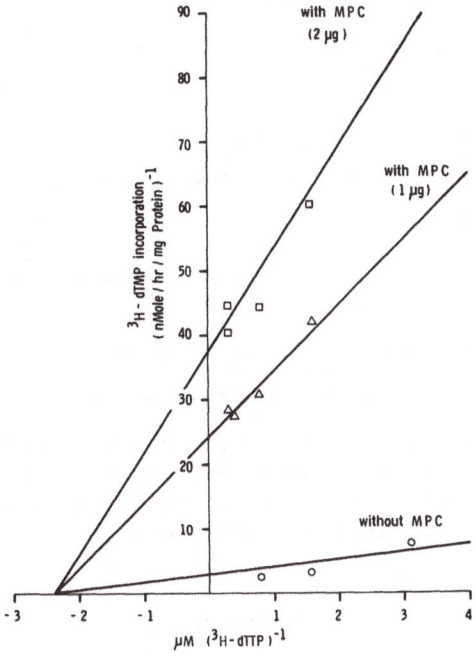

Figure 4: Lineweaver.Burk plot of the kinetics of FLV-DNA
          polymerase reaction in the absence(open circles)
          or presence of MPC (triangles= 1 µg/reaction
          mixture; squares= 2 µg).

     Of all the enzymatic studies carried out in our labo-
ratory so far, MPC has proved to be relatively specific
towards reverse transcriptase.

     The protective effect of MPC on leukemogenesis in
mice by FLV has been reported earlier (101). In these
experiments, the effect of MPC was investigated by pre-
incubating the virus containing suspensions with MPC.
These studies have been extended using MPC in-vitro and
in-vivo.

     The animals (mice:Groppel strain) were divided into
four groups of five each (donors): 1) Group 1 was injected
with a viral suspension (citrate plasma from FLV-infected
animals, dose= LD90) preincubated with Tris/HCl buffer,
pH 7.6 for 30 min. at 37 °C; 2) Group 2 was injected with

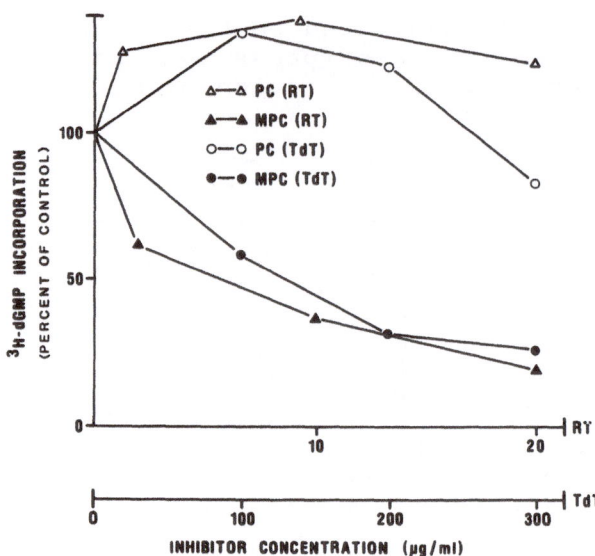

Figure 5: Effect of 5-mercapto polycytidylic acid (SH= 18.2%) on the activities of retroviral reverse transcriptase and on calf thymus terminal deoxynucleotidyl transferase.

TABLE 2

Assay for Leukemogenic Potential of Spleen Extracts from FLV-infected Mice after their in-Vitro/Vivo Treatment with 5-Mercapto Polycytidylic Acid.

TREATMENT OF DONOR MICE	LEUKEMOGENESIS IN RECEPIENT MICE AFTER INFECTION WITH SPLEEN EXTRACT[1]		
	NO. OF POSITIVE TOTAL NO. OF MICE	MEAN SURVIVAL TIME (DAYS)	MEAN SPLEEN WEIGHT(g)
Virus Suspension[2] (0.2 ml) + Tris buffer (37ºC, 30 min)	5/5	47	2. 41
Virus Suspension + 200 µg of MPC (37 ºC, 30 min.) + 50 µg MPC, i.p. (day 5 & 9)	2/5	123[3]	1. 05 (1.78, 2.10, 0.52,0.41,0.44)
Virus Suspension + Tris buffer (37 ºC, 2hr)	5/5	52.2	1. 80
Virus Suspension + 200 µg of MPC (37 ºC, 2 hr) + 50 µg MPC, i.p. (day 5 & 9)	1/5	110 4 (123)[3] 1 (97)	0. 38 (0.74,0.29,0.34 0.22, 0.31)

(1) Cell-free spleen extracts were prepared from spleens of donor mice on 10th day after virus infection, or other treatments as shown.(2) Citrate plasma(LD$_{90}$).
(3) Experiment terminated on 123rd day.

the viral suspension, as in 1, but preincubated with MPC
(200 ug per 0.2 ml suspension) at 37 °C for 30 min. These
animals received in addition, on day 5 and day 9 (post infection) 50, μg of MPC, injected intraperitoneally; 3)Group
3 was treated similar to group 1, except that the viral
suspensions were preincubated for 2hr; and 4) Group 4 was
treated in a similar manner as group 2, except that the
viral suspensions were preincubated for 2hr at 37 °C. On
the 10th day, animals were sacrificed and spleen extracts
were prepared, as described elsewhere (104).The spleen extract from each mouse was then analyzed individually with
respect to their leukemogenic potentiality. Each donor
spleen specimen was reinjected to a different "receipient"
mouse (20 in total), and the leukemogenesis was followed,
as shown in Table 2.

    All animals in group 1 and 3 developed splenomegally
and died between 40-60 days; whereas, in the MPC-treated
groups, of the 10 animals only 3 showed signs of splenomegally. In group 2, 2 animals had splenomegally but, in
spite of that, all animals survived till the 123rd day,
at which time our experiment was terminated. Similarly,
in the last group animals survived till the 123rd day;
one died on the 97th day. The spleen weights, shown in
the last column, also exhibit large differences between
the MPC-treated group,and the control group.In another
study we have analyzed the effect of MPC on normal mice
of the same strain. We failed to observe any effect of
MPC on the spleen weights of non-infected mice.

    In another study (98, 104) we have analyzed the in-
vivo effect of MPC on the leucocytes of mice infected
with the active Friend virus. Within 12-24 hrs. after
MPC injection (50 μg /mouse) a dramatic fall in the leucocyte count of animals infected with FLV was observed;
MPC failed to reduce the leucocyte number in mice not-
infected with the virus. It is interesting that, in one
of the infected animals MPC failed to suppress the leucocyte number; on the contrary, there was a gradual increase in leucocyte number. This animal died on the 13th
day of MPC treatment. Unfortunately, we were not able to
analyze the spleen of this animal. It is therefore difficult to interpret the reasons for failure of MPC effect
in this animal.

    Based on the biochemical selectivity of MPC action
and its biological effects, clinical trials of MPC in the
treatment of childhood leukemia were initiated. We began
these trials on cases with terminal phase of the disease.

TABLE 3

Clinical Data of Patients with ALL and AML submitted to MPC trials.

No.	Init.	Age	Sex	Diag.	Stage	Results
1	D. M.	8	♀	ALL	3rd rel.	PR , WBC ↓
2	T. I.	6 6/12	♀	ALL	2nd rel.	?
3	B. M.	6 7/12	♂	ALL	2nd rel.	?
4	M. M.	8 4/12	♂	ALL	2nd rel.	CR
5	J. O.	3 6/12	♂	ALL	2nd rel.	?
6	N. A.	10	♀	ALL	2nd rel.	CR
7	B. C.	5 8/12	♀	ALL	3rd rel.	CR
8	N. N.	7 11/12	♂	ALL	3rd rel.	?
9	M. A	12	♂	ALL	1st rel.	?
10	M. I.	8 6/12	♂	ALL	1st rel.	?
11	K. C.	11 3/12	♀	ALL	4th rel.	?
12	K. K.	10 9/12	♀	ALL	5th rel.	PR
13	L. J.	7	♂	ALL	3rd rel.	WBC ↓
14	F. D.	2 3/12	♂	ALL	init. ph.	WBC ↓
15	S. B.	7 11/12	♀	ALL	init. ph.	WBC ↓
16	H. B.	12 5/12	♀	ALL	init. ph.	WBC ↓
17	S. N.	4	♀	AML	init. ph.	WBC ↓
18	W. H.	5 6/12	♂	AML	init. ph.	WBC ↓
19	S. K.	4 9/12	♂	ALL	init. ph.	WBC ↓
20	F. A.	7 10/12	♀	AML	init. pH.	WBC ↓
21	L. F.	3 4/12	♂	ALL	init. p h.	WBC ↓
22	F. T.	4 9/12	♂	ALL	init. ph.	WBC ↓
23	C. E.	13 4/12	♂	ALL	init. ph.	WBC ↓

Clinical data of patients submitted to MPC trials are shown in Table 3. Of the 23 cases treated with MPC, 13 were in the terminal phase of the disease. These patients were resistent to all previous chemotherapeutic regimes which involved drugs such as prednisone, vincristin, daunorubicin, L-asparaginase, Ara-C, 6-mercaptopurine, methotrexate, cyclophosphamide and actinomycin D.

MPC preparation used in our clinical trials contained 15 percent of thiolated cytosine bases. The lyophilized product (MPC) was dissolved in 0.1M Tris/HCl buffer, pH 7.6, and diluted with 0.9% NaCl before use. This solution was sterilized by passing through a membrane filter (Milli-pore GmbH, Neu Isenburg, Germany). It was kept at 4 °C and used immediately, or within the next five days; solutions older than 5 days were reprecipitated, purified on the column and resterilized. In our clinical trials, MPC (sterile) was given intravenously at a dose 0.5 mg/kg body weight. The injections were given once a week.

Of the 13 terminal cases, complete remission was achieved in 3, and a partial remission achieved in 2 other cases. In the terminal cases, it is not possible to give an empirical evaluation of MPC-effectivity, since these children either lived for a short time only, or no remission could be achieved. The extent to which MPC affected the survival period in these cases can not be evaluated, since the patients were in a very heterogeneous state (1 to 5 relapses).Fever,occasionally accompanied by shivering was frequently observed under MPC treatment in the first hour after injection. However, these symptoms never lasted more than the first hour, and no other side effects could be observed.

On the basis of our experience with MPC on terminal cases, we were motivated to give MPC a clinical trial in the beginning of leukemia treatment. A monotherapy with MPC, as devised for terminal cases is, however, not possible. We therefore decided to introduce MPC therapy in the beginning of treatment of cases which at the time of diagnosis had leococytosis, The initial treatment, a single injection of MPC, was then followed up by the chemotherapy protocol, devised by Rhiem (106). Irrespective of ALL or AML, 24 hrs after MPC injection, there was a dramatic reduction of leukemic cells in all cases (Table 3).

The selective cytolysis of leukemic cells by MPC indicates that this compound binds to some "novel" component present in these cells, which may be the viral related reverse transcriptase. The reason for this selectivity may be the concentration range, which does not affect the normal cellular polymerases, as indicated by the enzymatic studies. Once the compound is bound to this"target" the cytotoxicity could then be a secondary event, probably by release of lysosomal enzymes.

2. Studies with Polyadenylic Acid Analogs: The fact that polyadenylic acid (Poly A) is easily transcribed by

the retroviral reverse transcriptase has led to the development of several poly A-analogs (110,111). The mode
of action of some of these analogs (Fig. 6) on various
DNA polymerases and viral reverse transcriptase has been
studied in our laboratory (109). These studies have been
recently reviewed (107,108) and therefore, will not be
described in detail.

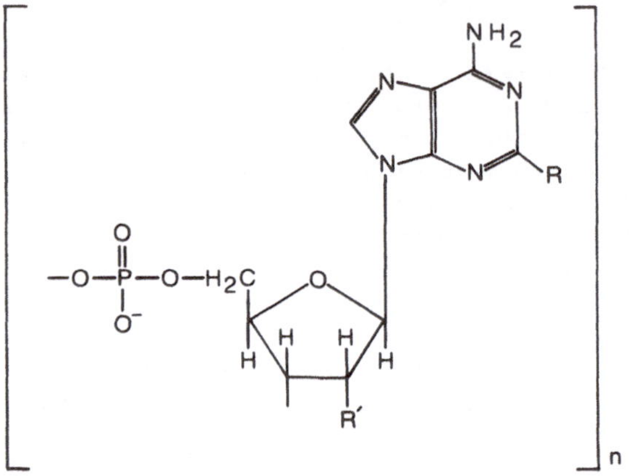

R = −S−CH$_3$, R′=−OH: poly(2-methylthioadenylic acid), (ms^{2}A)$_n$

R = −S−CH$_2$CH$_3$, R′=−OH: poly (2-ethylthioadenylic acid), (es^{2}A)$_n$

R = −H, R′=−F: poly (2′-fluoro-2′-deoxyadenylic acid), (dAfl)$_n$

Figure 6: Chemical Structures of Polyadenylic Acid Analogs.

        In summary, the polymer containing 2′-deoxy-2′-fluoro-
adenosine, (dAfl)$_n$, showed a concentration dependent stimulation of (rA).(dT)$_{12}$-catalyzed reverse transcriptase
reaction from Rauscher Leukemia Virus (RLV). A similar
stimulation of the (rA).(dT)$_{12}$-catalyzed DNA polymerase-
γ reaction was also observed. However, the (rC).(dG)$_{12}$-
dependent reverse transcriptase activity was inhibited
by (dAfl)n. The DNA polymerase-ß activity catalyzed by
(dA)$_n$.(dT)$_{12}$ was also inhibited by (dAfl)$_n$. We concluded
that (dAfl)$_n$ may have a secondary structure very similar
to (rA)$_n$, than to (dA)$_n$. This is supported by the physico-
chemical studies of Ikehara et al. (111). The interesting
observation is that (dAfl)$_n$ can be transcribed by the
retroviral reverse transcriptase even in the absence of

a complementary primer strand (Fig. 7). However, the K_m
of the reverse transcriptase reaction catalyzed by $(dAfl)_n$
is higher than the reaction catalyzed by $(rA)_n.(dT)_{12}$.

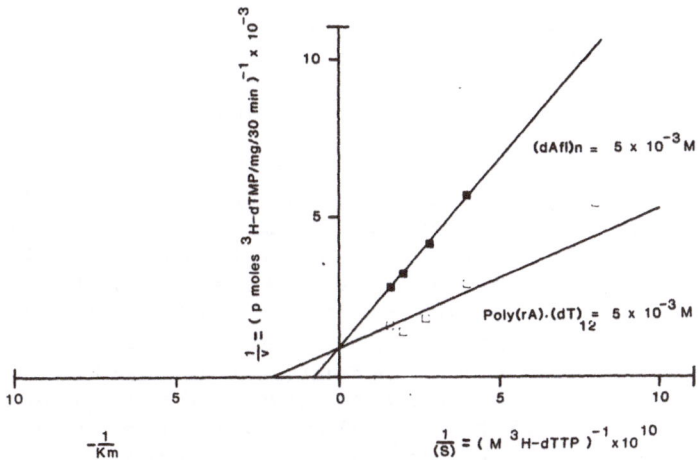

Figure   7: Lineweaver-Burk Plot of the Kinetics of Reverse
            transcriptase reaction, catalyzed by $(dAfl)_n$
            and $(rA)_n.(dT)_{12}$.

    In contrast, the 2-substituted derivatives, poly(2-
methylthio-adenylic acid) and poly(2-ethylthio-adenylic
acid), are not able to descriminate between the reactions
catalyzed by different template-primers. For example,
both the derivatives inhibit $(rA)_n.(dT)_{12}$- and $(rC)_n.(dG)_{12}$
catalyzed reverse transcriptase reaction to the same extent; though the methylthio derivative is a better inhibitor than the ethylthio analog. The DNA polymerase-α was
less sensitive to these inhibitors; whereas, the bacterial
DNA polymerase (Polymerase I) was completely resistent
to the action of all the derivatives used in this study.

CONCLUSION   AND   FUTURE   PROSPECTS

    The concept of antiviral chemotherapy emerges from
the basic knowledge about the biochemical differences
between the metabolic reactions operating in a normal
cell, and in a virus-infected cell.The knowledge about
the association of several RNA- and DNA viruses to human
cancer has been very rewarding in the exploitation of
such differences.We know of several DNA polymerizing activities, not present in the normal eukaryotic cell, are
present in virus-infected cells.These virus-specific DNA
polymerases are either induced by viruses, or in some

cases, associated as a structural entity of the virus. In both cases, these DNA polymerases are products of the viral genome. Recent advances in the biochemical methodology have contributed to the purification of these "novel" DNA polymerases from viruses, or virus-infected cells. The purified virus-specific DNA polymerases are biochemically and immunologically different to the constitutive cellular DNA polymerases α, β and γ.Thus, virus-specific DNA polymerases offer a unique molecular target for designing potential antiviral compounds.

In spite of a great variety of inhibitors which affect virus-induced DNA synthesis, an absolutely specific inhibitor of such viral polymerases is not yet available. The specificity is relative and often varies with the system studied. Thus, an inhibitor may have a Ki value of less than 10^{-5} M for the viral DNA-polymerase system, but a Ki of 10^{-3} M for the cellular DNA-polymerase system. As a result, it makes sense to evaluate the biological activity of such inhibitors under carefully controlled conditions. The question is: Is it possible to fractionate the antiviral and anticellular activities of such compounds in a particular dose range ? We know from our own studies that the cytotoxic and the antiviral activities of a compound can be influenced differently by structural alterations of the compound.

The future of the development of new DNA-polymerase inhibitors is dependent on two parameters: firstly, a good knowledge about the mode of action of the compond under investigation. This is important to localize the structural constraints of the molecule involved in exercising a particular function. On the basis of our present-day knowledge about the molecular processes involved in the replication of such viruses, it is possible to specify the possible target(s) for such inhibitors. This will facilitate the modification of the molecular structure in order to potentiate functional activity of the compound against that particular target reaction.Secondly, to develop new systems and procedures to identify certain binding interactions which occur during the in-vivo polymerization of DNA. The lysolecithin-permeabilized baby hamster kidney cell system, developed by Miller et al. (14,15) could contribute to study the interaction of such inhibitors with their molecular targets under semi-in vivo conditions.

ACKNOWLEDGEMENT

The experimental studies reported here were supported by Volkswagen Foundation (Grant No.14 03 05).

REFERENCES

1. BOLLUM,F.J.: J.Am.Chem.Soc. 80, 1766, 1958.

2. KORNBERG,T. and GEFTER,M.L.: Biochem.Biophys.Res.
   Commun. 40, 1348, 1970.

3. MOSES,R.E. and RICHARDSON,C.C.: Proc.Natl.Acad.Sci.
   U.S. 67, 674, 1970.

4. KNIPPERS,R.: Nature 228, 1050, 1970.

5. KORNBERG,T. and GEFTER,M.L.: Proc.Natl.Acad.Sci.U.S.
   68, 761, 1971.

6. WEISSBACH,A., BALTIMORE,D., BOLLUM,F.J., GALLO,R.C.
   and KORN,D.: Science 190, 401, 1975.

7. SARANGADHARAN,M.G., ROBERT.GUROFF,M. and GALLO,R.C.:
   Biochim. Biophys. Acta 516, 419, 1978.

8. LOEB,L.A.: In, The Enzymes (Edt. P.D. Boyer), vol 10,
   173, 1974. Academic Press, New York.

9. BOLLUM,F.J.: In, Progress in Nucleic Acid Research &
   Molecular Biology 10, 109, 1975.Academic Press,
   New York.

10. WEISSBACH,A.: Cell 5, 101, 1975.

11. BOLLUM,F.J.: In, Antiviral Mechanisms in the Control
    of Neoplasia (Edt. P. Chandra) pp.587-601, 1979.
    Plenum Press, New York.

12. FALASCHI,A. and SPADARI,S.: In, DNA-Synthesis(NATO
    Symposium) pp 487, 1978.Plenum Press,New York.

13. SUGINO,A. and NAKAYAMA,K.: Proc.Natl.Acad.Sci.U.S.
    77, 7049, 1980.

14. MILLER,M.R.,CASTELLOT,J.J. and PARDEE,A.B.: Bio-
    chemistry 17, 1073, 1978.

15. MILLER,M.R.,CASTELLOT,J.J. and PARDEE,A.B.: Exptl.
    Cell Res. 120, 421, 1979.

16. CASTELLOT,J.J., MILLER,M.R., LEHTOMAKI,D.M. and
    PARDEE,A.B.: J.Biol.Chem. 254, 6904, 1979.

17. BOLLUM,F.J.: J.Biol.Chem. 237, 1945, 1962.

18. McCAFFREY,R., SMOLER,D., and BALTIMORE,D.:Proc.Natl.
    Acad.Sci.U.S. 70, 521, 1973.

19. COLEMAN,M.S., HUTTON,J.J., SIMONE,P.D. and BOLLUM,
    F.J.: Proc.Natl.Acad.Sci.U.S. 71, 4404, 1974.

20. SARIN,P.S. and GALLO,R.C. :J.Biol.Chem. 249, 8051,
    1974.

21. SRIVASTAVA,B.I.S.: Cancer Res. 34,1015, 1974.

22. BERTAZZONI,U. and BOLLUM,F.J.: Terminal Transferase in Immunobiology and Leukemia 1982. Plenum Press, New York.

23. BERTAZZONI,U.,SCOVASSI,A.I., TORSELLO,S., BRUSAMOLINO, E., ISERNIA,P., BERNASCONI,C., GIENLLI,E. and SACCHI,N.: In, Biochemical and Biological Markers of Neoplastic Transformation (Edt. P. Chandra), pp 217-232, 1983. Plenum Press, New York.

24. SARIN,P.S., VIRMANI,M., PANTAZIS,P. and GALLO,R.C.: In, Biochemical and Biological Markers of Neoplastic Transformation (Edt. P. Chandra), pp 193-216, 1983. Plenum Press New York.

25. TEGTMEYER,P. and ANDERSEN,B.: Virol. 115, 67, 1981.

26. DULBECCO,R., AHRTWELL,L.H. and VOGT,M.: Proc.Natl.Acad Sci.U.S. 53, 403, 1965.

27. WEIL,R., MICHEL,M.R. and RUSCHMANN,G.K.. Proc.Natl. Acad.Sci.U.S. 53, 1468, 1965.

28. KIT,S., DUBBS,D.R. and FREARSON,P.M.: Cancer Res. 26, 262, 1966.

29. MARTIN,R.G.: Adv. Cancer Res. 34, 1, 1981.

30. ITO,K., ARENS,M. and GREEN,M.: J.Virol. 15,1507,1975.

31. KEIR,H.M., SUBAK-SHARPE,H., SHEDDEN,W.I.H., WATSON, D.H. and WILDY,P.: Virology 30, 154, 1966.

32. WEISSBACH,A., HONG,S.C.L., AUCKER,J. and MÜLLER,R.: J.Biol.Chem. 248, 6270, 1973.

33. ARON,G.M., PURIFOY,D.J.M. and SCHAFFER,P.A.: J.Virol. 16, 498, 1975.

34. HAY,J., MOSS,H. and HALLIBURTON,I.W.: Biochem.J. 124, 64p, 1971.

35. PURIFOY,D.J.M. and BENYESH-MELNICK,M.: Virology 68, 374, 1975.

36. KEMP,M.C., COHEN,J.C., O'CALLAGHAN,D.J. and RANDALL, C.C.: Virology 68, 467, 1975.

37. ALLEN,G.P., O'CALLAGHAN,D.J. and RANDALL,C.C.: Virology 76, 395, 1977.

38. MILLER,R.L. and RAPP,F.: J. Virol. 20, 564, 1976.

39. HUANG,E.S.: J. Virol. 16, 298, 1975.

40. HIRAI,K. and WATANABE,Y.: Biochim. Biophys. Acta 447, 328, 1976.

41. BOEZI,J.A., LEE,L.F., BLAKSLEY,R.W., KOENIG,M. and
    TOWLE,H.C.: J. Virol. 14, 1209, 1974.

42. MILLER,R.L., GLASER,R. and RAPP,F.: Virology 76,494,
    1977.

43. KNOPF,K.: Europ.J. Biochem. 98, 231, 1979.

44. POWELL,K.L. and PURIFOY,D.J.M.: J. Virol. 24, 618,1977

45. ERIKSSON,B., LARSSON,A., HELGSTRAND,E., JOHANSSON,N.G.
    and ÖBERG,B.: Biochim.Biophys.Acta 607,53, 1980.

46. FURMAN,P.A., St.CLAIR,M.H., FYFE,J.A., RIDEOT,J.L.,
    KELLER,P.M. and ELION,G.B.: J. Virol. 32,72, 1979.

47. MILLER,W.H. and MILLER,R.L.: J. Biol.Chem. 255, 7204,
    1980.

48. DARBY,G., FIELD,H.J. and SALISBURY,S.A.: Nature 289,
    81, 1981.

49. PEDRALI-NOY,G. and SPADARI,S.: J. Virol. 36, 457, 1980.

50. DERSE,D., CHENG,Y.C., FURMAN,P.À., St.CLAIR,M.H. and
    ELION,G.B.: J.Biol.Chem. 256, 11447, 1981.

51. LEINBACH,S.S., RENO,J.M., LEE,L.F., ISBELL,A.F. and
    BOEZI,J.A.: Biochemistry 15, 426, 1976.

52. KAPLAN,P.M., GREENMAN,R.L., GERIN,J.L., PURCELL,R.H.
    and ROBINSON, W.S.: J. Virol. 12, 995, 1973.

53. ROBINSON,W.S. and GREENMAN,R.L.: J.Virol.13, 1231,1974

54. TEMIN,H.M. and MIZUTANI,S.: Nature 226, 1211, 1970.

55. BALTIMORE,D.: Nature 226, 1209, 1970.

56. SHANK,P.R., COHEN,J.C., VARMUS,H.E., YAMAMOTO,K.R.,
    and RINGOLD,G.M.: Proc.Natl.Acad.Sci.U.S. 75, 2112,
    1978.

57. HSU,T.W., SABRAN,J.L., MARK,G.E., GUNTAKA,E.V. and
    TAYLOR,J.W.: J. Virol. 28, 810, 1978.

58. DHAR,R., McCLEMENTS,W.L., ENQUIST,L.W. and VANDE
    WOUDE,G.F.: Proc.Natl.Acad.Sci.U.S. 77,3971,1980.

59. SWANSTROM,R., DeLORBE,W., BISHOP,J.M. and VARMUS,H.E.:
    Proc.Natl.Acad.Sci.U.S. 78, 124, 1981.

60. YOSHIMURA,F.K. and WEINBERG,R.A.: Cell 16,323, 1979.

61. GUNTAKA,R.V., MAHY,B.W.J., BISHOP,J.M. and VARMUS,H.E.
    Nature 253, 507, 1975.

62. TEMIN,H.M.: Cell 21, 599, 1980.

63. MAJORS,J.E. and VARMUS,H.E.: Nature 289,253, 1981.

64. SHIMOTOHNO,K. and TEMIN,H.M.: Proc.Natl.Acad.Sci.U.S. 77, 7357, 1980.

65. HUGHES,S.H., MUTSCHLER,A., BISHOP,J.M. and VARMUS,H.E. Proc.Natl.Acad.Sci.U.S. 78, 4299, 1981.

66. RINGOLD,G.M.: Biochim. Biophys. Acta 560,487, 1979.

67. GALLO,R.C., YANG,S.S. and TING,R.C.: Nature 228,927, 1970.

68. GALLO,R.C., GALLAGHAR,P.E., MILLER,N.R., MONDAL,H., SAXINGER,W.C., MAYER,R.J., SMITH,R.G. and GILLES-PIE, D.H.: Cold Spring Harb.Symp.Quant.Biol. 39, 933, 1975.

69. CHANDRA,P.: In, Proceedings of the XIIth International Cancer Congress, Buenos Aires, Vol. 1,pp.29, 1978. Pergamon Press, Oxford.

70. CHANDRA,P.: Advances in Opthalmology 38, 234, 1979. Karger-Verlag, Basel.

71. CHANDRA,P.: In, Proceedings of the Federation of European Biochemical Societies, Dubrovnik (Edt. P. Mildner) Vol. 61,pp. 215, 1979. Pergamon Press.

72. CHANDRA,P. and STEEL,L.K.: Biochem. J. 167,513,1977.

73. CHANDRA,P., BALIKCIOGLU,S. and MILDNER,B.: Cancer Lett. 5, 299, 1978.

74. CHANDRA,P., STEEL,L.K. and CAVDAR,A.O.: In, Modern Trends in Human Leukemia (Edts. R. Neth et al.) Vol. III,pp.497,1979. Springer-Verlag,Berlin-Heidelberg-New York.

75. CHANDRA,P., STEEL,L.K., LAUBE,H. and KORNHUBER,B.: In, Antiviral Mechanisms in the Control of Neoplasia (Edt. P. Chandra) pp177, 1979. Plenum Press, New York.

76. CHANDRA,P., STEEL,L.K., LAUBE,H., BALIKCIOGLU,S., EBENER,U. and WELTE,K.: In, Proceedings of the 8th International Symposium on the Biological Characterization of Human Tumors, Athens. Vol. 5,pp 359, 1980. Excerpta Medica, Amsterdam.

77. WELTE,K., EBENER,U. and CHANDRA,P.:Cancer Lett. 7, 189, 1979.

78. CHANDRA,P.,STEEL,L.K.,LAUBE,H.,BALIKCIOGLU,S.,MILDNER, B.,EBENER,U.,WELTE,K. and VOGEL,A.: In, Cold Spring Harb. Conference on Cell Proliferation (Viruses in naturally occuring cancer) vol. 7, 775, 1980.

79. CHANDRA,P.,BALIKCIOGLU,S. and MILDNER,S.: Cell &
    Mol. Biol. 27, 239, 1981.

80. CHANDRA,P.: Editorial Review, Survey of Immunologic
    Res. (in press)

81. OHNO,T. and SPIEGELMAN,S.: Proc.Natl.Acad.Sci.U.S.
    74, 2144, 1977.

82. POIEZ,B.J.,RUSCETTI,F.W.,GAZDAR,A.F.,BUNN,P.A.,MINNA,
    J.D. and GALLO,R.C.: Proc.Natl.Acad.Sci.U.S 77,
    7415, 1980.

83. GALLO,R.C., ROBERT-GUROFF,M., KALYANARAMAN,V.S.,
    CECCHERINI,N.L, RUSCETTI,F.W., BRODER,S., SARANGA-
    DHARAN, ITO,Y.,M.MAEDA, M.WAINBERG and M.S. REITZ:
    In, Biochemical and Biological Markers of Neo-
    plastic Transformation (Edt. P. Chandra)pp.503,
    1983. Plenum Press, New York.

84. SARIN,P.S. and GALLO,R.C.: International Reviews in
    Science, Vol. 6 (Edt. K. Butron)pp 219,1974.
    Butterworths, London.

85. THOMPSON,F.M., LIBERTINI,L.J., JOSS,U.R. and CALVIN,
    M.: Science 178, 505, 1972.

86. TEMIN,H.M. and BALTIMORE,D.: Adv. Virus Res. 17,
    129, 1972.

87. LEVINSON,W., FARAS,A., WOODSON,B, JACKSON,J. and
    BISHOP,J.M.: Proc.Natl.Acad.Sci.U.S. 70, 164,1973.

88. CHANDRA,P.,EBENER,U.,STEEL,L.K.,LAUBE,H.,GERICKE,D.,
    MILDNER,B.,BARDOS,T.J.,HO,Y.K. and GÖTZ,A.: Ann.
    N.Y. Acad. Sci.: Third Conference on Antiviral
    Substances 284, 444, 1977.

89. CHANDRA,P.,ZUNINO,F.,GÖTZ,A.,WACKER,A.,GERICKE,D.,
    DI MARCO,A.,CASSAZA,A.M. and GIULIANI,F.: FEBS-
    Lett. 21, 154, 1972.

90. CHANDRA,P., DI MARCO,A.,ZUNINO,F.,CASAZZA,A.M.,GE-
    RICKE,D.,GUILIANI,F., SORANZO,C.,THORBECK,R.,GÖTZ,
    A.,ARCAMONE,F, and GHIONE,M.:Naturwissenschaften
    59, 448, 1972.

91. CHANDRA,P.,ZUNINO,F. and GÖTZ,A.:FEBS-Lett. 22,161,
    1972.

92. CHANDRA,P.:In, Medicinal Chemistry : Topics in current
    Chemistry Vol. 52, 99, 1974. Springer-Verlag.

93. CHANDRA,P.,WILL,G.,GERICKE,D. and GÖTZ,A.: Biochem.
    Pharmacol. 23, 3259, 1974.

94. CHANDRA,P. and BARDOS,T.J.: Res.Commun.Chem.Path.
     and Pharmacol. 4, 615, 1972.

95. CHANDRA,P., EBENER,U. and GÖTZ,A.: FEBS-Lett. 53,
     10, 1975.

96. CHANDRA,P., EBNER,U., STEEL,L.K., LAUBE,H., GERICKE,
     D., MILDNER,B., BARDOS,T.J., HO,Y.K. and GÖTZ,A.:
     Ann.N.Y.Acad.Sci. 284, 444, 1977.

97. CHANDRA,P.,STEEL,L.K.,EBENER,U.,WOLTERSDORF,M.,LAUBE,
     H., KORNHUBER,B., MILDNER,B. and GÖTZ,A.: Pharmacol.
     & Ther. A 1, 231, 1977.

98. CHANDRA,P., STEEL,L.K., EBENER,U., WOLTERSDORF,M.,
     LAUBE,H., KORNHUBER,B, MILDNER,B. and GÖTZ,A.:
     In, Inhibitors of DNA and RNA polymerases Edts.:
     P.S.Sarin and R.C.Gallo), pp47, Pergamon Press,
     Oxford, 1980.

99. CHANDRA,P.: In, Topics in current chemistry 52, 99,
     1974. Springer-Verlag, Heidelberg-New York.

100. CHANDRA,P.,EBENER,U.,BARDOS,T.J., GERICKE,D.,KORNHU-
     BER,B. and GÖTZ,A.: In, Modulation of host immune
     resustence in the prevention or treatment of cancer
     (Edt. M.A. Chirigos), Fogarty International Center
     Proceedings (USA) Vol. 28, 169, 1977.

101. CHANDRA,P.: In, Antimetabolites in Biochemistry, Bio-
     logy and Medicine (Edts. J. Skoda and P.Langen)
     pp.249, 1979. Pergamon Press, Oxford.

102. EBENER,U.: Doctoral Thesis (Frankfurt University),
     1977.

103. KOTLER,M. and BECKER,Y.: FEBS-Lett. 22, 222, 1972.

104. CHANDRA,P.,KORNHUBER,B., GERICKE,D., GÖTZ,A. and
     EBENER,U.: Z. Krebsforschg.u. Klin. Onkol. 83,
     239, 1975.

105. KORNHUBER,B. and CHANDRA,P.: In, Antiviral Mechani-
     sms in the control of Neoplasia (Edt. P. Chandra)
     pp. 577, 1979. Plenum Press, New York.

106. RHIEM,H., GARDNER,H.,JESSENBERGER,K. and TARIVERDIAN,
     G.: Proc.Amer.Assoc.Cancer Res. 15, 58, 1974.

107. CHANDRA,P., DEMIRHAN,I. and EBENER,U.: In, Terminal
     Transferase in Immunobiology and Leukemia (Edts.
     U.Bertazzoni and F.J. Bollum)pp 87, 1982. Plenum
     Press, New York.

108. CHANDRA,P.,DEMIRHAN,I.,EBENER,U.,KORNHUBER,B. and
     GERICKE,D.: In.Developments in Molecular Virology
     (Edt. Y. Becker) Vol. 3, (in press).Martinus
     Nijhoff/Dr.W.Junk Publishers, Holland.

109. CHANDRA,P., DEMIRHAN,I. and DE CLERCQ,E.: Cancer
      Lett. 12, 181, 1981.

110. DE CLERCQ,E., FUKUI,T.,KAKIUCHI,N.,IKEHARA,M.,HATTO-
      RI,M. and PFLEIDERER,W.: Cancer Lett.7,27,1979.

111. IKEHARA,M.,FUKUI,T. and KAKIUCHI,N.: Nucleic Acid
      Res. 5, 1877, 1978.

ANTIVIRAL AGENTS: WHY NOT A "PENICILLIN" FOR VIRAL INFECTIONS?

George J. Galasso

Chief, Development and Applications Branch
Microbiology and Infectious Diseases Program
National Institute of Allergy and Infectious Diseases
National Institutes of Health
Bethesda, Maryland   20205

The question asked most frequently by the lay person of someone involved in research on antiviral agents is "When are you going to do something about the common cold?"; the lay press contributes "Why isn't there something like penicillin for the treatment of viral infections?" The answer usually given is a simple explanation of virus replication and the difficulty of disrupting viral replication without affecting normal cellular functions. For many years it was believed that virus production depended exclusively on cellular metabolic processes. However, this pat answer is only a partial response; it certainly is no longer adequate for the virologist. We have seen from the preceding presentations that many viruses have specific targets which can be identified and characterized, and for which specific blocking agents can be synthesized. Why, then, have we not been more effective in developing new antiviral agents? Again, there is a simple answer – clinicians and virologists have been skeptical about the potential of antivirals, believing effective antivirals would also be toxic; further, we are only now learning sufficient details of the molecular biology of viruses to identify specific blocks. Thus, the effort to date to develop new agents has been minimal. To be successful, greater effort must be expended. All the compounds identified thus far are largely a result of serendipity. None are truly the result of a concentrated effort to develop targeted antiviral agents against a specific virus. Most have come through large screening programs with the production analogs of those substances which show some promise.

337

Targeted antivirals require extensive effort and considerable funds. The academic investigator is usually interested in a small segment of the overall developmental scheme. To properly pursue targeted antiviral development, a large coordinated effort of virologists, chemists, pharmacologists, toxicologists and clinicians is needed. Such an effort is affordable only by a dedicated pharmaceutical company or the government. Government funds are scarce, and becoming more so, with greater demands from increasing new health problems. Private industry considers this to be high-risk research with an unknown return. Taking the position of the research managers of a pharmaceutical firm, antiviral research must be viewed as a high-risk venture for research and development funds. The probability of success, even when following an estab-lished lead, is unknown. The cost of the research is high and getting higher, especially the clinical portion. Prior to deter-mining if the agent has clinical potential, necessary pre-clinical testing for safety is understandably extensive and expensive. Marketing departments also become involved in the decision-making process. If they determine that the potential market is insuffi-cient to guarantee a fair return for the investment, further work with the agent may be terminated. An effective drug for the treatment of viral encephalitis would never be developed by an organization concerned only with a return on the investment, at least not deliberately. Such organizations are interested primarily in more prevalent clinical problems such as the common cold and genital herpes. This means that research on diseases of relatively low frequency must be done by the government.

Therefore, it is difficult to foresee a single group taking on support of targeted antivirals. The solution is a strong collaboration among academicians, pharmaceutical firms and govern-ment toward the mutual goal of virus disease control through targeted antiviral agents. We have learned from the preceding presentations that the potential for effective antivirals is excellent. We have achieved considerable success already and, although I have presented a rather grim picture of the practical-ities, there is no question that the field is expanding and newer, more potent antivirals will be available in the near future. It is important however for investigators to be fully aware of what is involved. Toward that end, the remainder of this presentation will address drug development, testing and the current status of various agents.

STRATEGIES FOR ANTIVIRAL AGENT DEVELOPMENT

The preceding presentations have dealt with individual viruses and their specific targets for potential antiviral action; the following will identify some general areas for targeted

agents. This by no means is intended as an all inclusive list,
but serves only to illustrate that the problem is solvable. As
with all research, we need to maintain a "prepared mind" to
recognize the opportunities and be imaginative enough to overcome
the obstacles. We are now at the stage where we know enough about
viruses to be able to attempt targeted antiviral drug development.

Thus far, through serendipity, we have been surprised to
learn that some chemical moieties can be substituted in substrates
and become inhibitors; thymine can be replaced by a cytosine;
methyl groups by various alkyl groups, including 2-bromovinyl; and
the purine guanine substituted for thymine, as in acyclovir. We
also learned that deoxyribose can be substituted by an arabinoside
(ara-T), and that a sugar moiety is not essential for a substrate
in order for an enzyme (thymidine kinase) to induce phosphorylation.
Through continued vigilance, we will continue to be surprised; we
should be prepared to take advantage of these findings and translate
them into new antivirals.

There are several steps in the viral replication process
which can serve as targets: 1) prevention of viral attachment to
host cells by modification of virus or cell attachment sites, 2)
blocking penetration of cell membrane, possibly by modifying
receptor sites, 3) preventing viral uncoating, 4) inhibition of
RNA-dependent and DNA-dependent viral transcriptase, 5) interfering
or inhibiting viral nucleic acid synthesis, 6) interfering or
inhibiting viral protein synthesis, 7) inhibition of translational
processes of viral mRNA.

The key to any of the above is the identification of a
specific protein, or metabolic step, in viral replication which is
unique to the virus in question and which can be inhibited without
affecting, or at least not be toxic to, the host cell. Some
potential targets include:

1) <u>Selective phosphorylation</u>. Enzymes have been identified
which are induced in virus infected cells but not, or in
reduced amounts, in uninfected cells. These enzymes may be
useful in the selective phosphorylation of antiviral
compounds to the active form as is the case with ACV, BVDU,
etc. where the viral induced thymidine kinase activates the
agent.

2) <u>Viral enzyme inactivation</u>. Having identified viral
enzymes it may be possible to develop compounds which will
preferentially bind with them to inactivate them directly or
inactivate intermediates in the reaction between the enzyme
and its substrate.

3) <u>Inhibition of viral glycoprotein</u>. Many of the pathogenic
viruses possess a viral envelop which includes glycosylated
proteins, usually in the form of spikes which are involved in
initiation of infection. They include viral hemagglutinin,
neuraminidase and fusion protein. The biosynthesis of glycoproteins can be inhibited by interfering with intracellular
transport, with proteolytic cleavage, and with glycosylation
of the protein. Hemagglutination, neuraminidase, and fusion
activities of viruses depend on glycosylation of the viral
glycoproteins.

4) <u>Inhibition of viral DNA</u>. Compounds may be devised which
will act more readily to inhibit viral DNA than host DNA
synthesis. In addition, DNA inhibitors may have an advantage,
since DNA viruses usually induce a high rate of DNA synthesis
in infected cells. This is one of the reasons why deoxyribonucleoside analogs, such as 5-iododeoxyuridine, and bases
attached to arabinose instead of deoxyribose, such as vidarabine, are effective. All act on viral DNA to a greater
degree than host DNA. DNA inhibitors may also be effective
by a) affecting synthesis of precursors required for DNA
synthesis, b) interacting with the viral DNA molecules to
damage them, c) becoming incorporated into the viral DNA
causing a nonsense code.

5) <u>Inhibition of viral RNA</u>. Most single and double-stranded
RNA viruses are dependent on the RNA-dependent RNA polymerase
for replication. Inhibitors of this polymerase may prove
effective agents; such inhibitors have been found, but to
date they have proven to be too toxic or unable to penetrate
the host cell. The reverse transcriptase or RNA-dependent
DNA polymerase in oncogenic RNA viruses is a unique enzyme
which also offers a tempting target for drug design; inhibitory
compounds identified for such enzymes, have been insufficiently
potent or too toxic. Some viruses have been shown to have
segmented genomes, and the specific functions of the different
segments have been identified. It should be possible to
design inhibitors specific to such components.

6) <u>Inhibition of translational process</u>. Attempts should be
made to selectively prevent the translation of viral mRNA so
as to prevent the expression of the viral genomes. Although
it may be difficult to do so selectively, it is possible as
has been shown with interferon.

7) <u>Improved methodologies</u>. These include a) compounds
directed at different targets may be minimally effective
alone but may act synergistically when combined; b) better
delivery systems for getting the drug to the target site
should be developed including overcoming natural barriers

such as the blood-brain barrier, combining the drug with
highly specific monoclonal antibodies to get it to the target
organ and concentrate it, and techniques for localized slow
release of the drug (incorporation in liposomes or other slow
release implantation techniques, facilitation of skin penetration for topical drugs, adherence to mucous membranes; and,
c) prevention of rapid removal of drug, e.g. one problem with
intranasal instillation of drugs is their rapid removal by
mucous and cilliary action; antihistamines may be used to
slow down this action.

ANTIVIRAL DRUG EVALUATION

Identification of a potential antiviral agent in the laboratory is only the first of many steps on the way to the clinic.
There are various estimates of the cost of further development and
of the time involved. A fair estimate is $50 million and 7 years,
from the point of identifying a promising agent, provided all goes
well. In addition to proving efficacy the developer must satisfy
legal requirements. Up until 1962, the Food and Drug Administration (FDA) was responsible only for assuring that drugs were safe.
Following the passage of the Harris - Kefauver amendment a drug
could not be considered truly safe unless it were also effective.
Congress decreed that the FDA assure that developers prove their
products effective, as well as safe, before marketing.

Preclinical Testing

Prior to a trial in humans, extensive preclinical studies
must be performed. The positive in vitro effects must also be
obtained in animals. The tissue culture systems tell us about the
direct effect, but in vivo models give us essential information on
inactivation of the drug by body tissue, ability of the agent to
reach the target site, effects of the virus on the immune system
(e.g CMV and immunosuppression) which may be modified by the
agent, etc. Finding suitable animal-models is a problem. In many
instances, there are no adequate animal models, e.g. respiratory
syncytial virus infections. Even when models do exist they are
not truly representative of human disease. Even though a human
viral isolate can be used as a challenge and manifestations may
mimic the human condition, the portal of entry, pathogenesis,
details of viral replication and/or the role of the immune system
may differ. Ideally models should closely approximate the human
situation. The method of viral infection should approximate as
closely as possible the natural infection; further, the inoculum
must be selected so as not to overwhelm any beneficial effect of
the test compound, preferably an LD_{80}. Having shown in vivo

efficacy without excessive acute toxicity, further pharmacologic
and toxicity studies need to be performed.

Animals are used to establish the maximum tolerated dosage so
as to arrive at the therapeutic dosage (effective dose: maximal
tolerated dose). The metabolic disposition of the drug is examined
to determine adsorption, available tissue distribution, and excretion.
For example, phosphonoacetic acid was found to be very effective
against HSV infections in vitro and in vivo; however, it was found
to be retained by the bone and was abandoned as a potential clinical
drug. Animal studies are also performed to determine proper route
of administration (intramuscular, im; intravenous, iv; peroral, po).
The oral route is the one of choice for ease of administration
especially to children or patients with low platelet levels due to
immunosuppression. However, bioavailability of active drug is often
difficult to achieve by the oral route. In order to properly perform
these studies, the techniques to determine small quantities of the
active drug plus metabolites are necessary. Methodologies have
included high-pressure liquid chromatography, isotope techniques,
and radioimmunoassay.

In addition to the metabolic studies just mentioned, additional
toxicologic studies include acute and chronic toxicity, mutagenicity,
carcinogenicity, and effect on reproduction. Acute and chronic
toxicity studies are performed in at least two animal species (one
rodent and one non-rodent) to determine dose range and administration of high, non-lethal dosages for long term effects. Animals
are sacrificed and all tissues examined histologically. The chronic
toxicity studies involve drug administration for up to one year.
Animals are examined for clinical effects such as weight loss,
tremors, weakness. Clinical laboratory studies including hematology
and chemistry are performed. Following sacrifice, both gross and
microscopic studies are performed.

Teratogenicity studies are performed by administering the drug
to two species such as pregnant monkeys, rats, and rabbits during
the period of fetal organogenesis and examining the outcome of the
pregnancy (Segment II). Reproduction studies also include treating
pregnant rats during the second half of pregnancy and weaning and
observing the offspring for adverse perinatal effects (Segment
III). A third test involves treating male rats for 60 days (effect
on spermatogenesis) and female rats for 14 days (effect on ovogenesis)
and mating the animals to determine effect on fertility (Segment
I). Studies are customarily done in the order described here.
Segment II is done prior to starting human studies (IND), Segments
I and III prior to marketing (NDA).

Mutagenicity can be studied in various ways. Classic tests
are the dominant lethal and host-mediated assay. In the dominant
lethal test, male mice are given either the test compound, placebo

as a negative control, or triethylene melamine as the positive
control. These treated males are then mated with untreated
females to see if there are aborted implantations. The results
with the test compound are compared to the positive control. In
the host-mediated assay (Ames Test), mice are given the test
compound, placebo as the negative control, or N-methyl-N'-nitro-
N-nitrosoguanidine as the positive control. The animals are
injected with <u>Salmonella typhimurium</u> - a strain with labile
genetic nutritional requirements. The bacteria are recovered and
plated on various media to see if their nutritional requirements
have been genetically altered. The bacteria recovered from the
animals, which have recieved the positive control, should show
mutations. The bacteria from the animals that received placebo or
test drug should show no alterations. Another test involves
mammalian bacterial and mammalian cell systems with and without
metabolic cell transplantation. Transformation can de determined
<u>in vitro</u> by observing growth patterns when exposed to drug-organized
growth is replaced by disorganized growth. When these cells are
then placed back into the animals they were derived from, tumors
may result. The field of mutagenicity testing is rapidly changing.
Newer tests coming into routine use include the sister-chromatid
exchange in Chinese hamster ovary cell culture, <u>in vivo</u> rat bone
marrow, and Chinese hamster lung culture.

Although tumorogenesis studies need not be done prior to
initiation of clinical investigation, they are required prior to
or as a condition of drug approval. Subacute toxicity studies
determine the maximally tolerated dose that can be administered to
animals for 21-24 months. Administering excessive doses could
preclude animals surviving their usual life-span, resulting in
incomplete studies. The test and control animals are then sacrificed and examined for incidence and type of tumors.

The unsatisfactory nature of tests in animals to predict what
will happen in humans is well appreciated. Every species metabolizes
drugs somewhat differently, therefore results cannot be readily
extrapolated to humans. It is equally difficult to devise a test
or find a model for unsuspected toxic manifestations. Thus,
preclinical studies serve only as indicators of what can be
expected from a drug. If a compound passes all tests, it does not
mean it will be safe when given to humans, nor is the converse
true. The ultimate test is the human. Drugs which fail the
preclinical screens should not be used in the clinic; except in
instances where the potential benefit far outweighs the risk.

Clinical Studies

If a drug has demonstrated therapeutic effects and acceptable
toxicity in the preclinical tests, an Investigational New Drug
(IND) application can be submitted to the FDA. This will include

all the results of the preclinical tests as well as manufacturing
methods, the clinical protocol of the proposed study, and the
qualifications of the clinical investigator. The FDA will review
the IND within 30 days of receipt; if found acceptable, clinical
studies can be initiated. However, even if proven safe in animals,
continued surveillance in humans is mandatory even after the drug
is licensed. There are many instances where toxicities were not
observed until the drug had been on the market for some time.

Clinical studies are divided into three phases. Phase I
studies are performed in small numbers of patients or volunteers
(20-50), preferably in individuals who stand to receive some
benefit of the drug. The purpose is to determine bioavailability,
pharmacokinetics and metabolic disposition. The primary purpose
is to lay the groundwork for rational dosage regimens in subsequent
clinical efficacy evaluation. Phase II testing again involves
comparatively small, well-controlled studies in up to 200 patients
with the disease for which the drug is proposed. The primary
purpose here is to determine early efficacy, dose-response,
toxicity and therapeutic levels in various biologic fluids.
Careful monitoring of all parameters is required, particularly the
principle target organs for drug toxicity: the eye, central
nervous system, skin, gastrointestinal tract, liver, kidneys and
hematopoietic system. Phase III studies are conducted in sufficient
numbers of patients (up to 2000) to prove the efficacy of the
drug; with further monitoring for safety, tolerance and definition
of any adverse effects. It is of interest that about 90% of
investigational drugs do not go beyond Phase II testing because
they do not show enough therapeutic promise.

In all clinical testing, the FDA requires that Institutional
Review Boards be established in the Institutions where the clinical
study is to be conducted. These Boards must review, approve and
monitor the proposed study. Further, the protocol must include a
proper consent form indicating that the volunteer has been fully
informed of the study to be performed, including all the possible
risks and alternatives.

Once the developer is convinced that efficacy has been
demonstrated and that the drug is safe, a New Drug Application
(NDA) may be submitted to the FDA. The NDA includes: 1) full
reports of investigations, including animal and clinical studies
as outlined above, 2) a statement of the drug's composition
including manufacturing, processing, and packaging details, as
well as a description of the facilities and control measures for
drug uniformity, 3) samples of the drug and components, 4) a copy
of the proposed labeling.

The IND/NDA reviews are done by at least four FDA staff
members; a medical officer (the team leader), a pharmacologist,

statistician and chemist (or microbiologist). The NDA review is usually completed within 180 days. Once approved the drug may be marketed. Approval may be withdrawn if: 1) new information shows the drug unsafe, 2) not effective, 3) the NDA contains an untrue statement. Therefore, drug monitoring must be continued. The FDA also monitors to assure that advertisements are consistent with the package insert and not misleading.

The FDA is often criticized for withholding important drugs from U.S. clinics, either by refusing approval or by taking a long time to approve the New Drug Application. Its charge is to approve only those drugs of proven efficacy and relatively no toxicity - this latter aspect is dependent on the benefit-risk ratio. I, for one, feel that it does an excellent job of protecting the public's interest. This view is shared by many nations that use the FDA as a guide in their own deliberations. When some of us who deal with the FDA get frustrated, we should take some time to put the shoe on the other foot and try to understand their responsibilities in fulfilling their scientific mandate while coping with public attitudes and legal restrictions.

CURRENT STATUS OF ANTIVIRAL AGENTS

There are currently few antivirals generally accepted as efficacious. These include iododeoxyuridine, adenine arabinoside and trifluorothymidine ointments for herpetic keratitis; acyclovir ointment for genital herpes; oral amantadine for influenza A; intravenous adenine arabinoside for herpes encephalitis and zoster in the immunosuppressed patient; and intravenous acyclovir for mucosal and cutaneous herpes infections in the immunocomprised patient and for very severe cases of initial genital disease in the normal host.

Topical agents for herpes keratoconjunctivitis

This is the first viral infection found amenable to treatment with an antiviral agent. Iododeoxyuridine (IDU) was found to be effective against this serious herpes infection but too toxic to use systemically for other indications. It soon became apparent that more effective drugs were needed because IDU was not always clinically effective, due to viral resistance and some ocular toxicity. In 1977, the FDA approved adenine arabinoside (Vira-A) ointment which was shown to be more effective than IDU; subsequently trifluorothymidine (F_3T) also was found to be effective and was approved for general use. Other compounds such as 5-ethyl-2'deoxyuridine, tromantadine HCl, and iododeoxycytidine are available in foreign countries; but their efficacy is controversial and they are not as widely accepted as the former three compounds.

New treatments currently under study include acyclovir (ACV),
bromovinyl deoxyuridine (BVDU), and combination of thermomechanical
debridement of the diseased corneal epithelium and either F_3T or
interferon. The combined treatment regimen appears to be the most
effective to date. Stromal and anterior chamber disease (uveitis)
are still areas where current antiherpetic therapy is ineffective.

Amantadine/Rimantadine

Although amantadine has been available for prevention and
treatment of influenza since 1966, it has not received the acceptance it deserves. It was not until clinical studies were repeated
in the late 70's together with a substantial data on side effects
that the drug received wider acceptance (1,2). Careful investigation showed that the side effects were no more severe than those
experienced with many of the antihistamines that are readily
available. The withdrawal rate due to amantadine side effects in
these clinical studies was 6.2%.

Although the side effects are negligible, concern still
exists among some physicians resulting in increasing interest in
an amantadine analogue, rimantadine. Clinical studies have
demonstrated that it is equally effective as amantadine and is
associated with fewer side effects. Studies comparing the two
drugs have demonstrated prophylactic efficacy of 85% for rimantadine
and 91% for amantadine when compared to placebo (3). Therapeutic
comparison trials also showed the two drugs efficacious, with a
slight edge for amantadine (4). Further definitive trials have
been completed but have not been analyzed for publication.

Adenine arabinoside

Vidarabine was the first antiviral agent to be approved by
the FDA for the systemic treatment of a life-threatening illness.
The effecacy of this drug against brain biopsy proven herpes
encephalitis has been unequivocally demonstrated. This is a
disease with a high mortality rate (70%) and with severe residual
effects in the survivors. Vidarabine was able to reduce mortality
to 28%, with a substantial increase in survivors able to return to
normal activity. The drug is most effective when used early in
the infection and in younger patients (5,6). Vidarabine will
eliminate the virus but can not repair the damaged nervous system,
an important point to remember.

With the increasing incidence of genital herpes, neonatal
herpes is becoming of greater concern worldwide. The disease in
the newborn may be a) disseminated, with multiple organ involvement,
often including the brain; b) localized to the central nervous
system; or c) localized to the skin, eye, and/or month. In a well
controlled trial mortality in infants with disseminated disease

was reduced by vidarabine (15 mg/kg/day) from 86% to 57%; with
localized CNS disease it was reduced from 50% to 10%; the combined
mortality was reduced from 74% to 38%, similar to that described
above for herpes encephalitis. There was no mortality associated
with localized skin, eye or mouth infections, but severe sequelae
occurred in 38% of placebo recipients as compared to none in the
drug recipients (7).

Vidarabine has also been proven effective in treatment of
zoster in immunocompromised patients. In a double-blind placebo-
controlled study the drug proved effective in accelerating healing
of cutaneous lesions, decreased cutaneous dissemination and zoster
related complications. It has also been reported that the duration
of postherpetic neuralgia was reduced and no serious drug toxicity
was observed. Treatment was most effective when started within
the first three days (8). A similar double-blind placebo-controlled
study was performed in immunocompromised patients with varicella
with comparable results. Drug therapy accelerated cessation of
new lesion formation, more rapid resolution of fever, and reduced
varicella-related complications. Vidarabine therefore, showed a
good therapeutic index for treatment of chickenpox in immunocompromised children when given early in infection (9).

Since vidarabine is relatively insoluble and must be administered by slow drip with considerable fluid, studies are now
underway to determine if the vidarabine monophosphate, which is
almost completely soluble, can be substituted with equal beneficial
effects.

Acyclovir

ACV (Zovirax) appears to be one of the more interesting
anti-herpes compounds currently under extensive clinical study.
It is highly specific being activated by viral induced thymidine
kinase. As previously stated, it is already approved by the FDA
for topical treatment of the initial episode genital herpes (10)
and for systemic treatment of serious herpes infections in the
immunosuppressed patient (11). The drug did not prevent the
establishment of recurrent infection and had no effect other than
reduced viral shedding in treatment of recurrences.

ACV also proved effective prophylactically when given to bone
marrow transplant patients with a history of recurrences. Seven
of 10 patients on placebo developed herpes infections compared to
none of the patients on prophylactic drug (12).

Currently, studies are being done using oral ACV. In a
double-blind placebo-controlled study, oral drug (200 mg per dose)
administered five times daily for 10 days appeared to be an
effective treatment of first episodes of genital herpes similar

an topical treatment. Systemic treatment of the initial episode
may be preferred for its potential to reduce new lesion formation
and systemic symptoms such as malaise and fever. No effect on
recurrence rate was observed.

Clinical studies to compare ACV and vidarabine in herpes
encephalitis and in neonatal herpes are currently underway.

Fluoropyrimidine compounds

Interesting newly synthesized pyrimidine analogues, 2'fluoro-
5'iodoaracytosine (FIAC), 2'fluoro-5-iodo-1-Beta-arabino-furanozyl-
uracil (FIAU) and 2'-fluoro-5-methylarauracil (FMAU) appear to
have good antiviral activity in vitro against herpes viruses in
comparison with the other agents. There have been some reports on
the clinical efficacy of these compounds, mainly FIAC. Preclinical
studies now underway indicate good clinical potential. Animal
models show effectiveness even when the drugs are administered
orally 72-96 hours post-infection. Although all three compounds
appear attractive for clinical evaluation, the developers are
currently comparing them prior to deciding which of the three to
pursue. There are also some toxicity issues which remain to be
clarified.

Bromovinyldeoxyuridine

An equally exciting new drug in previous presentations is
BVDU. As indicated, it is a very specific antiviral that is more
effective against HSV-1 than HSV-2, and that also appears to be
effective against zoster. The pre-clinical studies have been
completed, and phase 1 clinical evaluation is underway. Animal
studies are impressive in that the progression of disease can be
halted even after symptoms have appeared when the drug is administered orally. Preliminary clinical trials in Europe indicate good
results against zoster. However, basic studies on the pharmacokinetics, metabolism and toxicity still need to be completed. If
it passes all the necessary tests, BVDU has excellent clinical
potential.

Ribavirin (virazole)

This drug appears to have a broad spectrum of activity
against RNA and DNA viruses in tissue culture and animal model
systems. One of the most promising aspects of this agent is its
potential against such arenaviruses as the agents of Lassa fever
and Bolivian hemorrhagic fever (Machupo virus), and against
Pichinde virus. Early studies in Sierra Leone with oral drug
proved ineffective in therapy of Lassa fever. Subsequent monkey
studies demonstrated that intravenous or intramuscular administration at higher dosages was more effective in saving the animals,

even seven or eight days after infection with Lassa fever virus.
The results were even better when the drug was given in conjunction
with immune plasma. In new randomized study underway in Sierra
Leone patients are being given ribavirin or ribavirin plus Lassa
immune plasma. These studies are nearing completion. It is too
early to predict their outcome, although the overall results with
combined treatment appear encouraging.

The initial clinical trials with ribavirin were directed
against influenza. Although oral treatment proved to be ineffective,
studies are continuing using ribavirin as an aerosol in influenza
pneumonia patients (14). Preliminary experiments have also been
done using aerosolized ribavirin in infants with respiratory
syncytial virus infection. Early results show some promise of
efficacy (15).

Phosphonoformic Acid (PFA Foscarnet)

Early excitement was generated in the mid 70's when phosphonoacetic acid proved to be effective against herpes infections in
animal models. However, as mentioned earlier, when preclinical
studies showed the drug to be retained by bone, studies were
terminated. Attempts to find equally effective, less toxic
analogs were initiated. The result was PFA.

Clinical studies with PFA topical ointment have been performed
in Europe against labial herpes. The results indicate that if
treatment is initiated very early, the vesicular period can be
shortened and there is a tendency for fewer new vesicles in the
treated patients. However, there was no difference in time to
crusting and healing. Unfortunately, virological studies were not
performed. It would have been useful to have determined if the
drug had a beneficial effect on amount and duration of viral
shedding. Hopefully, these data can be accumulated from the
ongoing studies. However, the problem of bone retention also
exists with PFA, but not to the extent as with PAA.

Studies in genital herpes are currently underway in Sweden;
others are planned for the U.K., Finland and Canada. Data from
the Swedish study should be available soon.

Arildone

(4-[6-(2-chloro-4-methoxy)phenoryhexyl]-3,5-heptanedione) and
chemically related compounds appear to have antiviral activity
against both RNA and DNA viruses including herpesviruses, polioviruses, rhinoviruses, picornaviruses, and respiratory syncytial
viruses. These compounds seem to inhibit viral uncoating. They
appear most effective when administered at the same time or prior
to the virus challenge. Although some clinical evaluations

against herpes are in progress, it does not appear to be as
promising as the other compounds under study.

Enviroxime

2-amino-1-(isopropylsulfonyl)-6-benzimidazole phenylketone
oxime is one of the few drugs undergoing clinical trial for a
virus infection other than herpes. It appears to be a specific
agent against rhinoviruses. A double-blind placebo-controlled
study was done using the drug as a nasal spray. Volunteers were
studied with a rhinovirus type 4 challenge for prophylaxis or
therapy. Although no prophylactic efficacy was demonstrated, some
reduction of symptoms and viral shedding was observed in subjects
with higher drug levels (16). Drug administered as a spray is
rapidly removed by nasal mucous and ciliary action; therefore,
lack of efficacy may be due to a mechanical problem of drug
retention. This is obvious from the inconsistency of drug levels
in nasal washes following drug administration. Further studies
are warranted to maximize the positive effect, specifically
through drug formulation and delivery.

Other Chemical agents of interest

A nucleoside analog similar in structure to acyclovir,
currently under study by at least three groups, has been reported
to have potent antiviral effect against HSV-1 and HSV-2 as well as
human CMV. Although assigned different names, the more commonly
used name is DHPG; 9-(1,3-dihydroxy-2-propoxymethyl) guanine. It
is effective when given orally and may prove useful in the treatment
of CMV infections unlike other agents currently under study. A
drug found to be effective against a broad range of picornaviruses
but not other viruses is 2-(3,4-dichlorophenoxy)-5-nitrobenzo-
nitrile. The suitability of this compound for the clinic remains
to be shown but its high therapeutic ratio and specificity in
vitro make it a promising antiviral agent. Several other compounds
that are in limited use and/or under study should be mentioned.
Aedurid, Cebe-Viran and Viru-Merz are available in Europe for the
treatment of ocular and other herpes infections; their clinical
efficacy has not been well established. Based on the ability of
2-deoxy-D-glucose to inhibit some enveloped viruses in vitro,
clinical studies were performed to determine its role against
genital herpes. There has been one report claiming efficacy, but
it is open to criticism and has not been substantiated. One
compound which has received considerable publicity over the years
and for which there is no conclusive demonstration of efficacy is
methisoprinol (isoprinosine). It was originally presented as an
antiviral agent and more recently has been advocated as an immune
potentiator. Although this compound is used in several parts of
the world, a careful review of the literature does not show any
clear evidence of clinical efficacy for any viral disease.

PROBLEMS OF RESISTANCE

As more antiviral agents are developed and approved for general use, there is no question but that they will be over prescribed and used for indications other than those for which the drug was approved. There will also be instances where drugs will be used prophylactically for indefinite periods as, for example for genital herpes. This raises the concern that resistant viral strains may develop. The largest amount of data in this area has been accumulated with herpes viruses. It is not difficult to develop resistant strains by growing the virus in vitro with increasing concentrations of drug; this has been done with PAA, Vidarabine, BVDU, and ACV (17). Resistance may occur by various mechanisms such as a) alteration, deletion or mutation of the gene for thymidine kinase, with this virus not producing the activating enzyme for some of the antivirals. These viruses are termed TK^- strains. b) selection of TK^- viruses through inhibition of the TK^+ viruses which may permit the emergence of the TK^- viruses present in the wild virus natural population, c) alteration of substrate specificity - through mutation of the genome, a thymidine kinase of reduced specificity or activity may result which is not as effective in activating the antiviral.

Studies on the physical mapping of the DNA sequences of HSV responsible for mutations conferring drug resistance have been done. One study (17) reported overlapping regions for Vira-A, PAA and ACV resistance, but BVDU resistance could be transferred separately. Since BVDU possesses a 3'-OH group in the ribosyl configuration, which the others do not, this suggests resistance to BVDU may be based on a structural difference from the other three drugs. This report suggests that resistance to these compounds is mediated by mutations within the DNA polymerase locus of HSV and that multiple resistance to antiviral drugs having similar mechanism of action may be encountered in clinical use.

Resistant strains have appeared in patients treated with drugs (16, 17) and cross-resistance has been observed. Thus, drug resistance is a real concern, not just a laboratory exercise; however, the resistant strains have been considerably less virulent and are characterized by low grade lesions in patients. Extensive animal studies have been done comparing the resistant strains to the parent wild type and in all instances a 100-1000 fold decrease in virulence has been observed. The animals have considerably lower lesion scores and viral shedding, and the infection is no longer lethal. These animals are also protected from subsequent challenge with the wild type. Therefore, although the development of resistance is possible and something to be concerned about, there is currently no evidence that resistant strains are more virulent. Nonetheless, it is important that caution be exercised in long term prophylactic use and that the indiscriminate use of antivirals for anything other than the approved indications be avoided. We should learn from our experiences with antibiotics.

ANTIVIRAL AGENT COMBINATION THERAPY

As stated previously, one approach to improve antiviral effect is combination therapy. Use of combined antivirals may result in: 1) reduction of the development of resistance - if two drugs with different mechanisms of inhibition are used, the likelihood of drug resistance is greatly reduced; 2) enhanced efficacy - by acting at different or even the same site a synergistic effect may be seen; 3) reduced toxicity - by using reduced amounts of drugs one may achieve a similar or better effect with fewer toxic side effects; 4) increase clinical potential of agents - some antivirals may not have favorable therapeutic ratios when used alone but subtoxic amounts may be sufficient when used in combination.

Thus, the resultant effective combination must have at least additive effects, no interference or antagonism between compounds, no increased toxicity and reduced incidence of resistant viruses. Several studies have been done demonstrating such beneficial combinations (20, 21, 22); unfortunately, in some instances increased toxicity has also been observed both in vitro, in animal models and in the clinic.

In doing preliminary studies for comparison of antiviral agents for relative efficacy, it is important to do the comparisons using the most effective dosage in the system used. Some studies have been done using the same molecular level or equal dosage. This can result in erroneous conclusions since compound A may seem more effective than compound B at comparable dosage but compound B should be given at dose 10x and may have a much better therapeutic ratio than compound A. Subsequent combination studies should start at the optimal dosages for each drug and work downward.

In order for combination therapy to be accepted by the FDA, the combination must undergo pre-clinical toxicity evaluation and the combination must, of course, be more effective than either component alone.

The patients who stand the greatest chance of benefit from combined therapy are the immunosuppressed and those with liver or kidney dysfunctions. Some of the available antivirals cause bone marrow suppression, renal and liver toxicity, and accumulate in patients with altered renal function. Reduction in dosage through combined therapy may circumvent these problems.

INTERFERON

No volume on antiviral agents is complete without some
discussion of interferon. Interferon is the oldest and still the
best broad spectrum antiviral, but it has not yet shown sufficient
efficacy to be available to the general clinician. Unfortunately,
the publicity generated over the past three years has led to
considerable disappointment on the part of the layperson as well
as some scientists because "it still hasn't cured anything."
However, this is an unfair attitude, the blame for which rests
largely with the laypress that, in 1980, wrote about the "miracle
drug."

Slowness with its application in the clinic was due for many
years to the fact that we were unable to produce sufficient
quantities for proper clinical trials. It was through Kari
Cantell's untiring efforts on production of leucocyte interferon
that some small trials were initiated in 1973. It was shown that
exogenous interferon, applied locally prior to and following
rhinovirus challenge, could provide complete protection from
clinical illness. However, large amounts of interferon were
needed and it was not a practical approach to the common cold
problem. Clinical studies in the United States continued using
interferon systemically against infectious viral diseases primarily
zoster, varicella and subsequently chronic hepatitis. Studies in
Europe concentrated on cancer, mainly osteogenic sarcoma. These
studies using the Cantell material demonstrated some efficacy
against infectious diseases and cancer, but one of their major
contributions was the acquisition of considerable data on the
pharmacology/toxicology of interferon. Important data on tissue
distribution, effective levels, half-life, and side effects was
accumulated.

Based on these results, oncologists took note of interferon
in the late seventies, and several new clinical studies were
initiated. The promise shown by these studies attracted the
attention of government authorities, the laypress and industry,
and more funds became available. Simultaneously, new technologies
such as DNA recombinant and monoclonal antibody techniques were
developed. Interferon seemed to be an ideal substance to which
these techniques for production and purification can be applied.
In very short order the supply of interferon was no longer a
problem. However, new problems developed. Originally, it was
believed that interferon was a glycoprotein with antiviral properties, produced by cells in response to various stimuli. We have
now learned that we are not dealing with one substance but with
many. Not only are we dealing with three major types i.e. alpha
(leucocyte), beta (fibroblast) and gamma (immune), but there are
several subtypes as well. Further, the interferons have several
immune potentiating and cell modifying properties as well. They

can inhibit the development of delayed type hypersensitivity, accelerate graft rejection in small dosages or slow it in large doses; dependent on dosage, they can either stimulate or depress antibody production; they can stimulate phagocytosis, enhance action of sensitized T lymphocytes, natural killer lymphocytes and K lymphocytes, as well as other properties important in cyto-pathology.

With the emergence of different types of interferon with a multitude of properties, the selection of which interferon to use is becoming difficult. This is particularly the case with alpha interferon since the cloned materials each represent one of the 12 or more subtypes. It is unreasonable to assume that they are all equally effective. What is needed is to develop some laboratory means of differentiating the several types and understanding their respective properties. This unfortunately will take considerable time. Because of the urgency to do clinical studies, we will probably learn more in the clinic than in the laboratory, where these studies should be done.

Fortunately, the recombinant interferons appear to be similar to the natural product in their biological properties. It appears likely that eventually the recombinant interferons will be the ones used in the clinic, even more likely we may end up with new hybrids created in the laboratory which may prove more effective than any of the natural products. However, care must be taken to avoid producing materials which may prove antigenic.

Clinical studies to date have shown considerable potential. There are some studies showing that topical application together with trifluorothymidine or debridement for the treatment of herpetic keratitis is the most effective regimen (23,24,25). Interferon has also been reported to be effective against adeno-virus keratoconjunctivitis.

There is currently considerable activity aimed at demonstrating the effectiveness of interferon against the common cold. Since availability is no longer a problem, interest in this area has been revised. However, the practicality of this prophylactic methodology may be questioned. Interferon is not without toxicity and application of high doses locally to the nose over prolonged periods may not prove feasible. A more likely approach is pro-phylaxis in the family setting when a family member develops a cold.

Systemic application has been shown to be effective for zoster and varicella (26, 27) as well as for prevention of herpes reactivation (27); however, with the availability of other drugs, there has not been a great interest in obtaining licensure for this treatment. There are currently ongoing trials to determine whether interferon has a role in the treatment of active herpes genitalis in the initial episode, and if recurrences in such treated patients occur less frequently. Studies are also in progress to see if cytomegalovirus infections can be prevented in kidney transplant patients, where this disease is very important.

The role of interferon in chronic hepatitis B is also under study. Although it does seem to affect several parameters of chronicity, interferon does not appear to eradicate the disease. More recently, studies treating patients sequentially with interferon and adenine arabinoside monophosphate, which also has a beneficial effect, are being performed by Merigan. However, some toxicity has been observed which may preclude this approach.

There have been a number of other miscellaneous clinical studies including rubella, Epstein-Barr, Ebola, Lassa fever and dengue viruses. However, the most promising area for interferon efficacy appears to be warts. There are several anecdotal reports demonstrating efficacy against laryngeal papilloma and genital warts. There are currently some definitive studies underway to demonstrate whether interferon does indeed have efficacy.

Interferon may have a role in the treatment of several other viral diseases or diseases in which viruses may be implicated. For example, rabies and chronic infections such as Creutzfeldt-Jakob encephalopathy and multiple sclerosis, and in rare but devastating infections such as induced by Ebola virus. EBV and CMV mononucleosis may also be candidates, as well as arboviruses (some studies have been done treating dengue patients with interferon), arenaviruses, and other exotic agents.The recently reported syndromes of Kaposi's sarcoma and/or pneumocystis pneumonia with severe immunodeficiency in young homosexual men, drug abusers, hemophiliacs, and Haitians may represent another case where interferon treatment could be beneficial.

Interferon is not without toxicity and side effects. The toxicity may also be more severe when administered at very high dosages and in certain conditions. The need for carefully controlled, preferably placebo-controlled, studies prior to any claim for efficacy cannot be overly stressed. Open and preliminary studies are often misleading.

What has not been discussed here are the relative merits of alpha, beta, or gamma interferon. There is considerably more experience with the alpha and beta types; they do have some differences, such as tissue tropism, but the differences are slight. The third type, gamma-interferon, may have considerably different clinical potential, but this has yet to be demonstrated. Additional questions concern the possibly deleterious effects of interferon in bronchial asthma, rheumatoid arthritis, autoimmune

disease, etc. However, at the current rate of progress, many of
these questions will be resolved in the near future.

ADDITIONAL CONCERNS

     If antiviral agents are to be of clinical value, a greater
effort must be exerted in the development of rapid viral diagnostic
techniques. In the past, if a patient had an apparent infectious
disease which was not bacterial, physicians were content to give a
clinical diagnosis of a "virus infection." This was mainly due to
the fact that there was no treatment for viral diseases other than
symptomatic relief and there was no real need to know the specific
virus. However, we are now at the stage where we must know if the
patient has influenza, particularly influenza A, as opposed to
other respiratory viruses, whether he/she has herpes encephalitis
as opposed to other causes etc. because there are approved drugs
for their treatment and others well on the way. In all the
efficacy studies done to date, the drugs have been effective only
if administered early in the infection. If viral pathogenesis has
progressed too far, there is little hope of useful intervention.
Therefore, as we strive to find new antivirals against a wide
range of viruses, there must be a concomitant effort in the
development of new, rapid, accurate, and simple viral diagnostic
techniques.

     Another factor in the proper development of antiviral agents
is the need for carefully done, randomized double-blind placebo-
controlled studies. Having identified clinical potential through
phase 1 testing, it is imperative that studies be done in this
fashion with the only exception being when it is ethically
impossible to do so. It is only in this manner that reliable and
acceptable data can be obtained. In most instances, this is the
most expedient and effective way of showing efficacy because
smaller numbers of volunteers can be used.

     Too often anecdotal studies are put into the literature and
there is a tendency to accept them as proof of efficacy. In some
instances, patients can be harmed by using ineffective and
sometimes toxic drugs as has been demonstrated by the use of
cytosine arabinoside against zoster and IDU against herpes
encephalitis. Great care must be exercised in reviewing data to
ascertain that claims of efficacy are not made prematurely and
that clinical efficacy has indeed been demonstrated.

     An important by-product of these studies is the acquisition
of additional information about the natural history and pathogenesis
of the disease under study. It is important to perform these
studies in such a fashion that we gain greater insight in the
pathogenesis of the disease. In some instances, we have learned

that a viral infection is not as serious as originally believed
e.g. zoster in the immunocompromised, or that the incidence of a
specific agent may not be what was expected e.g. incidence of
HSV-1 vs HSV-2 in initial genital herpes. This is of value not
only in determining proper use of the drug but even if the drug,
does not work we have obtained considerable information about the
disease.

FUTURE PROSPECTS

Just as new technologies, such as DNA recombinant technology and
monoclonal antibodies, have resulted in tremendous activity in the
interferon field; the demonstration of effective antivirals such
as vidarabine and ACV in the treatment of serious acute and
chronic infections has led to an increased interest in the development and testing of new antiviral agents. For the successful
progress of targeted antiviral agent development, continuation of
research on the molecular biology of viral replication is paramount.
We must fully understand all aspects of the chemistry of viral
replication, identify all the virus-specific intermediates so as
to be able to effectively develop modalities of inhibiting viral
production without affecting the normal cell. Once this inhibition
has been accomplished in vitro, we must be able to successfully
translate it to the in vivo systems. This transition is not an
easy one.

       The drug must be able to: 1) reach the target organ, if it is
to treat encephalitis it must penetrate the blood-brain barrier,
if it is applied topically it must penetrate the dermis; 2) be
stable in the form administered and not be readily inactivated by
the normal body fluids; it should be stable in the presence of
enzymes etc. encountered when administered in the tissues or
digestive system as appropriate; 3) be cleared by the tissue
within a reasonable time; phosphoacetic acid proved to be a
dramatic herpes virus inhibitor in vitro and in vivo systems but
was found to be retained in the bone and abandoned as a possible
clinical drug; 4) resist and not adversely, affect the immune
system; substances such as novel interferons could induce antibody
and be ineffective in subsequent administration, or they may
inhibit antibody formation negating the immediate beneficial
antiviral effect; 5) not interfere with the normal metabolic
processes of uninfected cells; it should not be mutagenic or
teratogenic; 6) be an effective viral inhibitor; and 7) not induce
viral resistance.

       Viral infections are one of the leading causes of morbidity
and mortality worldwide with growing evidence that they may also
be the causative agents of many chronic diseases. The need for
increased research in antiviral agent development is obvious. The

important targets are diseases of high incidence and those with
considerable morbidity and mortality, even if relatively uncommon,
i.e., the common cold with an average of 2/person/year and viral
encephalitis approximately 1/100,000 with 70% mortality. Some of
the viruses which deserve our early attention are the agents of
the common cold, diarrheal disease, hepatitis, respiratory syncytial,
parainfluenza, toga, arena, rabies viruses, with even more effective
agents needed for herpes and influenza viruses.

The agents already approved for general use are listed above
and the list of potential new agents is increasing at a satisfying
rate. If one were to gaze into the crystal ball, one would
predict that rimantadine will be approved for prophylaxis and
therapy of influenza A, vidarabine will be approved for the
treatment of neonatal herpes, ACV will also be shown effective for
the control of genital herpes, zoster and neonatal herpes when
given systemically, BVDU will eventually be shown to be effective
in zoster. The fluoropyrimidine compounds will be shown to be
effective in zoster as well as other herpes infections. DHPG has
considerable potential for the clinical treatment of CMV infections
and ribavirin aerosol will most likely be available for treatment
of influenza pneumonia and RSV infections of infants and for
arenavirus infections such as Lassa fever, this latter treated in
conjunction with immune plasma. It is my guess that interferon
will prove clinically effective and be approved for the treatment
of warts (laryngeal papilloma and genital warts) and for the
prevention of the common cold. However, the practicality of
interferon use for the common cold may prove difficult. It will
be interesting to review this prognostication in five years to see
which of the above are available in the pharmacy. There is also
little question that additional agents will be available for
respiratory diseases and genital herpes. Although the currently
available agents and those under study are adequate, more potent,
less toxic compounds are needed.

The establishment of a facility dedicated to the effective
and efficient development of new targeted antiviral agents would
be ideal. Such a facility would bring together the needed
expertise of the molecular virologist, chemist, biologist,
pharmacologist, manufacturing chemist, biostatistician, and
clinician. Their goal would be the development of effective
clinically useful agents unrestricted by funds or the profit
motive. Unfortunately, this is a utopian dream not available in
the real world. Therefore, we must depend on the willing
collaboration of government, academia and industry. Such a goal
is obtainable but it requires a much closer working relationship
and willingness than is currently in practice.

There is also the need for a closer collaboration between

nations on the development of more uniform requirements for safety, toxicity and efficacy. This would make it considerably easier and more palatable for the developer to expend the required effort. Currently the demands of each country are dissimilar and usually require a clinical efficacy study to be done in the country where approval is requested. The WHO has taken the initial steps in becoming aware of the latest developments in antiviral agents so that they may be applied more expeditiously, perhaps they might also take the lead in bringing some reasonable order to the testing required prior to marketing.

CONCLUSION

The success we have achieved thus far is merely the prelude, it should serve only as the model for the development of more effective agents. We have demonstrated that antiviral agents can be used in the clinic, many of the new agents under study are amenable to modification to improve their efficacy. The molecular virologist has also provided us with a great deal of information about viruses and the possible targets needed to devise more potent and less toxic agents. We need the continued input of the molecular virologist and the chemist; there is little question that this union together with the pharmacologist and investigative clinician will produce many clinically useful compounds in the near future.

REFERENCES

1. National Institute of Allergy and Infectious Diseases, Amantadine: Does if have a role in the prevention and treatment of influenza? Annals of Internal Med. 92:256 (1980).

2. S.W, Younkin, R.F Betts, F.A Roth and R.G Douglas, Jr., Reduction in fever and symptoms in young adults with influenza A/Brazil/78 H1N1 infection after treatment with aspirin or amantadine, Antimicrob. Ag. Chemother. 23:577 (1983).

3. R. Dolin, R.C. Reichman, H.P. Madore, R. Maynard, P.N. Linton, and J. Webber-Jones, A controlled trial of amantadine and rimantadine in the prophylaxis of influenza A infection, NEJM. 307:580 (1982).

4. L.P. Van Voris, R.F. Betts, F.G. Hayden, W.A. Christmas, and R.G. Douglas, Jr., Successful treatment of naturally occurring influenza A/USSR/77 H1N1, JAMA. 245:1128 (1981).

5.   R.J. Whitley, S-J. Soong, R. Dolin, G.J. Galasso, L.T.
     Ch'ien, C.A. Alford, Jr. and the Collaborative Antiviral
     Study Group, Adenine Arabinoside therapy of biopsy-proven
     herpes simplex encephalitis, NEJM. 297:289 (1977).

6.   R.J. Whitley, S-J. Soong, M.S. Hirsch, A.W. Karchmer, R.
     Dolin, G. Galasso, J.K. Dunnick, C.A. Alford, and the NIAID
     Collaborative Antiviral Study Group,Herpes simplex
     encephalitis: vidarabine therapy and diagostic problems,
     NEJM. 304:313 (1981).

7.   R.J. Whitley, A.J. Nahmias, S-J. Soong, G.J. Galasso, C.L.
     Fleming, C.A. Alford, Jr. and the NIAID Collaborative
     Antiviral Study Group, Vidarabine therapy of neonatal herpes
     simples virus infection, Pediatrics. 66:495 (1980).

8.   R.J. Whitley, S-J Soong, R. Dolin, R. Betts, C. Linneman,
     Jr., C.A. Alford, Jr. and the NIAID Collaborative Antiviral
     Studies Group, Early therapy to control the complications of
     herpes zoster in immunosuppressed patients, NEJM. 307:971
     (1982).

9.   R.J. Whitley, M. Hilty, R. Haynes, Y. Bryson, J.D. Connor,
     S-J. Soong, C.A. Alford, Jr. and the NIAID Collaborative
     Study Group, Vidarabine therapy of varicella in
     immunosuppressed patients, J. Ped. 101:125 (1982).

10. L.Corey, A.J. Nahmias, M.E. Guinan, J.K. Benedetti, C.W.
     Critchlow and K.K. Holmes, A trial of topical acyclovir in
     genital herpes simplex virus infections, NEJM. 305:1313
     (1982).

11. J.C.Wade, B. Newton, C. McLaren, N. Flournay, R.E. Keeney and
     J.D. Meyers, Intravenous acyclovir to treat mucocutaneous
     herpes simples infection after marrow transplantation: a
     double blind trial, Ann Intern Med, 97:265 (1982).

12. R.Saral, W.H. Burns, D.L. Laskin, G.W. Santos, and P.S.
     Lietman, Acyclovir prophylaxis of herpes-simplex-virus
     infections: a randomized double-blind controlled trial in
     bone-marrow-transplant patients, NEJM. 305:63 (1981).

13. Y.J. Bryson, M. Dillon, M. Lovett, G. Acuna, S. Taylor, J.D.
     Cherry, L. Johnson, E. Weismeier, W. Growdon, T. Creagh-Kirk,
     and R. Keeney, Treatment of first episodes of genital herpes
     simplex virus infection with oral acyclovir, NEJM. 308:916
     (1983).

14. H.W. McClung, V. Knight, B.E. Gilbert, S.Z. Wilson, J.M.
     Quarles and G.W. Devine, Ribavirin aerosol treatment of

influenza B virus infection, <u>JAMA</u>. 249:2671 (1983).

15. C.B. Hall, E.E. Walsh, J.F. Hruska, R.F. Betts and W.J. Hall,
     Ribavirin treatment of experimental respiratory syncytial
     viral infection, <u>JAMA</u>. 249:2666 (1983).

16. R.A. Levandowski, C.T. Pachucki, M. Rubenis and G.G. Jackson,
     Topical Enviroxime against rhinovirus infection, <u>Antimicrob.</u>
     <u>Ag</u>. <u>Chemother</u>. 22:1004 (1982).

17. C.S. Crumpacker, L.E. Schnipper, P.N. Kowalsky and D.M.
     Sherman, Resistance of herpes simplex virus to adenine
     arabinoside an E-5-(2-bromovinyl)-2'-deoxyuridine: a physical
     analysis, <u>JID</u>. 146:167 (1982).

18. C.D. Sibrack, L.T. Gutman, C.M. Wilfert, C. McLaren, M.H. St.
     Clair, P.M. Keller, and D.W. Barry, Pathology of acyclovir-
     resistant herpes simplex virus type 1 from an
     immuno-deficient child, <u>JID</u>. 146:673 (1982).

19. H.H. Balfour, Jr., Resistance of herpes simplex to acyclovir,
     <u>Annals</u> <u>of</u> <u>Intern</u>. <u>Med</u>. 98:404 (1983).

20. R.F. Schinazi and A.J. Nahmias, Different in vitro effects of
     dual combinations of anti-herpes simplex virus combination,
     <u>Amer</u> <u>J</u>. <u>Med</u>. 73:40 (1982).

21. R.F. Schinazi, J. Peters, C.C. Williams, D. Chance and A.J.
     Nahmias, Effect of combinations of acyclovir and vidarabine
     or its 5' monophosphate on herpes simplex viruses in cell
     culture and in mice, <u>Antimicrob</u>. <u>Ag</u>. <u>Chemother</u>. 22:499
     (1982).

22. S.A. Spector, M. Tyndall and E. Kelley, Inhibition of human
     cytomegalovirus by trifluorothymidine, <u>Antimicrob</u>. <u>Ag</u>.
     <u>Chemother</u>. 23:113 (1983).

23. R. Sundmacher, D. Neuman-Haefelin and K. Cantell, Successful
     treatment of dendritic keratitis with human leucocyte inter-
     feron, <u>Albrecht</u> <u>von</u> <u>Graefes</u> <u>Archiv</u> <u>fur</u> <u>Klinische</u> <u>und</u> Experi-
     <u>mentell</u> <u>Ophthal</u>. 20:39 (1976).

24. B.R. Jones, D.J. Coster, M.G. Falcon, and K. Cantell, Topical
     therapy of ulcerative herpetic keratitis with interferon,
     <u>Lancet</u>. 2:128 (1976).

25. E.W.J. DeKoning, O.P. van Bijsterveld and K. Cantell,
     Combination therapy for dendritic keratitis with human
     leucocyte interferon and trifluorothymidine, <u>Brit</u>. <u>J</u>.
     <u>Ophthal</u>. 66:509 (1982).

26. T.C.Merigan, R.H. Rand, R.B. Pollard, P.S. Abdallah, G.W.
     Jordan and R.P. Fried, Human leucocyte interferon for the
     treatment of herpes zoster in patients with cancer, NEJM.
     298:981 (1978).

27. A.M. Arvin, J.H. Kushner, S. Feldman, R.L. Bachner, D.
     Hammong and T.C. Merigan, Human leucocyte interferon for the
     treatment of varicella in children with cancer, NEJM. 306:761
     (1982).

28. G.J. Pazin, J.A. Armstrong, M.T. Lam, G.C. Tarr, P.J. Janetta
     and M. Ho, Prevention of reactivated herpes simplex infection
     by human leucocyte interferon after operation on the
     trigeminal root, NEJM. 301:225 (1979).

PARTICIPANTS

ALLAUDEEN, H.S., Philadelphia, U.S.A.
ANDRIES, K., Beerse, Belgium
BABA, M., Fukushima, Japan
BARWOLFF, D., Berlin-Buch, DDR
BATES, D.K., Michigan, U.S.A.
BATTISTINI, A., Roma, Italy
BEAUCHAMP, L., Research Triangle Park, U.S.A.
BERGSTROM, D.E., Grand Forks, U.S.A.
BERNAERTS, R., Leuven, Belgium
BISHOP, D.H.L., Birmingham, U.S.A.
BOBEK, M., Buffalo, U.S.A.
BOVY, P., Brussels, Belgium
BRAIG, H.R., Groningen, The Netherlands
BREUER, E., Jerusalem, Israel
BROWN, D.M., Cambridge, United Kingdom
BROWN, P., Bethesda, U.S.A.
BUFFEL, D., Heverlee, Belgium
CANTONI, G.L., Bethesda, U.S.A.
CARRASCO, L., Madrid, Spain
CHANDRA, P., Frankfurt am Main, W. Germany
CHEN, M.S. Syracuse, U.S.A.
CHENG, Y.-C., Chapel Hill, U.S.A.
CHIANG, P.K., Washington D.C., U.S.A.
CHOPPIN, P., New York, U.S.A.
CLEMENTI, M., Ancona, Italy
CONDE, S., Madrid, Spain
CRUMPACKER, C., Boston, U.S.A.
DATEMA, R., Sodertalje, Sweden
DE CLERCQ, E., Leuven, Belgium
DE LAS HERAS, F.G., Madrid, Spain
DO ROSARIO VASCONCELOS, M., S. Pedro do Estoril, Portugal
DRACH, J.C., Michigan, U.S.A.
EGGERS, H.J., Koln, W. Germany
ELLAMES, G., High Wycombe, United Kingdom
ERICSON, A.-C., Sodertalje, Sweden
FERREIRA, W., Lisbon, Portugal
FIELD, H.J., Cambridge, United Kingdom

FOX, J.J., Rye, U.S.A.
FYFE, J.A., Research Triangle Park, U.S.A.
GALASSO, G.J., Bethesda, U.S.A.
GARCIA LOPEZ, T., Madrid, Spain
GHAZZOULI, I., Wien, Austria
GOSSELIN, G., Montpellier, France
GUSCHLBAUER, W., Gif-sur-Yvette, France
HARNDEN, M.R., Epsom, United Kingdom
HAYATSU, H., Okayama, Japan
HUTCHINSON, D.W., Coventry, United Kingdom
ILSE, D., Don Mills, Canada
IMAI, J., Bethesda, U.S.A.
JOHANSSON, N.G., Sodertalje, Sweden
KALAYCI, C., Ankara, Turkey
KATZ, E., Jerusalem, Israel
KERN, E.R., Salt Lake City, U.S.A.
KORANT, B.D., Wilmington, U.S.A.
KOZARICH, J.W., New Haven, U.S.A.
KRAMER, M.J., Nutley, U.S.A.
LA COLLA, P., Cagliari, Italy
LANGEN, P., Berlin-Buch, DDR
LANKINEN, H., Stockholm, Sweden
LARDER, B., Cambridge, United Kingdom
LOWE, G., Oxford, United Kingdom
MARTIN, J.A., Welwyn Garden City, United Kingdom
MAUDGAL, P.C., Leuven, Belgium
MOFFATT, J.G., Palo Alto, U.S.A.
MONTGOMERY, J.A., Birmingham, U.S.A.
MOORE, B., Sandwich, United Kingdom
OXFORD, J., London, United Kingdom
OEDIGER, H., Leverkusen, W. Germany
OTTO, M.J., New Haven, U.S.A.
PFLEIDERER, W., Konstanz, W. Germany
PRUSOFF, W.H., New Haven, U.S.A.
RAPP, F., Hershey, U.S.A.
RINEHART, K.L., Urbana, U.S.A.
ROBINS, M.J., Edmonton, Canada
ROSENTHAL, H.A., Berlin, DDR
ROSENWIRTH, B., Wien, Austria
SACKS, S.L., Vancouver, Canada
SALISBURY, S.A., Cambridge, United Kingdom
SARIH, L., Bordeaux, France
SCHELLEKENS, H., Rijswijk, The Netherlands
SCHRODER, H.C., Mainz, W. Germany
SHANNON, W.M., Birmingham, U.S.A.
SHIGETA, S., Fukushima, Japan
SIM, I.S., High Wycombe, United Kingdom
SKEHEL, J.J., London, United Kingdom
STAWINSKI, J., Poznan, Poland
STREISSLE, G., Wuppertal, W. Germany

SUHADOLNIK, J., Philadelphia, U.S.A.
SWALLOW, D.L., Macclesfield, United Kingdom
TISDALE, S.M., Beckenham, United Kingdom
TORRENCE, P.F., Bethesda, U.S.A.
TYRRELL, D.A.J., Salisbury, United Kingdom
TZOTZOS, G.T., Athens, Greece
VERHEYDEN, J.P.H., Palo Alto, U.S.A.
WALKER, R.T., Birmingham, United Kingdom
WARING, M.J., Cambridge, United Kingdom
WHITLEY, R.J., Birmingham, U.S.A.
WIEWIOROWSKI, M., Poznan, Poland
WIGDAHL, B.L., Hershey, U.S.A.
WYATT, P., Birmingham, United Kingdom
ZERIAL, A., Vitry-sur-Seine, France
ZUR HAUSEN, H., Heidelberg, W. Germany